Sustainable Cementitious Materials for Civil and Transportation Engineering

Sustainable Cementitious Materials for Civil and Transportation Engineering

Guest Editors

Junjie Wang
Jianhe Xie
Yongliang Liu

MDPI

Basel • Beijing • Wuhan • Barcelona • Belgrade • Novi Sad • Cluj • Manchester

Guest Editors

Junjie Wang
Department of Civil Engineering
Tsinghua University
Beijing
China

Jianhe Xie
School of Civil and Transportation Engineering
Guangdong University of Technology
Guangzhou
China

Yongliang Liu
Department of Engineering Mechanics
Jilin University
Changchun
China

Editorial Office
MDPI AG
Grosspeteranlage 5
4052 Basel, Switzerland

This is a reprint of the Special Issue, published open access by the journal *Materials* (ISSN 1996-1944), freely accessible at: https://www.mdpi.com/journal/materials/special_issues/ceme_mat_civil_transport_eng.

For citation purposes, cite each article independently as indicated on the article page online and as indicated below:

Lastname, A.A.; Lastname, B.B. Article Title. *Journal Name* **Year**, *Volume Number*, Page Range.

ISBN 978-3-7258-5665-7 (Hbk)
ISBN 978-3-7258-5666-4 (PDF)
https://doi.org/10.3390/books978-3-7258-5666-4

Cover image courtesy of Junjie Wang

Contents

About the Editors vii

Preface ix

Junjie Wang, Jianhe Xie and Yongliang Liu
Sustainable Cementitious Materials for Civil and Transportation Engineering
Reprinted from: *Materials* **2023**, *16*, 6290, https://doi.org/10.3390/ma16186290 1

Rong Huang, Lei Xu, Zihang Xu, Qihang Zhang and Junjie Wang
A Review on Concrete Superplasticizers and Their Potential Applications for Enhancing the Performance of Thermally Activated Recycled Cement
Reprinted from: *Materials* **2024**, *17*, 4170, https://doi.org/10.3390/ma17174170 3

Lifan Zheng, Junjie Wang, Kefei Li, Mingyu Wang, Shimeng Li and Lin Yuan
Advances in the Experiments of Leaching in Cement-Based Materials and Dissolution in Rocks
Reprinted from: *Materials* **2023**, *16*, 7697, https://doi.org/10.3390/ma16247697 18

Agnieszka Lal and Joanna Fronczyk
Does Current Knowledge Give a Variety of Possibilities for the Stabilization/Solidification of Soil Contaminated with Heavy Metals?—A Review
Reprinted from: *Materials* **2022**, *15*, 8491, https://doi.org/10.3390/ma15238491 37

Muhammad Imran Khan, Muslich Hartadi Sutanto, Nur Izzi Md. Yusoff, Salah E. Zoorob, Waqas Rafiq, Mujahid Ali, et al.
Cementitious Grouts for Semi-Flexible Pavement Surfaces—A Review
Reprinted from: *Materials* **2022**, *15*, 5466, https://doi.org/10.3390/ma15155466 64

Shuhua Zhang, Guangrong Tan, Zhiqiang Qi, Bin Tian, Jijun Cao and Bofu Chen
Relationship Between the Carbonation Depth and Microstructure of Concrete Under Freeze–Thaw Conditions
Reprinted from: *Materials* **2024**, *17*, 6191, https://doi.org/10.3390/ma17246191 93

Waiching Tang, Ali M. Onaizi, Sagheer A. Onaizi, Umer Sajjad and Yanju Liu
Integrated Use of Furnace Bottom Ash as Fine Aggregate and Cement Replacement for Sustainable Mortar Production
Reprinted from: *Materials* **2024**, *17*, 3834, https://doi.org/10.3390/ma17153834 118

Wentao Li, Ke Huang, Feng Chen, Lihua Li, Yang Cheng and Kang Yang
Engineering Properties and Microstructure of Soils Stabilized by Red-Mud-Based Cementitious Material
Reprinted from: *Materials* **2024**, *17*, 2340, https://doi.org/10.3390/ma17102340 142

Young-Jin Kim, Sang-Rak Sim and Dong-Woo Ryu
Experimental Study on Effects of CO_2 Curing Conditions on Mechanical Properties of Cement Paste Containing CO_2 Reactive Hardening Calcium Silicate Cement
Reprinted from: *Materials* **2023**, *16*, 7107, https://doi.org/10.3390/ma16227107 163

Lei Xu, Yong Lai, Daoxun Ma, Junjie Wang, Molan Li, Le Li, et al.
Effects of Fiber and Surface Treatment on Airport Pavement Concrete against Freeze–Thawing and Salt Freezing
Reprinted from: *Materials* **2022**, *15*, 7528, https://doi.org/10.3390/ma15217528 177

Yangsen Cao, Xinzhou Li, Zhuangzhuang Liu, Jiarong Li, Fan Zhang and Baozeng Shan
Laboratory Study and Field Validation of the Performance of Salt-Storage Asphalt Mixtures
Reprinted from: *Materials* **2022**, *15*, 6720, https://doi.org/10.3390/ma15196720 . . . . **194**

Wenxiu Jiao, Aimin Sha, Zhuangzhuang Liu and Shuo Li
Influence of Fineness Levels and Dosages of Light-Burned Dolomite on Portland Cement Performance
Reprinted from: *Materials* **2022**, *15*, 5798, https://doi.org/10.3390/ma15165798 . . . . **214**

Peng Cao, Xuebing Hu, Enlong Liu, Jianzhong Chen, Shouchao Jiang and Hao Ding
Research on Mechanical Behavior of the Steel–Concrete–Steel Composite Structures Subjected to High Temperature of Fire
Reprinted from: *Materials* **2022**, *15*, 4872, https://doi.org/10.3390/ma15144872 . . . . **229**

Ke-Xian Zhuo, Guo-Tao Liu, Xue-Wei Lan, Dong-Ping Zheng, Si-Quan Wu, Pei-Zong Wu, et al.
Fracture Behavior of Steel Slag Powder-Cement-Based Concrete with Different Steel-Slag-Powder Replacement Ratios
Reprinted from: *Materials* **2022**, *15*, 2243, https://doi.org/10.3390/ma15062243 . . . . **244**

About the Editors

Junjie Wang

Junjie Wang is currently an Assistant Professor in the Department of Civil Engineering, Tsinghua University in China. He has been focusing on the research of low-carbon recycled cement and recycled concrete. As the PI, he presided over a National Key R&D Young Scientist Project (2023YFB3711800), a National Natural Science Foundation (Youth) project (52008232) of China, a sub-project of the National Key R&D Program (2020YFC1522403-1), and a "Chunhui Plan" Project of the Ministry of Education of China. He participated as Co-PI in a State Key Program of National Natural Science of China. He presided over the international consulting project Saudi SABIC villa phase I maintenance consulting project and a high-plateau airport pavement concrete durability and control technology research project. He is the Chief Young Scientist of National Key R&D of China who has continuously been placed on the list of Elsevier's top 2% of scientists in the world since 2023, he is an excellent master's thesis supervisor at Tsinghua University (2024), and he is a director of the Concrete and Cement Products Branch of the Chinese Ceramic Society. He has published more than 80 papers in peer-reviewed scientific journals, among which he is the first/corresponding author of 60 papers (including 12 IF > 10 articles, one ESI hot paper, and two ESI highly cited papers) in *Cement and Concrete Research* (IF = 11.4), *Cement and Concrete Composites* (IF = 10.5), and other top journals. Junjie's H-index is 36, his i10-index is 57, his total number of citations in Scopus is >3600, and the highest number of citations of a single article of his is >440. Since 2023, he has been listed as one of the top 2% of the world's top scientists.

Jianhe Xie

Jianhe Xie received his Ph.D. from the South China University of Technology and is currently a Professor at Guangdong University of Technology and Guangdong AIB Polytechnic. His research focuses on advanced construction materials and sustainable infrastructure, with expertise in fiber-reinforced polymer (FRP)-strengthened concrete structures, geopolymer recycled concrete, ultra-high-performance concrete, and seawater sea-sand concrete. He has led more than 20 national and provincial research projects, including four funded by the National Natural Science Foundation of China (NSFC) and the Guangdong Natural Science Fund for Distinguished Young Scholars. He has published over 100 academic papers, among which more than 80 are SCI-indexed where his is the first or corresponding author, achieving an H-index of 45. He also holds over 30 authorized Chinese patents and has contributed to the development of several provincial and municipal technical standards. Prof. Xie's achievements have been recognized with multiple awards, including the First Prize of Science and Technology Progress from the Ministry of Education, the Guangdong Science and Technology Award, and the Guangdong Outstanding Young Scientist Award. He has also been listed among the World's Top 2% of Scientists for consecutive years. As an active member of professional societies and a reviewer for more than 20 international journals, Prof. Xie remains committed to advancing both theoretical research and engineering practice in civil and transportation engineering.

Yongliang Liu

Yongliang Liu graduated from University College London (UCL) in 2017 with a PhD in civil engineering materials, and has been working at the department of engineering mechanics as Assistant Professor under the School of Mechanical and Aerospace Engineering (SMAE), Jilin University, China, since 2018. With a combined background of mechanics, civil engineering and physics of polymers, he is very interested in the testing and evaluation of organic–inorganic composites including static, dynamic mechanical and fatigue properties, viscoelastic constitutive equations of engineering polymer materials, the design and assessment of different slab tracks in high-speed railways, vibration control and modelling, etc. In addition, he is interested in cold-in-place recycling, from technologies, methodologies, to engineering applications. He is a member of ICE, ACI and ASCE, and has published more than 20 papers.

Preface

As global infrastructure demands surge, the imperative to drastically reduce the environmental footprint of construction materials, particularly cementitious binders that are central to civil and transportation engineering, has never been more urgent. Cement production alone accounts for a significant portion of global CO_2 emissions, driving the need for transformative research and innovation towards sustainable alternatives and optimized performance. This Reprint, "Sustainable Cementitious Materials for Civil and Transportation Engineering", emerges as a timely response to this global challenge, assembling cutting-edge research that pushes the boundaries of material science for a greener future.

This collection delves into the multifaceted realms of sustainable cementitious materials. Its scope encompasses innovative approaches to material design, performance enhancement under harsh environmental conditions, and the valorization of industrial by-products. Key themes include the following: Material Innovation: Development and characterization of novel binders (e.g., CO_2-reactive calcium silicate cement, red-mud-based materials) and optimized conventional systems (e.g., using light-burned dolomite); Waste Valorization: High-volume utilization of industrial residues like furnace bottom ash, steel slag powder, and thermally activated recycled cement as replacements for original cement; Durability: Investigating the resilience of concrete and stabilized soils against freeze–thaw cycles, salt scaling, carbonation, high temperatures, and leaching phenomena; Sustainable Processes: Exploring CO_2 curing as a carbon sequestration and strengthening mechanism; Performance Enhancement: Strategies for improving concrete for demanding applications like airport pavements and semi-flexible surfaces, including fiber reinforcement and specialized grouts; Critical Reviews: Synthesizing knowledge on superplasticizers, leaching/dissolution mechanisms, and stabilization/solidification techniques for contaminated soils.

The primary aim of this Reprint is to disseminate significant advances in the science and engineering of sustainable cementitious materials. It seeks to provide researchers, engineers, and practitioners with a comprehensive resource detailing the following: Innovative Solutions: Practical pathways for reducing the carbon footprint and resource consumption of cementitious materials; Enhanced Understanding: Deep insights into the microstructure–property relationships governing material performance, especially under environmental actions that are crucial for long-term durability in infrastructure; Recycled Materals: Robust experimental and field data supporting the viability and performance of new materials and waste-incorporated systems.

The impetus for compiling this Reprint stems from the undeniable urgency of the climate crisis and the construction sector's pivotal role in mitigating it. The research presented here is motivated by the need to do the following:

- Transition away from resource-intensive, high-emission conventional practices;
- Develop materials that ensure infrastructure durability and resilience in the face of changing climates and extreme events;
- Create economically viable circular economy models within the construction industry by effectively utilizing abundant waste streams;
- Provide scientifically sound data to inform standards, specifications, and policy decisions promoting sustainable construction.

The diverse contributions within this Reprint collectively represent a significant stride forward. They illuminate pathways for reducing environmental impact, enhancing material performance and durability, and unlocking the potential of waste resources. It is our hope that the knowledge shared herein will inspire further innovation and accelerate the adoption of truly sustainable cementitious materials, paving the way for a more resilient and environmentally responsible built environment for generations to come.

Junjie Wang, Jianhe Xie, and Yongliang Liu
Guest Editors

materials MDPI

Editorial

Sustainable Cementitious Materials for Civil and Transportation Engineering

Junjie Wang [1,*], Jianhe Xie [2] and Yongliang Liu [3]

1 Department of Civil Engineering, Tsinghua University, Beijing 100084, China
2 School of Civil and Transportation Engineering, Guangdong University of Technology, Guangzhou 510006, China; jhxie@gdut.edu.cn
3 School of Mechanical and Aerospace Engineering, Jilin University, Changchun 130025, China; yliu2018@jlu.edu.cn
* Correspondence: junjiewang@tsinghua.edu.cn

The current Special Issue entitled "Sustainable Cementitious Materials for Civil and Transportation Engineering" aims to discuss current research on the preparation, characterization, and application of sustainable cementitious materials for civil and transportation engineering, with a special focus on the development of low-carbon construction materials.

Concrete is the most widely used material for construction and building globally. The traditional cement and concrete industry is a main consumer of natural resources and contributes approximately 5~8% to total CO_2 emissions. To address issues pertaining to sustainability and reduce the carbon footprint, it has become a new trend to research and develop novel sustainable cementitious materials, especially materials with low carbon emissions. In this context, on the one hand, construction and demolition waste (CDW) obtained from civil and transportation engineering, recycled concrete aggregates (RCAs) and recycled asphalt pavements (RAPs), for instance, could be treated and reused in the manufacturing of cementitious materials without compromising their properties, which would promote the sustainability of raw materials.

The recycling of CDW [1,2] is a complex process, which usually includes the shredding, separation and processing of waste. First, the collected CDW is transported to crushing equipment for crushing. At present, common crushing equipment include jaw crushers, impact crushers, cone crushers, etc. During the crushing process, large pieces of the construction waste are broken into smaller blocks for subsequent separation and processing. Crushed construction wastes need to meet certain particle size distribution requirements in order to produce recycled aggregates, recycled sand and recycled powders that meet the requirements. After crushing, construction waste needs to be separated from different substances through a separation link in order to obtain purer recycled aggregates, recycled sand and recycled powder. Usually, separation is carried out via equipment such as vibrating screens and winnowing machines. A vibrating screen passes the crushed construction waste through a sieve with different pore sizes and divides it into aggregates and sand with different particle sizes. The winnowing machine uses the principle of aerodynamics to separate light substances such as plastic and paper from heavy substances such as bricks and concrete blocks. After separation, some impurities may remain in the construction waste, which need to be further processed. For example, magnetic separation equipment can be used to separate ferromagnetic substances such as steel bars, nails, etc.; selective flocculation equipment can separate suspended solids and soil in water.

After further processing, the construction waste could be converted into recycled aggregates, recycled sand and recycled powder. In the recycling process, it is necessary to control the particle size distribution, particle shape, impurity content and other indicators of recycled aggregates, recycled sand and recycled powder to ensure that their quality and performance meet the requirements of concrete manufacturing.

Portland cement is a main source of carbon emissions; on the other hand, Portland cement could be partially or completely replaced with recycled cement [3,4], industrial solid waste, clay minerals, limestone, etc., to produce low-carbon binders such as low-clinker cement (i.e., limestone calcined clay cement (LC^3), calcium aluminate cement, and calcium sulfoaluminate cement) and alkali (acid)-activated binders (i.e., geopolymer) [5,6]. For example, 50 wt.% clinker can be replaced with limestone and calcined clay to prepare LC^3 of comparable strength, but it can reduce CO_2 emissions 20–40%, costs by 15–25%, and thermal energy requirements by 20%. Moreover, geopolymer is an alkali-activated binder with excellent properties and is a promising alternative to Portland cement. During the preparation of geopolymer concrete, solid waste and an alkali activator that is derived from the waste are frequently used, and therefore the carbon footprint can be further reduced. Successful applications of these sustainable cementitious materials in civil and transportation engineering are also significant.

The research interests of this Special Issue include, but are not limited to, the following: low-carbon cementitious binders; carbonation-enhanced concrete; low-carbon cement and concrete technology based on non-Portland cement systems, such as alkali-activated materials or geopolymeric materials; recycled aggregate concrete; green admixtures for cement and concrete; the durability of low-carbon concrete; case studies on applications of sustainable cementitious materials in specific regions for civil and transportation engineering.

Funding: This work was supported by National Natural Science Foundation of China (grant nos.: 52008232, 12372180, and 12072078).

Conflicts of Interest: The authors declare no conflict of interest.

References

1. Xu, L.; Wang, J.; Li, K.; Lin, S.; Li, M.; Hao, T.; Ling, Z.; Xiang, D.; Wang, T. A systematic review of factors affecting properties of thermal-activated recycled cement. *Resour. Conserv. Recycl.* **2022**, *185*, 106432. [CrossRef]
2. Wang, J.; Xu, L.; Li, M.; Wang, Y.; He, H.; Xiang, D.; Li, K.; Hao, T. Investigations on factors influencing physical properties of recycled cement and the related carbon emissions and energy consumptions. *J. Clean. Prod.* **2023**, *414*, 137715. [CrossRef]
3. Xu, L.; Wang, J.; Li, K.; Li, M.; Lin, S.; Hao, T.; Wang, T.; Guo, Y.; Zheng, L. Investigations on the rehydration of recycled blended SCMs cement. *Cem. Concr. Res.* **2023**, *163*, 107036. [CrossRef]
4. Xu, L.; Wang, J.; Li, K.; Hao, T.; Li, Z.; Li, L.; Ran, B.; Du, H. New insights on dehydration at elevated temperature and rehydration of GGBS blended cement. *Cem. Concr. Compos.* **2023**, *139*, 105068. [CrossRef]
5. Xie, J.; Zhao, J.; Wang, J.; Wang, C.; Huang, P.; Fang, C. Sulfate resistance of recycled aggregate concrete with GGBS and fly ash-based geopolymer. *Materials* **2019**, *12*, 1247. [CrossRef] [PubMed]
6. Xie, J.; Wang, J.; Rao, R.; Wang, C.; Fang, C. Effects of combined usage of GGBS and fly ash on workability and mechanical properties of alkali activated geopolymer concrete with recycled aggregate. *Compos. Eng.* **2019**, *164*, 179–190. [CrossRef]

materials

MDPI

Review

A Review on Concrete Superplasticizers and Their Potential Applications for Enhancing the Performance of Thermally Activated Recycled Cement

Rong Huang †, Lei Xu †, Zihang Xu, Qihang Zhang and Junjie Wang *

Department of Civil Engineering, Tsinghua University, Beijing 100084, China; hr23@mails.tsinghua.edu.cn (R.H.); xulei.2021@tsinghua.org.cn (L.X.); zh-xu23@mails.tsinghua.edu.cn (Z.X.); zqh1522525607@163.com (Q.Z.)

* Correspondence: junjiewang@tsinghua.edu.cn; Tel.: +86-(0)1-783-723-8638

† These authors contributed equally to this work.

Abstract: With the rapid development of the construction industry worldwide, a large amount of waste concrete is generated each year, which has caused serious environmental problems. As a green and sustainable building material, thermally activated recycled cement (RC) has received widespread attention. However, the unique properties of RC, such as the high water demand and short setting time, necessitate the use of specialized superplasticizers that are different from those used in ordinary Portland cement. As an important component for the application of RC, superplasticizer has an important impact on the performance modification of RC. This article summarizes the recent research progress of potential superplasticizers for RC, with a view to providing a reference for the research and application of superplasticizers for RC. Based on the differences between ordinary Portland cement and RC, the paper discusses potential superplasticizers that may be suitable for RC, and points out that future development of potential modified superplasticizers can include altering the molecular structure to improve adsorption onto the surfaces of RC or to enhance the durability of concrete with RC.

Keywords: superplasticizers; thermal-activated recycled cement; mechanisms

1. Introduction

Due to the swift growth of the global industry, there has been a substantial surge in the need for cement. According to statistical data, China's cement production was projected to reach 2.02 billion tons by 2023 [1]. The production of cement not only depletes a significant quantity of natural resources but also releases a substantial amount of greenhouse gases [2–4]. China generates over 2 billion tons of garbage each year as a result of building restoration and destruction. Of this waste, around 65% is comprised of waste concrete. Consequently, the efficient usage of waste concrete has emerged as a prominent study area in recent years [5,6]. Thermally activated recycled cement (RC) is a new type of building material with cementitious properties that is obtained by heat treatment after recycling waste concrete or hardened cement paste [7]. Initially, it was discovered that concrete can regain some of its strength after being exposed to fire, as observed in studies on the fire resistance of concrete [8,9]. Numerous studies have demonstrated that thermal activation can enable the hardened cement paste to undergo rehydration. Bogas et al. found that thermally activated RC at 650 °C has a compressive strength 2.8 times greater than that of unactivated RC [10]. The temperature required for thermal activation is lower than the temperature needed for the calcination of clinker in ordinary Portland cement (OPC). This process significantly reduces CO_2 emissions, making it an important method for recycling waste concrete through thermal activation [11]. The production of RC not only reduces the environmental impact of concrete production but also saves natural resources [12,13]. The quality of thermally activated RC is connected to being subjected to the highest treatment temperature, which is usually between 300 °C and 900 °C [14]. Shui et al. discovered the best compressive strength at 28 d at 800 °C, but Bogas et al. selected 650 °C as the optimal thermal- activation temperature, while Xu et al. identified 750 °C as the optimal thermal-activation temperature [10,14–18]. The difference in thermal-activation temperature may be due to the different precursor sources of RC resulting in complex composition and different mechanical properties and durability performance [19,20]. RC has a large specific surface area and high total porosity, which leads to a high water requirement for rehydration, and thus the chloride migration coefficient of RC concrete is higher than that of OPC concrete [21,22]. The water–cement ratio and porosity have a significant effect on the compressive strength of RC, so it is necessary to find suitable methods to modify the properties of RC pastes or RC concretes [20,23].

Superplasticizers are additives that minimize water demand per unit while improving building performance without increasing the volume of entrained air [24]. Its composition features include surface-active substances with hydrophilic ion groups on the carbon chain [25]. The first commonly employed superplasticizers were lignosulfonates, although their water reduction was limited and had a considerable influence on the flowability of concrete [26]. With the gradual enrichment of the types of concrete, the types of superplasticizers have also increased, and naphthalene superplasticizers and polycarboxylic acid superplasticizers are beginning to be widely used in concrete [27,28]. Although there have been many types of superplasticizers, which are all based on OPC, the component morphology of RC and OPC is distinct. As a significant component of modified RC, there is a lack of superplasticizers specialized to RC, hence it is vital to explore the potential superplasticizers for RC [29].

This review aims to provide an overview of the potential superplasticizers that can be used in thermally activated RC. It explores the mechanisms by which these superplasticizers interact with the cementitious materials and the effect of different types of superplasticizers on the properties of RC. The review also discusses the challenges and opportunities associated with the use of superplasticizers in RC, providing insights into the future directions of research in this field. By understanding the potential of superplasticizers to enhance the properties of RC, researchers and industry professionals can develop more sustainable and efficient superplasticizers for RC, contributing to excellent mechanical properties, durability, and environmental protection performance. RC is an important development direction for future building materials [30].

2. Classification and Mechanisms of Superplasticizers

Superplasticizers are mainly divided into organic and inorganic superplasticizers. Organic superplasticizers contain organic polymer compounds as their key components, which may be separated into amine sulfonate superplasticizers, water-soluble resin-type superplasticizers, naphthalene sulfonate formaldehyde condensates, aliphatic superplasticizers, polyacrylate-type superplasticizers, and polycarboxylate superplasticizers, depending on their composition. Amine sulfonate superplasticizers can achieve a water reduction rate of more than 25%, which can effectively enhance the workability and durability of concrete, but their production cost is high, and their molecular weight is too large or too small, easily leading to seepage, segregation, or sloughing of concrete [31,32]. As a typical type of water-soluble resin-based superplasticizers, melamine formaldehyde superplasticizers not only have a high water reduction rate but also maintain a good slump in addition to reducing the porosity, thereby improving the permeability and durability of concrete [33–35]. Melamine formaldehyde superplasticizers may have compatibility concerns with specific cement types and need to be tested to establish the optimum cement type and dosage [36]. Naphthalene series superplasticizers include highly efficient water-reducing agents and not only have obvious water-reducing effects but are also inexpensive and have a relatively mature process; however, the slump loss is large [37]. Aliphatic superplasticizers have a wide source of raw materials for production, fast growth of concrete strength, and show no crystallization and precipitation in winter, but because the raw materials contain flammable and explosive chemicals, these superplasticizers are dangerous to produce [38–40]. Polyacrylate-type superplasticizers have high-efficiency water-reduction properties, improve concrete microstructure, and are more compatible with different types of cement, but they have problems with compatibility with other chemical admixtures [41]. Plank [42] classified polycarboxylate polymer superplasticizers (PCEs) according to different chemical structures into methyl methacrylate/acrylate emulsion type PCE, acrylate emulsion type PCE, amide/imidazole type PCE, and amphoteric type PCE. PCEs have good-water reduction properties, good dispersion to the cement matrix, and good compatibility with other admixtures, but have significant temperature sensitivity and poor compatibility with diverse varieties of cement [43]. Inorganic superplasticizers contain inorganic salts as the major components, including silicate, phosphate, borate, etc. Inorganic superplasticizers are used to increase the fluidity of concrete while reducing water consumption by modifying the composition and structure of cement and reducing its degree of gelatinization. Inorganic water-reducing agents are usually derived from nature or industrial by-products, which are environmentally friendly, but the water-reducing effect may not be as good as that of organic water-reducing agents; therefore, they must be used in conjunction with organic water-reducing agents and thus are not used in mainstream operations [44]. Table 1 outlines the mechanisms of the involvement of organic superplasticizers in OPC cementitious materials and the current status of their application.

Table 1. Classification and corresponding properties of different types of superplasticizers.

Type	Characteristic	Mechanisms	Dosage	Application	Reference
Polycarboxylate superplasticizers (PCEs)	This type of superplasticizer is currently the most advanced in technology and has the best application prospects. It has characteristics such as low dosage, high water-reducing efficiency, low slump loss, significant enhancing effect, and is a green and environmentally friendly high-efficiency superplasticizer.	PCEs are adsorbed onto the surface of cement particles through their negatively charged anchoring groups (such as carboxyl groups) to form a thicker adsorption layer, generating electrostatic repulsion and steric hindrance, thereby enhancing the flowability of the cement slurry.	The amount added is low, with a high water-reduction efficiency, typically at 0.5–2.0% of the total weight of the cementitious material.	Suitable for preparing high-durability, high-fluidity, high-slump-retention, and high-strength concrete.	[25,45–47]
Naphthalene sulfonate formaldehyde condensates	A higher water-reducing efficiency (15–25%) can be achieved, no air-entrainment is induced, there is minimal impact on setting time, and it exhibits good compatibility with cement, as well as various other admixtures. Furthermore, it is relatively cost-effective.	By forming complexes with calcium ions on the surface of cement particles through its sulfonate groups, the flowability of the cement slurry is increased.	The amount added is typically 0.2–2.0% of the total weight of the cementitious material, commonly used at 0.2–0.5%.	Applicable for prestressed concrete engineering; can enhance the early strength and later strength of concrete.	[48,49]

Table 1. *Cont.*

Type	Characteristic	Mechanisms	Dosage	Application	Reference
Amine sulfonate superplasticizers	The molecule has a complex structure, containing a large number of hydrophilic functional groups such as sulfonate, amino, hydroxyl, etc., and has a very good water-reducing effect and improves the durability of concrete.	Because its amino and sulfonate groups interact with cement particles, the dispersion and flowability of the cement slurry are improved, while the cement dosage and water requirement are reduced.	The amount added is relatively high, typically at 1.0–3.0% of the total weight of the cementitious material.	Suitable for improving the durability of concrete.	[50,51]
Aliphatic superplasticizers	The strengthening effect on concrete is obvious, with minimal slump loss.	Having longer carbon chains, it can form a protective film on the surface of cement particles, reducing inter-particle friction, thus improving flowability and reducing water requirements.	The amount added typically ranged from 1.5 to 2.0%.	Applicable for situations where the reinforcement effect on concrete is significant, and the slump loss is minimal.	[52]
Melamine formaldehyde superplasticizers	The appearance is a white powder, soluble in water, with good dispersibility for powdery materials, a high water-reduction efficiency, and good fluidity and self-healing properties.	By its melamine resin structure interacting with the surface of cement particles, the flowability of the cement slurry is improved, and the cement dosage is reduced.	The amount added varies depending on the product, but is typically 0.5–1.5% of the total weight of the cementitious material.	Applicable for improving poor workability of concrete caused by poor aggregate quality.	[53,54]
Polyacrylate-type superplasticizers	This type is not only highly efficient in reducing water and improving the concrete structure but can also control the slump loss, is compatible with a variety of cement types, works at a low dosage, and still maintains high mobility.	Increasing the dispersion of cement particles is mainly due to improving the spatial exclusion between the particles and the polyacrylate-type superplasticizers' air-entraining isolation "ball" effect.	The water decrease rate is as high as 21.3% at a dosage of 0.35% of the total weight of the cementitious material.	Can be applied to many kinds of cement-based concrete.	[55–62]

The acting mechanism of superplasticizers is mainly to adsorb on the surface of cement particles and change the charge characteristics of cement particles, thereby improving the dispersibility of cement particles, reducing the friction resistance between cement particles, accelerating the rate of cement hydration reaction, and improving the fluidity of cement paste and concrete [50,63]. Based on the advantages and disadvantages of the water-reducing agents listed above and after a detailed comparison of the differences between OPC and RC in Section 3, several superplasticizers potentially suitable for RC are listed in Section 4.

3. Comparison of Physical and Chemical Properties between RC and OPC

OPC is mainly composed of calcium silicates (C_3S and C_2S) and calcium aluminates (C_3A, C_4AF). The main products of its hydration are C–S–H gel and $Ca(OH)_2$ [64]. These hydration products will undergo dehydration and decomposition reactions under high temperatures. Compared with OPC, RC has very complex chemical compositions and usually contains a higher amount of calcium oxide and many polycrystalline dicalcium silicates, while it lacks tricalcium silicate [20]. In addition to the main hydration products in OPC, carboaluminate often occurs during the rehydration of RC [4,65]. The particle size of thermally activated RC is usually less than 150 μm, and the density ranges from 2650 to 2950 kg/m^3 [7]. As expected, RC also requires a large amount of water. The water–cement ratio to achieve standard consistency is usually higher than 0.6 [19]. Research has shown that due to the ball-bearing effect of slag, mixing slag with RC particles can improve flowable efficiency of RC paste, promote better particle distribution, and prevent particle agglomeration [66].

In addition, our previous studies [16,67] found that the particle size distribution of RC obtained at different thermal-activation temperatures was similar to that of OPC, but the specific surface area differed significantly after thermal activation and was 15–45 times that of OPC. This was mainly caused by a large amount of dehydrated amorphous phases (Figure 1), which was also the most important factor for high water demand and high initial hydration heat. From the results of isothermal calorimetry, it was found that at the initial stage of hydration, the heat flow of RC was over 10 times that of OPC, which is consistent with the high specific surface area. Except for the initial wetting heat peak, there is an additional heat peak from rehydration after about 8 h, similar to the acceleration period of OPC, which illustrates that the rehydration mechanism of RC not only includes the wetting mechanism but also the dissolution–precipitation mechanisms. The reactant causing the second heat flow peak was not the polycrystalline dicalcium silicate, but rather the dehydrated amorphous nesosilicates. With the increase in the activation temperature of RC from 550 to 750 °C, the content of the dehydrated amorphous phase decreased, the content of α'_H-C_2S increased, and the peak height of the second heat flow was consistent with the content of amorphous phases [67].

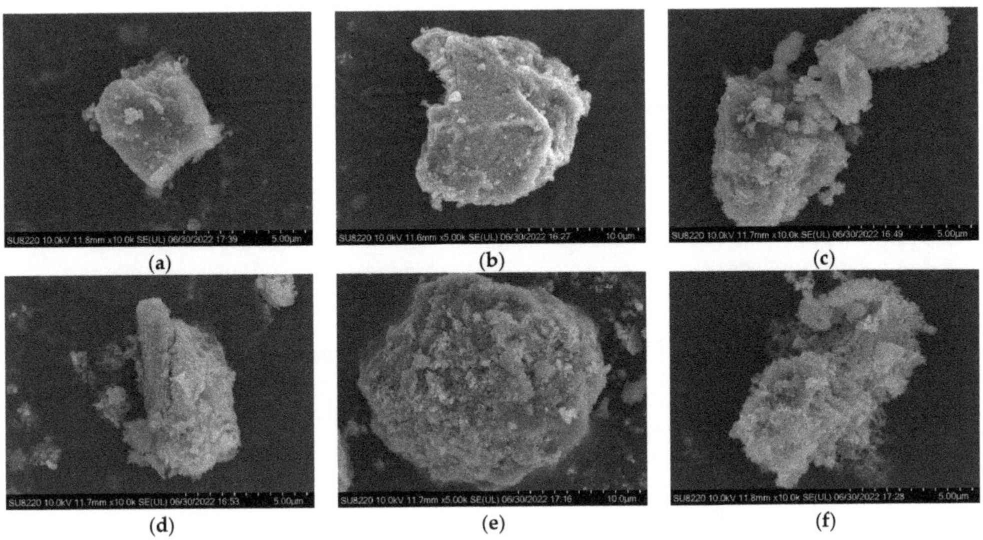

Figure 1. Scanning electron microscopy (SEM) images of untreated and recycled cement (RC) particles [16]. (**a**) Untreated; (**b**) RC450 (RC was treated at 450 °C for 3 h); (**c**) RC550; (**d**) RC650; (**e**) RC750; (**f**) RC850. Reproduced with permission from [16].

Table 2 summarizes the physical attributes and chemical compositions of OPC and RC for comparison. Due to the lack of C_3S of OPC in RC, the CH content after rehydration is lower than that after OPC hydration. The hydration mechanism of RC is similar to that of OPC, but with high f-CaO content, quick hydration, high exothermicity, short condensation time, low initial strength, and facile cracking [7,68].

Table 2. Comparison of physical properties and chemical composition of ordinary Portland cement (OPC) and RC.

	Standard	OPC	RC	Reference
Particle size (μm)	GB175-2023 [69]	<45	<75–150	[69–72]
Blaine specific surface (m^2/kg)	EN 196-6 [73]	300–450	800–4400	[20,69,74,75]
Chemical composition		C_3S, C_2S, C_4AF, C_3A, gypsum, limestone	Contains more f-CaO and polycrystalline C_2S, amorphous AFm phase, but lacks C_3S.	[20,69]
w/b (water-cement ratio)	EN 196-3 [76]	0.25–0.35	0.5–0.75	[21,66,69,77–81]
28-day compressive strength (MPa)	EN 196-1 [82] GB175-2023 [69]	32.5–62.5	3–32	[7,10,14,69,70,77]

4. Potential Superplasticizers for Thermally Activated RC

4.1. Polyacrylate-Type Superplasticizers

Polyacrylate-type superplasticizers have good dispersion and stability, and they are currently the most studied organic superplasticizers. In recent years, researchers have synthesized polyacrylate-type superplasticizers with different branch structures, different molecular weights, and different functional groups through molecular structure design, to improve their performance in the application to different types of cementitious materials. Studies have shown that the different structure layout of polyacrylate-type superplasticizers has a significant impact on the performance of cementitious materials [83,84]. Polyacrylate-type superplasticizers are more compatible with all types of cements, and therefore may also be more compatible with RC.

The theoretical basis for the design of polyacrylate superplasticizers mainly includes Deryagin–Landau–Verwey–Overbeek theory, steric hindrance and the double-layer model of dispersed particles, and degradation theory of macromolecules. Based on these theories, the elements that determine the molecular structure are adjusted to design superplasticizers with the required performance. For example, Yamada et al. [56] studied the influence of chemical structure on polyacrylate-type superplasticizers and the influence of polyacrylate-type superplasticizers on cement particle dispersion from aspects such as purity of polymers, functional groups including carboxyl, sulfonic acid, and side chains of polyoxyethylene, and polymerization degree of the main chain. They found that superplasticizers with long side chains, short main chains, and high sulfonic acid density structures have good dispersion properties, and that the presence of high-density anionic functional groups in the structure can prolong the setting time of cement paste. The research of Ferrari et al. [57] found that the molar ratio of macromonomer to carboxylic acid monomer in polyacrylate superplasticizers is a key factor influencing the water-reduction effect of polyacrylate superplasticizers, with the optimal ratio being 1:3. Increasing this ratio can increase the adsorption amount of polyacrylate superplasticizers on the surface of cement particles. Winnefeld et al. [58] studied the influence of different structures of polyacrylate-type superplasticizers on the workability and early hydration of cement mortar. The results showed that reducing the density of poly (ethylene oxide) side chains can improve workability, while the length and molecular weight of the main chain have little effect on superplasticizers.

In recent years, the characteristics of polyacrylate superplasticizers, namely the density and sequence distribution of branches in polyacrylate superplasticizers, have received widespread attention. Puertas et al. [59] used infrared spectroscopy, proton nuclear magnetic resonance, and ultraviolet–visible spectroscopy to analyze the structure of superplasticizers, identifying the main functional groups as esters, carboxylates, and ethers. At the same time, they used a rotational rheometer to measure the viscosity of superplasticizer solution and obtained a high viscosity with the superplasticizer molecules having a large number of long branches. Borget et al. [60] used proton nuclear magnetic resonance technology to characterize the graft degree, which is the number of branches, and applied ^{13}C nuclear magnetic resonance technology to characterize the sequence distribution of branches in the polyacrylate superplasticizer molecule by identifying the carbons in the esters and carboxylic acid bonds. Experimental results showed that branches are randomly distributed in the molecule structure. Liu et al. [61] used a technique of copolymerization of large monomers and small monomers, applying free radical-initiated solution polymerization to synthesize a series of water-soluble comb-shaped polymers—poly acrylic acid grafted with polyethylene glycol monomethyl ether (PAA-g-mPEG). They characterized the structure using Fourier transform infrared spectroscopy (FTIR) and nuclear magnetic resonance hydrogen spectrum (^{1}H-NMR) and studied their side chain crystallization behaviors. They used differential scanning calorimetry (DSC) to characterize and analyze the thermal properties and crystallization of different side chain lengths of mPEG. They used phase contrast microscopy and atomic force microscopy (AFM) to observe film crystal morphology, with the result indicating that the crystal morphology of comb-shaped polymers is highly branched under limited conditions. Preliminary analysis showed that mPEG chain length and its weight percentage in the copolymer had an impact on crystal morphology. Wang et al. used a copolymer P(MMA-co-MAh) with homemade methyl methacrylate and maleic anhydride as reactants and polyethylene glycol monomethyl ether (PEGME) as the grafting monomer, and synthesized the comb-shaped P(MMA-co-MAh) copolymer polyethylene glycol diester (P(MMA-co-MAh)-g-PEGME). They characterized the structure of the synthesized comb-shaped copolymer using FTIR, $^{13}C/^{1}H$ nuclear magnetic resonance (NMR) spectrum, and H,C-correlation spectroscopy (H,C-COSY) spectrum; they also analyzed the physical properties of the synthesized copolymer using thermal gravimetric analysis (TGA) and DSC [62].

4.2. Polymer Polycarboxylate Superplasticizers

RC hydration will quickly release a large amount of heat, which may not be suitable for temperature-sensitive polycarboxylate-type superplasticizers, but Xue et al. found that low-heat cements show good compatibility and were not weaker than OPC, low-heat cements for C_2S-dominated, and RC with a relatively similar chemical composition [85].

The performance of polymer polycarboxylate superplasticizer is closely related to its adsorption in the water–cement particle interface, so it is necessary to understand the conformation and self-assembly behavior of the superplasticizer molecules in a cement system. Borget et al. [60] used static and dynamic light scattering to measure the rotational radius and hydrodynamic radius of the superplasticizer molecules in a simulated cement pore solution with a certain pH and ion strength. They reported the hydrodynamic radius was around 4.4–10 nm, and within the range of $8 < pH < 12.8$ and $3 \times 10^{-2} <$ ion strength I (mol/L) $< 4 \times 10^{-1}$, the hydrodynamic radius of the superplasticizer molecules remained essentially unchanged. Using Gay and Raphael's average field model [84] (Figure 2), it can be seen that the polymer polycarboxylate superplasticizer molecule shows a worm-like conformation with a flexible main chain.

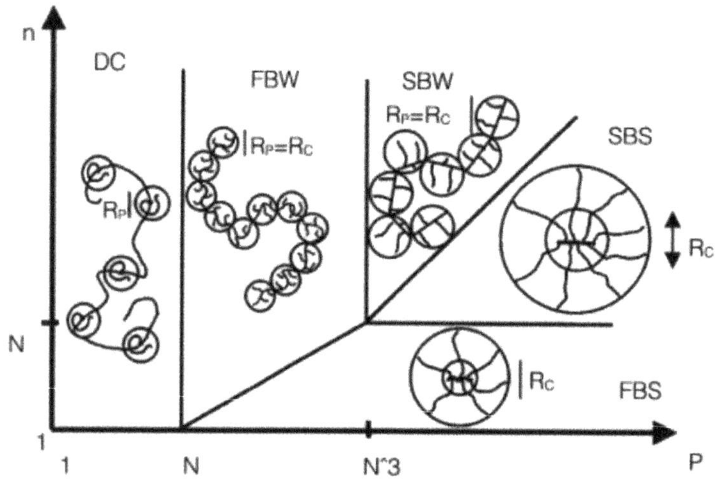

Figure 2. Conformational behavior of comb polymers with different structures in good solvents (made of n segments, each containing N monomers along the backbone and P monomers in a side-chain) (DC: Decorated Chain; FBW: Flexible Backbone Worm; SBW: Stretched Backbone Worm; SBS: Stretched Backbone Star; FBS: Flexible Backbone Star; RP: Radius of Gyration of Side Chains; RC: Radius of Gyration of Core) [86]. Reproduced with permission from [86].

With the progress of cement hydration, the superplasticizer molecules may exist in various bonding modes in the cement system. Flatt and Houst [87] divided the superplasticizer molecule systems into three categories: (1) those consumed by reactions, such as ettringite (AFt) and C–S–H formation that are encapsulated or absorbed by reaction products to form intercalation structures, forming the so-called organic–mineral phase (OMP), also including co-precipitation or formation of micelles; (2) those adsorbed on the surface of cement particles; (3) additional molecules dissolved in water when adsorbed superplasticizer molecules reach saturation (reach the saturated adsorption amount, adsorption area, etc.). The superplasticizer molecules that play a dispersing role are mainly the second type [88].

Plank et al. [86] used a method of hydrating C_3A in a water solution of superplasticizer to prepare an organic-layered Ca-Al-(PC)-LDH (layered double hydroxide) nanocomposite, and found that the superplasticizer molecules intercalated into the layered double hydroxide, and the length of the molecular side chains determined the layer spacing. Therefore, when the superplasticizer is added to the cement system, there may be a competition between the adsorption and hydration of the mineral layers [89].

Giraudeau et al. [89] prepared and investigated the formation, structure, and stability of a composite of polycarboxylic acid superplasticizer and calcium aluminate dihydrate using a coprecipitation method. Based on TRAPDOR (Transfer of Populations in Double Resonance) nuclear magnetic resonance, small-angle neutron scattering (SANS), and small-angle X-ray diffraction test results, it was proposed that the conformation of the superplasticizer molecules in the composite is not a mushroom or comb-like layout adsorbed on the surface, but rather a hemispherical chain conformation, where the superplasticizer's flexible worm-like main chain adheres to the surface of the layered composite structure, while PEO branches form a hemispherical shape that separates between layers, with a layer spacing that is around twice the radius of the hemisphere.

In a published paper in 2010, Plank et al. [90] found that the content of dissolved sulfates in cement pore solutions determines whether intercalation complexes can be formed. Superplasticizer molecules enter the interlayers through exchange with the layer-bonded OH^-, while highly negatively charged sulfates are more likely to replace superplasticizer molecules and fill the layer space to form various moisture-content monosulfide-type calcium aluminate. Therefore, cement with longer side chain superplasticizer molecules or high sulfate content (SO_4^{2-}/C_3A molar ratio of 0.75) containing easily soluble sulfates (sulfates of alkali metals or calcium) can prevent the consumption of superplasticizer by intercalation. The high molecular weight of uncapped polyacrylic acid superplasticizer is due to the presence of certain negative charges on the main chain, which adsorb onto the surfaces of positively charged cement particles (or blended particles) under the electrostatic interactions, thereby exerting a certain degree of interaction. As cement is composed of different mineral phases, different minerals have different interactions with the superplasticizer. Yoshioka et al. [91] investigated the adsorption characteristics of cement components with superplasticizers and found that C_3A and C_4AF adsorb greater amounts of superplasticizers than C_3S and C_2S, mainly due to their hydration products having a positive zeta potential. Plank and Hirsch [92] measured the zeta voltage of several products during early hydration of cement and found that calcium aluminates precipitated from solution and monosulfide-type aluminates (AFm) have positive zeta potentials, while those of hydroxide, potassium gypsum, and slaked lime have close to zero or negative values. Therefore, calcium aluminates and monosulfide-type aluminates (AFm) are more likely to adsorb superplasticizer molecules. Zingg et al. [93] measured the zeta voltage of C_3A in a variety of solutions including KOH, K_2SO_4, and simulated pore solutions, and found that the zeta potential of C_3A in simulated pore solution changed from positive to negative, indicating that C_3A is prone to adsorbing negative charges. At this time, the solution undergoes significant flocculation and precipitation, but adding 1 wt.% of superplasticizer does not result in precipitation, which further proves the interaction between C_3A and superplasticizer.

Kawa et al. [94] used In-lens field-emission scanning electron microscopy (FESEM), energy-dispersive X-ray spectroscopy (EDS), and Auger electron spectroscopy (AES) to analyze the surface status and element concentration distribution of cement particles after adding superplasticizer. They confirmed that the superplasticizer forms a three-dimensional network-like adsorption layer on the surface of cement particles. Due to the complexity of the surface structure of cement, it is difficult to directly characterize the adsorption layer thickness of superplasticizer molecules on the surface of cement particles. Greczynski and Hultman [95] used Ar^+ etching-assisted X-ray photoelectron spectroscopy technology to measure the bonding energy and energy spectrum intensity of carbon at different etching depths before and after the adsorption of superplasticizer molecules on gypsum surfaces. The adsorption thickness of polycarboxylic acid superplasticizer molecules on gypsum particles was found to be 7.5 nm. Kauppi et al. [96] used inactive MgO planes as an inert template and characterized the range of steric repulsion using spherical MgO colloid probes, thereby estimating the characteristic thickness of superplasticizer molecules to be between 1.5–5 nm.

Using the AFM method, Houst et al. [97] found that the relationship between the adsorbed layer thickness (L_{AFM}) and the hydrodynamic radius (Rh) of the superplasticizer molecule in solution was approximately linear, and the adsorbed layer thickness was approximately 30% to 50% of the hydrodynamic radius. They inferred that the adsorption conformation of the superplasticizer molecule was a relatively extended conformation of the main chain, with negatively charged groups on the main chain anchored to the cement particle surface, while the solvent-swollen side chains were extended in water.

4.3. Sulfate-Based Superplasticizers

Sulfate-based superplasticizers, which are also extensively studied organic superplasticizers, have been investigated by modifying their molecular structure, such as introducing different functional groups and changing the molecular chain length. The appropriate molecular structure of a sulfate-based superplasticizer can improve the fluidity and mechanical properties of specified cementitious materials.

4.4. Inorganic Superplasticizers

Inorganic superplasticizers have advantages of environmental protection and low-cost advantages and are an important direction for research on superplasticizers for thermal-activated RC. In a sense, silica fumes can be used as an inorganic superplasticizer, mainly by changing the surface charge properties of cement particles to improve their dispersion [98]. In the future, the influence of silica fumes on thermally activated RC can be studied by changing its composition and structure. Phosphate-based superplasticizers and borate-based superplasticizers are also important directions for research. It has been found that phosphate-based superplasticizers and borate-based superplasticizers can significantly improve the fluidity and mechanical properties of thermally activated RC and are expected to become potential superplasticizers for thermally activated RC [99,100].

5. Conclusions

Superplasticizers are chemical additives that can significantly improve the fluidity of concrete. They work by reducing the surface tension between cement particles and water, thereby reducing the amount of water required to improve the workability of concrete. The potential superplasticizers for RC have been reviewed, discussed, and compared in this paper. Because of the different physical and chemical properties between RC and ordinary cement, the superplasticizers suitable for RC should differ from those of OPC, and the mechanisms might also be different. The following conclusions can be drawn.

- The differences between OPC and RC include:
    (1) Chemical composition: RC contains more CaO than OPC and lacks C_3S.
    (2) Specific surface area: The specific surface area of RC is larger than that of OPC.
    (3) Hydration rate: RC releases a lot of heat in the early stage of hydration, and the hydration rate of RC is faster than that of OPC.

    These differences lead to higher water demand of RC than OPC.
- The different mechanisms of superplasticizers when applied to ordinary cement and RC include:
    (1) Surface property disparity: More rougher surfaces and impurities may be contained in RC particles, which necessitate the use of superplasticizers with stronger dispersive capabilities to enhance their workability.
    (2) Chemical reactivity: The potential presence of any residual chemical additives or aging products in RC necessitates that the superplasticizer possesses chemical reactivity that is compatible with these components to prevent performance degradation caused by chemical reactions.
    (3) Adsorption behavior: The adsorption behavior of the superplasticizer can be altered by the different mineral components in RC, which affects its water-reducing effectiveness and stability.
- Characteristics of superplasticizers that may be suitable for RC include:
    (1) With high-efficiency dispersing capabilities, dispersion effects at the molecular level can be optimized through the design of long-chain structures and side chains; this method is suitable for addressing the irregular and diverse particle size distribution in RC and can effectively cope with any potential impurities.
    (2) The ability to adapt to complex cementitious matrices can be enhanced by altering the molecular structure, such as the introduction of functional groups and the adjustment of molecular chain lengths.
    (3) The performance of polyacrylate superplasticizers in thermally activated RC could be significantly impacted by their molecular structure. By designing and adjusting the molecular structure, the performance of thermally activated RC-based materials can be improved by altering the surface charge properties.
    (4) Temperature sensitivity of superplasticizers is another important factor. The initial hydration heat release of RC is large and can easily have an impact on the

high temperature sensitivity of the superplasticizers, such as polycarboxylate superplasticizers. They can be considered when supplemented together with plasticizers based on lignosulfonates.

- Future research and development directions:
    (1) The impact of different components in RC from different sources on the adsorption of superplasticizers should be investigated. This includes assessing the surface charge characteristics of various mineral phases and how they are affected by the adsorption and dispersing performance of superplasticizers.
    (2) An in-depth study of the adsorption and self-assembly behavior of polyacrylate-based superplasticizers within the RC system should be conducted to better understand their mechanisms of action in the dispersion of RC particles. This involves exploring the structure and function of the superplasticizer molecular layer on the surface of RC particles.
    (3) Compounding technology can be utilized to blend two or more high-efficiency superplasticizers in specific proportions, thereby altering some of their individual properties while internally coordinating to produce a synergistic effect for RC.
    (4) Furthermore, further systematical research is needed to investigate the adsorption and lubrication behaviors of different types of superplasticizers and their possible participation in RC rehydration in RC pastes, in order to gain a deeper understanding of their mechanisms of action in RC paste.

Author Contributions: Conceptualization, R.H., L.X. and J.W.; methodology, R.H., L.X. and J.W.; investigation, R.H., L.X. and Z.X.; resources, R.H. and L.X.; data curation, R.H., L.X. and J.W.; writing—original draft preparation, R.H., L.X., Z.X., Q.Z. and J.W.; writing—review and editing, R.H. L.X., Q.Z. and J.W.; visualization, R.H., L.X. and Z.X.; supervision, J.W.; project administration, J.W.; funding acquisition, J.W. All authors have read and agreed to the published version of the manuscript.

Funding: This research was funded by National Natural Science Foundation of China grant number 52038004, 52008232.

Institutional Review Board Statement: Not applicable.

Informed Consent Statement: Not applicable.

Data Availability Statement: Dataset available on request from the authors.

Conflicts of Interest: The authors declare no conflicts of interest.

References

1. China Cement Association. Cement Industry Analysis Report 2023. Available online: http://lwzb.stats.gov.cn/pub/lwzb/fbjd/202405/W020240527578179123377.pdf (accessed on 24 May 2024).
2. Ali, N.; Abbas, J.; Anwer, M.; Alwi, S.K.K.; Anjum, M.N.; Author, C.; Jaffar, A. The greenhouse gas emissions produced by cement production and its impact on environment: A review of global cement processing. *Int. J. Res. Sci.* **2015**, *2*, 488–500.
3. CBS News. Cement Industry Accounts for about 8% of CO_2 Emissions. One Startup Seeks to Change That. Available online: https://www.cbsnews.com/news/cement-industry-co2-emissions-climate-change-brimstone/ (accessed on 4 March 2024).
4. Wang, J.J.; Wang, Y.L.; Yu, J.; Xu, L.; Li, M.L.; Cheng, J.H.; Li, Z. Effects of sodium sulfate and potassium sulfate on the properties of calcium sulfoaluminate (CSA) cement based grouting materials. *Constr. Build. Mater.* **2022**, *353*, 129045. [CrossRef]
5. Liu, X.Y.; Liu, L.; Lyu, K.; Li, T.Y.; Zhao, P.Z.; Liu, R.D.; Zuo, J.Q.; Fu, F.; Shah, S.P. Enhanced early hydration and mechanical properties of cement-based materials with recycled concrete powder modified by nano-silica. *J. Build. Eng.* **2022**, *50*, 104175. [CrossRef]
6. El-Hawary, M.; Al-Sulily, A. Internal curing of recycled aggregates concrete. *J. Clean. Prod.* **2020**, *275*, 122911. [CrossRef]
7. Xu, L.; Wang, J.; Li, K.; Lin, S.; Li, M.; Hao, T.; Ling, Z.; Xiang, D.; Wang, T. A systematic review of factors affecting properties of thermal-activated recycled cement. *Resour. Conserv. Recycl.* **2022**, *185*, 106432. [CrossRef]
8. Crook, D.N.; Murray, M.J. Regain of strength after firing of concrete. *Mag. Concr. Res.* **1970**, *22*, 149–154. [CrossRef]
9. Alonso, C.; Fernandez, L. Dehydration and rehydration processes of cement paste exposed to high temperature environments. *J. Mater. Sci.* **2004**, *39*, 3015–3024. [CrossRef]
10. Bogas, J.A.; Carriço, A.; Pereira, M.F.C. Mechanical characterization of thermal activated low-carbon recycled cement mortars. *J. Clean. Prod.* **2019**, *218*, 377–389. [CrossRef]

11. Sousa, V.; Bogas, J.A. Comparison of energy consumption and carbon emissions from clinker and recycled cement production. *J. Clean. Prod.* **2021**, *306*, 127277. [CrossRef]
12. Xiao, J.Z.; Ma, Z.M.; Sui, T.B.; Akbarnezhad, A.; Duan, Z.H. Mechanical properties of concrete mixed with recycled powder produced from construction and demolition waste. *J. Clean. Prod.* **2018**, *188*, 720–731. [CrossRef]
13. Zheng, Y.X.; Xi, X.Y.; Liu, H.; Du, C.W.; Lu, H.B. A review: Enhanced performance of recycled cement and CO_2 emission reduction effects through thermal activation and nanosilica incorporation. *Constr. Build. Mater.* **2024**, *422*, 135763. [CrossRef]
14. Shui, Z.; Xuan, D.; Chen, W.; Yu, R.; Zhang, R. Cementitious characteristics of hydrated cement paste subjected to various dehydration temperatures. *Constr. Build. Mater.* **2009**, *23*, 531–537. [CrossRef]
15. Xu, L.; Wang, J.J.; Li, K.F.; Li, M.L.; Lin, S.Y.; Hao, T.Y.; Wang, T.Y.; Guo, Y.P.; Ling, Z. Investigations on the rehydration of recycled blended SCMs cement. *Cem. Concr. Res.* **2023**, *163*, 107036. [CrossRef]
16. Xu, L.; Wang, J.J.; Li, K.F.; Hao, T.Y.; Li, Z.; Li, L.; Ran, B.; Du, H. New insights on dehydration at elevated temperature and rehydration of GGBS blended cement. *Cem. Concr. Compos.* **2023**, *139*, 105068. [CrossRef]
17. Xu, L.; Hu, X.C.; Yang, Q.R.; Ran, B.; Li, K.F.; Wang, J.J.; Zhang, X. Insight into multi-ionic adsorption behavior of recycled cement paste exposed to chloride solutions. *Constr. Build. Mater.* **2024**, *426*, 136142. [CrossRef]
18. Xu, L.; Wang, J.J.; Hu, X.C.; Ran, B.; Huang, R.; Tang, H.Y.; Li, Z.; Li, B.W.; Wu, S.H. Modification of recycled cement with phosphogypsum and ground granulated blast furnace slag. *Constr. Build. Mater.* **2024**, *426*, 136241. [CrossRef]
19. Wang, J.J.; Xu, L.; Li, M.L.; Wang, Y.L.; He, H.; Xiang, D.; Li, K.F.; Hao, T.Y. Investigations on factors influencing physical properties of recycled cement and the related carbon emissions and energy consumptions. *J. Clean. Prod.* **2023**, *414*, 137715. [CrossRef]
20. Wang, J.; Xu, L.; Li, M.L.; He, H.; Wang, Y.; Xiang, D.; Lin, S.Y.; Zhong, Y.; Zhao, H.Y. Effect of pre-carbonation on the properties of cement paste subjected to high temperatures. *J. Build. Eng.* **2022**, *51*, 104337. [CrossRef]
21. Carriço, A.; Real, S.; Bogas, J.A. Durability performance of thermoactivated recycled cement concrete. *Cem. Concr. Compos.* **2021**, *124*, 104270. [CrossRef]
22. Xie, J.H.; Wang, J.J.; Li, M.L.; Xu, L.; Xiang, D.; Wang, Y.L.; He, H.; Zhu, Y.; Zhao, J.X. Estimation of chloride diffusion coefficient from water permeability test of cementitious materials. *Constr. Build. Mater.* **2022**, *340*, 127816. [CrossRef]
23. Real, S.; Bogas, J.A.; Carriço, A.; Hu, S. Mechanical Characterisation and Shrinkage of Thermoactivated Recycled Cement Concrete. *Appl. Sci.* **2021**, *11*, 2454. [CrossRef]
24. Liu, J.; Yu, C.; Shu, X.; Ran, Q.; Yang, Y. Recent advance of chemical admixtures in concrete. *Cem. Concr. Res.* **2019**, *124*, 105834. [CrossRef]
25. Xia, Y.; Shi, W.; Xiang, S.; Yang, X.; Yuan, M.; Zhou, H.; Yu, H.; Zheng, T.; Zhang, J.; Jiang, Z.; et al. Synthesis and Modification of Polycarboxylate Superplasticizers—A Review. *Materials* **2024**, *17*, 1092. [CrossRef]
26. Zhou, B.W.; Ha, C.Y.; Mo, J.Q.; Deng, L.L.; Shen, M.M. Process of Researches and Applications on Lignosulfonate Surfactants. *Polym. Bull.* **2013**, *5*, 76–82. [CrossRef]
27. Jiang, Y.M. Synthetic Naphthalene Superplasticizer and Performance Study. Master's Thesis, Chongqing Normal University, Chongqing, China, 2017. Available online: https://kns.cnki.net/kcms2/article/abstract?v=sxrP1m9hSI9aIt1D1R2HaQdbTAIziEFPGNDR0xn2eHOPc_07CowB8EDchZgR8Hgu7rG6naA7x_XFuhPY-saflx6vqMhJ_tl0vrYygkvcP7z8zL851-ucIfRU8frgBjcKsFWDEbOreoTWwo3nKcquTFFjvimVixa274L1A9u9ENc7HHhr4IvyHxs-lUHDcwhrxFN31jNPSw=&uniplatform=NZKPT&language=CHS (accessed on 1 May 2017).
28. Li, X.G.; Wang, L.Y. Research Progress of Polycarboxylate Superplasticizer. *Res. Appl. Build. Mater.* **2021**, *3*, 24–26+12. [CrossRef]
29. Wu, H.X.; Gao, J.M.; Liu, C.; Luo, X.; Chen, G.F. Combine use of 100% thermoactivated recycled cement and recycled aggregate for fully recycled mortar: Properties evaluation and modification. *J. Clean. Prod.* **2024**, *450*, 141841. [CrossRef]
30. Wang, J.J.; Liu, E. The relationship between steady-state chloride diffusion and migration coefficients in cementitious materials. *Mag. Concr. Res.* **2020**, *72*, 1016–1026. [CrossRef]
31. Li, Q.; Zhao, M.Z.; Li, Q.Z. Synthesis and behavior analysis of amino-sulfonic acid-based high performance water-reducer. *Concrete* **2001**, *11*, 25–28.
32. Zhu, B.S.; Qiao, M.; Wu, J.Z.; Ran, Q.P. The factors affecting the properties of the aminosulfonate superplasticizer and the exploration of structural modification. *New Build. Mater.* **2019**, *46*, 5–8+18. [CrossRef]
33. Cao, S.Y.; Dong, T.N.; Cao, Y.C.; Lei, J.B.; Zhang, S.Y.; Hang, Z.S.; Huang, Y. Monodisperse sulfonated melamine formaldehyde resin microspheres: Synthesis by dispersion polymerization and light-diffusing performance characterization. *AIP Adv.* **2024**, *14*, 035325. [CrossRef]
34. Pei, M.S.; Wang, D.J.; Hu, X.B.; Xu, D.F. Synthesis of sodium sulfanilate-phenol-formaldehyde condensate and its application as a superplasticizer in concrete. *Cem. Concr. Res.* **2000**, *30*, 1841–1845. [CrossRef]
35. Yilmaz, V.T.; Glasser, F.P. Influence of sulphonated melamine formaldehyde superplasticizer on cement hydration and microstructure. *Adv. Cem. Res.* **1989**, *2*, 111–119. [CrossRef]
36. Yan, R.F.; Yin, S.H.; Zhang, H.S.; Wang, L.M.; Chen, D.P. Effect of superplasticizer on the setting behaviors and mechanical properties of tailings-waste rock cemented paste backfills. *Case Stud. Constr. Mater.* **2022**, *18*, e01714. [CrossRef]
37. Jiang, Y.M.; Liu, J.T.; Li, R. Research Progress of Naphthalene Series Highly Efficient Water Reducing Agent. *Chem. World* **2017**, *58*, 124–128. [CrossRef]

38. Han, Y.L. Comparative analysis of batch and continuous preparation processes for synthetic aliphatic water reducers. *Process Manag.* **2015**, *03*, 202–214.
39. Zhou, M.S.; Qiu, X.Q.; Yang, D.J.; Wang, W.X. Synthesis and evaluation of sulphonated acetone–formaldehyde resin applied as dispersant of coal–water slurry. *Energy Convers. Manag.* **2007**, *48*, 204–209. [CrossRef]
40. Wang, X. Study on Selection and Dosage of Polycarboxylate and Aliphatic Series Superplasticizer. *J. Taiyuan Univ.* **2019**, *37*, 1–5. [CrossRef]
41. Wei, A.M.; Wang, L.J.; Niu, L.Q. Mechanism of action and research progress of polyacrylic acid high efficient water reducing agents. *Creat. Living* **2008**, 36–38.
42. Plank, J.; Sakai, E.; Miao, C.W.; Yu, C.; Hong, J.X. Chemical admixtures—Chemistry, applications and their impact on concrete microstructure and durability. *Cem. Concr. Res.* **2015**, *78 Part A*, 81–99. [CrossRef]
43. Yuan, J.T.; Liu, B.; Lu, C.F. Research Review of Water Reducing Agents for Concrete. *Sichuan Build. Mater.* **2015**, *41*, 9–10+12.
44. He, T.S.; Li, P.; Xu, Y.L.; Qian, Q.; Dan, W. Effect of Compound Use of Micro Inorganic Salt and Different Efficient Water-reducing Agents on the Performance of Concrete. *Bull. Chin. Ceram. Soc.* **2016**, *35*, 753–757. [CrossRef]
45. Feng, H.; Feng, Z.J.; Wang, W.S.; Deng, Z.L.; Zheng, B.C. Impact of polycarboxylate superplasticizers (PCEs) with novel molecular structures on fluidity, rheological behavior and adsorption properties of cement mortar. *Constr. Build. Mater.* **2021**, *292*, 123285. [CrossRef]
46. Ma, R.; Wang, Y.B.; Li, H.; Bai, Y. Progress in the polycarboxylate superplasticizer and their structure-activity relationship—A review. *Mater. Today Commun.* **2023**, *35*, 105838. [CrossRef]
47. Xiong, G.Y.; Guo, X.L. Effects and mechanism of superplasticizers and precursor proportions on the fresh properties of fly ash-slag powder based geopolymers. *Constr. Build. Mater.* **2022**, *350*, 128734. [CrossRef]
48. Ruckstuhl, S.; Suter, M.J.F.; Giger, W. Sorption and mass fluxes of sulfonated naphthalene formaldehyde condensates in aquifers. *J. Contam. Hydrol.* **2003**, *67*, 1–12. [CrossRef]
49. Huynh, L.; Feiler, A.; Jenkins, P. The effect of adsorbing naphthalene sulfonate formaldehyde condensates upon the interactions between metal oxides. *Colloids Surf. A* **2001**, *181*, 79–89. [CrossRef]
50. Fan, Z.R.; Kong, L.J.; Lu, J.T.; Wang, X.B. Mechanism study of effect of superplasticizers on the fluidity of alkali-activated materials. *Mater. Struct.* **2023**, *56*, 29. [CrossRef]
51. Ji, X.P.; Pan, T.H.; Zhao, W.H.; Liu, J.Z.; Sha, J.F.; Han, F.Y. Interaction of superplasticizers with C_3A: Understanding the superplasticizer compatibility with cement. *J. Mater. Civ. Eng.* **2023**, *35*, 04023276. [CrossRef]
52. Du, K.L. Synthesis and application of aliphatic superplasticizer by two-stage acetone process. *Earth Environ. Sci.* **2020**, *571*, 012159. [CrossRef]
53. Martin, K.; Joerg, S.; Alfred, T.; Andreas, K. Drying and curing behaviour of melamine formaldehyde resin impregnated papers. *J. Appl. Polym. Sci.* **2014**, *131*, 39860. [CrossRef]
54. Mangane, M.B.C.; Argane, R.; Trauchessec, R.; Lecomte, A.; Benzaazoua, M. Influence of superplasticizers on mechanical properties and workability of cemented paste backfill. *Miner. Eng.* **2018**, *116*, 3–14. [CrossRef]
55. Zhang, R.G.; Li, Q.; Liu, Y.; Guo, H.L.; Lei, X.J. The synthesis of polyacrylate type superplasticizer and its application. *Chem. Technol. Mark.* **2006**, *09*, 42–43+58.
56. Yamada, K.; Takahashi, T. Effects of the chemical structure on the properties of polycarboxylate-type superplasticizer. *Cem. Concr. Res.* **2000**, *30*, 197–207. [CrossRef]
57. Ferrari, G.; Cerulli, T.; Clemente, P.; Dragoni, M.; Gamba, M.; Surico, F. Influence of carboxylic acid carboxylic ester ratio of carboxylic acid ester superplasticizer on characteristics of cement mixtures. *Spec. Publ.* **2000**, *195*, 505–520.
58. Winnefeld, F.; Becker, S.; Gtz, J.P.T. Effects of the molecular architecture of comb-shaped superplasticizers on their performance in cementitious systems. *Cem. Concr. Compos.* **2007**, *29*, 251–262. [CrossRef]
59. Puertas, F.; Santos, H.; Palacios, M.; Martnez-Ramrez, S. Polycarboxylate superplasticiser admixtures: Effect on hydration, microstructure and rheological behaviour in cement pastes. *Adv. Cem. Res.* **2005**, *17*, 77–89. [CrossRef]
60. Borget, P.; Galmiche, L.; Meins, J.F.L.; Lafuma, F. Microstructural characterisation and behaviour in different salt solutions of sodium polymethacrylate-g-PEO comb copolymers. *Colloids Surf. A-Physicochem. Eng. Asp.* **2005**, *260*, 173–182. [CrossRef]
61. Liu, Z.; Wang, W.R.; Gao, Y.F. Synthesis of a water-soluble comb polymer containing mPEG side chains and the restricted crystallization behavior of their side chains. *Acta Polym. Sin.* **2006**, *1*, 26–31. [CrossRef]
62. Wang, S.H.; Yan, H.X.; Ma, X.Y.; Huang, Y.; Zhang, Q.L. Synthesis, structure and properties of comb-like copolymer P(MMA-co-MAh)-g-PEGME. *Acta Polym. Sin.* **2008**, *9*, 880–886. [CrossRef]
63. Chen, S.L.; Zhang, J.L.; Sun, S.M.; Zhong, K.H.; Shao, Q.; Xu, H.J.; Huang, H.L.; Wei, J.X. Dispersion. fluidity retention and retardation effect of polyacrylate-based ether superplasticizer nanomicelles in Portland cement. *Constr. Build. Mater.* **2021**, *290*, 123149. [CrossRef]
64. Taylor, H.F.W. *Cement Chemistry*; Thomas Telford Publishing: London, UK, 1997. [CrossRef]
65. Wang, Y.L.; Yu, J.; Wang, J.J.; Xiang, D.; Gu, H.; Cheng, J.H. Effects of sodium aluminate and quicklime on the properties of CSA grouting materials. *J. Build. Eng.* **2022**, *58*, 105060. [CrossRef]
66. Zhang, L.L.; Ji, Y.S.; Huang, G.D.; Li, J.; Hu, Y.J. Modification and enhancement of mechanical properties of dehydrated cement paste using ground granulated blast-furnace slag. *Constr. Build. Mater.* **2018**, *164*, 525–534. [CrossRef]

67. Xu, L.; Wang, J.J.; Li, K.F.; Li, Z.; Li, L. The role of $C_{12}A_7$, α'_H-C_2S and dehydrated amorphous nesosilicate in rehydration of recycled cement. In Proceedings of the 16th International Congress on the Chemistry of Cement, Bangkok, Thailand, 18–22 September 2023.
68. Kwon, E.; Ahn, J.; Cho, B.; Park, D. A study on development of recycled cement made from waste cementitious powder. *Constr. Build. Mater.* **2015**, *83*, 174–180. [CrossRef]
69. *GB175-2023*; General Portland Cement, China. China National Standardization Administration: Beijing, China, 2023.
70. Wang, J.J.; Mu, M.L.; Liu, Y.L. Recycled cement. *Constr. Build. Mater.* **2018**, *190*, 1124–1132. [CrossRef]
71. Rui, Y.; Shui, Z. Effect of dehydrated cement paste (DCP) adidition on cementitious of cement in early age. In Proceedings of the International Conference on Waste Engineering and Management, Shanghai, China, 3–15 October 2010.
72. Wu, A.X.; Li, H.; Cheng, H.Y.; Wang, Y.M.; Li, C.P.; Ruan, Z.E. Status and prospects of researches on rheology of paste backfill using unclassified-tailings (Part 1): Concepts, characteristics and models. *Chin. J. Eng.* **2020**, *42*, 803–813. [CrossRef]
73. *EN 196-6*; Methods of Testing Cement—Part 6: Determination of Fineness. European Committee for Standardization (CEN): Brussels, Belgium, 2018.
74. Bogas, J.A.; Carriço, A.; Tenza-Abril, A.J. Microstructure of thermoactivated recycled cement pastes. *Cem. Concr. Res.* **2020**, *138*, 106226. [CrossRef]
75. Guo, H.; Gao, R.; Liu, S.H.; Feng, C.H.; Qin, M.J.; Sun, G.L. Effect of ultra-low dosage graphene oxide on the properties of recycled cement-based materials. *J. Build. Eng.* **2024**, *91*, 109637. [CrossRef]
76. *EN 196-3*; Methods of Testing Cement—Part 3: Determination of Setting Times and Soundness. European Committee for Standardization (CEN): Brussels, Belgium, 2016.
77. Bogas, J.A.; Real, S.; Cruz, R.; Azevedo, B. Mechanical performance and shrinkage of compressed earth blocks stabilised with thermoactivated recycled cement. *J. Build. Eng.* **2023**, *79*, 107892. [CrossRef]
78. Bogas, J.A.; Real, S.; Carriço, A.; Abrantes, J.C.C.; Guedes, M. Hydration and phase development of recycled cement. *Cem. Concr. Compos.* **2022**, *127*, 104405. [CrossRef]
79. Yu, R.; Shui, Z.H. Influence of agglomeration of a recycled cement additive on the hydration and microstructure development of cement based materials. *Constr. Build. Mater.* **2013**, *49*, 841–851. [CrossRef]
80. Serpell, R.; Lopez, M. Properties of mortars produced with reactivated cementitious materials. *Cem. Concr. Compos.* **2015**, *64*, 16–26. [CrossRef]
81. Zhang, X.; Ding, X.Z.; Ong, C.K.; Tan, B.T.G.; Yang, J. Dielectric and electrical properties of ordinary Portland cement and slag cement in the early hydration period. *J. Mater. Sci.* **1996**, *31*, 1345–1352. [CrossRef]
82. *EN 196-1*; Methods of Testing Cement—Part 1: Determination of Strength. European Committee for Standardization (CEN): Brussels, Belgium, 2016.
83. Abile, R.; Russo, A.; Limone, C.; Montagnaro, F. Impact of the charge density on the behaviour of polycarboxylate ethers as cement dispersants. *Constr. Build. Mater.* **2018**, *180*, 477–490. [CrossRef]
84. Wen, X.D.; Feng, L.; Hu, D.Y.; Wang, K.; Zhang, Z.Y. Effect of side-chain length in polycarboxylic superplasticizer on the early-age performance of cement-based materials. *Constr. Build. Mater.* **2019**, *211*, 26–32. [CrossRef]
85. Xue, H.; Tian, Y.; Wang, Z.D. Research on Compatibility of Low-heat Cement and Water Reducing Agent. *China Concr.* **2024**, *5*, 32–35.
86. Gay, C.; Raphael, E. Comb-like polymers inside nanoscale pores. *Adv. Colloid Interface Sci.* **2001**, *94*, 229–236. [CrossRef]
87. Flatt, R.J.; Houst, Y.F. A simplified view on chemical effects perturbing the action of superplasticizers. *Cem. Concr. Res.* **2001**, *31*, 1169–1176. [CrossRef]
88. Plank, J.; Dai, Z.; Andres, P.R. Preparation and characterization of new Ca–Al–polycarboxylate layered double hydroxides. *Mater. Lett.* **2006**, *60*, 3614–3617. [CrossRef]
89. Giraudeau, C.; Lacaillerie, J.B.E.; Souguir, Z.; Nonat, A.; Flatt, R.J. Surface and intercalation chemistry of polycarboxylate copolymers in cementitious systems. *J. Am. Ceram. Soc.* **2009**, *92*, 2471–2488. [CrossRef]
90. Plank, J.; Dai, Z.M.; Keller, H.; Hossle, F.; Seidl, W. Fundamental mechanisms for polycarboxylate intercalation into C_3A hydrate phases and the role of sulfate present in cement. *Cem. Concr. Res.* **2010**, *40*, 45–57. [CrossRef]
91. Yoshioka, K.; Tazawa, E.; Kawai, K.; Enohat, T. Adsorption characteristics of superplasticizers on cement component minerals. *Cem. Concr. Res.* **2002**, *32*, 1507–1513. [CrossRef]
92. Plank, J.; Hirsch, C. Impact of zeta potential of early cement hydration phases on superplasticizer adsorption. *Cem. Concr. Res.* **2007**, *37*, 537–542. [CrossRef]
93. Zingg, A.; Winnefeld, F.; Holzer, L.; Pakusch, J.; Becker, S.; Gauckler, L.; Zingg, A.; Winnefeld, F.; Holzer, L.; Pakusch, J.; et al. Adsorption of polyelectrolytes and its influence on the rheology, zeta potential, and microstructure of various cement and hydrate phases. *J. Colloid Interface Sci.* **2008**, *323*, 301–312. [CrossRef] [PubMed]
94. Kawa, H.U.; Hanehara, S.; Rasaka, T.K.S.; Sawaki, D. Effect of admixture on hydration of cement, adsorptive behavior of admixture and fluidity and setting of fresh cement paste. *Cem. Concr. Res.* **1992**, *22*, 1115–1129. [CrossRef]
95. Greczynski, G.; Hultman, L. Towards reliable X-ray photoelectron spectroscopy: Sputter-damage effects in transition metal borides, carbides, nitrides, and oxides. *Appl. Surf. Sci.* **2021**, *542*, 148599. [CrossRef]
96. Kauppi, A.; Andersson, K.M.; Bergstriim, L. Probing the effect of superplasticizer adsorption on the surface forces using the colloidal probe AFM technique. *Cem. Concr. Res.* **2005**, *35*, 133–140. [CrossRef]

97. Houst, Y.F.; Bowen, P.; Perche, F.; Kauppi, A.; Borget, P.; Galmiche, L.; Meins, J.F.L.; Lafuma, F.; Flatt, R.J.; Schober, I.; et al. Design and function of novel superplasticizers for more durable high-performance concrete (superplast project). *Cem. Concr. Res.* **2008**, *38*, 1197–1209. [CrossRef]
98. Hamada, H.M.; Abed, F.; Katman, H.Y.B.; Humada, A.M.; Jawahery, M.S.A.; Majdi, A.; Yousif, S.T.; Thomas, B.S. Effect of silica fume on the properties of sustainable cement concrete. *J. Mater. Res. Technol.* **2023**, *24*, 8887–8908. [CrossRef]
99. Sun, T.; Shui, Z.H.; Huo, T. Rehydration performance of binary binders made with dehydrated cement paste and phosphogypsum. *Key Eng. Mater.* **2011**, *474–476*, 1238–1242. [CrossRef]
100. Zhang, L.L.; Ji, Y.S.; Li, J.; Gao, F.R.; Huang, G.D. Effect of retarders on the early hydration and mechanical properties of reactivated cementitious material. *Constr. Build. Mater.* **2019**, *212*, 192–201. [CrossRef]

materials

MDPI

Review

Advances in the Experiments of Leaching in Cement-Based Materials and Dissolution in Rocks

Lifan Zheng, Junjie Wang *, Kefei Li, Mingyu Wang, Shimeng Li and Lin Yuan

Department of Civil Engineering, Tsinghua University, Beijing 100084, China; zlf21@mails.tsinghua.edu.cn (L.Z.); likefei@tsinghua.edu.cn (K.L.); wangmy21@mails.tsinghua.edu.cn (M.W.); sm-li21@mails.tsinghua.edu.cn (S.L.); yuan21@mails.tsinghua.edu.cn (L.Y.)
* Correspondence: junjiewang@tsinghua.edu.cn

Abstract: Leaching in cement-based materials and dissolution in rocks are important problems in civil engineering. In the past century, concrete damage caused by leaching have occurred worldwide. And, rock dissolution is usually the main cause of karst rock erosions. This paper provides a review of the causes, influencing factors, and effects on engineering properties of dissolution of rocks and leaching of cement-based materials. The applied experimental methods for leaching and dissolution have been sorted out and discussed. In situ field experiments can be used to study dissolution under natural conditions, while the laboratory experiments can effectively shorten the experiment time length (by changing pH, temperature, pressure or other factors that affect the leaching or dissolution) to quickly investigate the mechanism of dissolution and leaching. Micro tests including XRD, SEM, EDS, and other testing methods can obtain the changes in material properties and microstructures under leaching and dissolution. In addition, with the advances in technologies and updated instruments, more and more new testing methods are being used. The factors affecting the leaching and dissolution include environmental factors, materials, and solvent parameters. The mechanisms and deterioration processes of leaching and dissolution varies according to the types of material and the compositions.

Keywords: leaching; rocks; cement-based materials; leaching mechanism; dissolution

1. Introduction

Cement-based materials represented by concrete have a long history of thousands of years. For example, concrete was widely used in Roman structures founded from 300 B.C. to 476 A.D., which are still relatively intact today [1]. In 1824, Portland cement, mainly composed of silicate, was patented and made concrete an important material [1,2]. In civil engineering, leaching is often used to describe the corrosion of soluble components in concrete caused by water or other media [3]. People's understanding of leaching has a long-time developing process. In the early 20th century, researchers recognized the risk of leaching and later Moskovin of the Soviet Union conducted a more systematic study [4]. Some scholars have pointed out that any concrete structure made of Portland cement must inevitably undergo the leaching of lime and lose all bonding properties within a certain period of time [5], and studies on leaching of concrete under different working conditions were conducted [6]. In the past 100 years, leaching of concrete has occurred frequently around the world, leading to an increasing emphasis on the leaching mechanism of concrete.

Dissolution in rocks refers to the process that soluble substances of rocks dissolve under the long-term action of solvents like rainwater, and the dissolved substances are lost, resulting in the formation of dissolution pores and cracks on the surface or inside of rocks, which commonly leads to karstification. Karstification is widely distributed worldwide, with a total karst area of 3.63 million kilometers in China. Among them, there are 2.03 million kilometers of exposed surface on carbonate rock, which is the largest contiguous exposed karst area in the world [7–9]. Karst and erosion not only brings many engineering problems [10] but also plays an important role in the actual projects and global carbon cycles [11–13]; in the area of oil and gas exploration in deep-buried layers, dissolution of rocks is also an urgent research issue [14–16]. During the 20th century, with the continuous increase in human engineering activities, karst disasters, as an engineering geological problem, were increasingly valued by the engineering community. In 1973, the International Association of Engineering Geosciences held an international conference in Hanover, Germany, titled "Karst Collapse and Settlement—Engineering Geological Issues Related to Soluble Rocks". This was also the first international symposium on karst issues held by the engineering community [17]. In the 1990s, Chinese scholars compiled a map of soluble rock types in China and measured the distribution area of carbonate rocks in various provinces [18]. Afterwards, scholars from various countries around the world conducted extensive research on the causes, mechanisms, impacts, and establish mathematical models for dissolution, such as the process and mechanism of carbonate rock dissolution [19–27] and evaporites [28]. Some researchers also analyze and summarize karst phenomena under different types of materials or environmental conditions [29,30].

In this paper, we provide a review of experimental methods, mechanisms, and impact on engineering properties for dissolution in rocks and leaching in cement-based materials, and we analyze and compare the advantages of various methods in order to provide ideas and references for future research on experimental methods and conditions.

2. Experiment Methods for Leaching and Dissolution

2.1. Macroscopic Experiments

According to the external environment in which the experiment is conducted, macroscopic methods can be divided into two types: in situ field experiments and laboratory experiments.

2.1.1. In Situ Field Experiments

In situ field experiments are often conducted using rock tablets. The sample is cut into standard-sized thin sheets, and is placed under different types of natural conditions, and researchers calculate the changes of various properties of the tablets for a period of time to characterize the dissolution [31,32]. The demonstration of rock tablets can be seen in Figure 1. Due to the difference in mineral composition between standard tablets and rocks excavated from local rock formations, rocks from the test site are usually used to produce tablets to obtain more accurate results [33]. But, the preparing conditions of the samples can also affect the results of the experiment [34]. Smooth tablets have lower dissolution rates than rough ones under the same conditions, and the lithology and altitude of the experiment area also have impacts. The totally weathered area of the tablets is larger than the surface area, as the dissolution occurs not only on the surface but also in the voids and cavities of the limestone tablet [35].

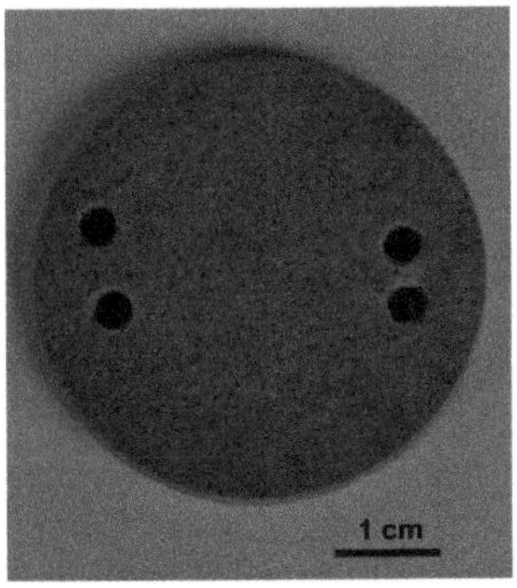

Figure 1. Limestone tablet [32].

This method can be used in various karst experiments. For instance, Li. et al. cut the carbonate rocks in a certain area of Guangxi into thin sections and buried them in forest, shrubland, and farmland environments for a period of six months for natural dissolution, and they quantitatively studied the impact of drought of different durations on karst in three environments [36]. Yan Wei et al. buried the dolomite samples in the soil layers of dry land, paddy fields, and other areas for one year, obtained the dissolution rates of the samples under different land-use conditions, and analyzed the corresponding carbon sink intensity [37]. Wu Jianqiang et al. used the in situ field method to study the relationship between the dissolution rate of carbonate rocks and the content of calcium oxide in the southern Jiangsu region [38].

2.1.2. Laboratory Simulation Experiment

Leaching and dissolution are relatively long processes, which usually take several years or decades under natural conditions to obtain obvious changes. Accelerated experiments are methods that accelerate leaching and dissolution by strengthening one or several influencing factors, which can obtain results in a short period of time to explore the mechanism. Experiments are usually conducted in groups, and some use equipment that can provide coupling experimental conditions [39,40]. A group of rock specimens used in a dissolution experiment is shown in Figure 2 as an example. Acidic reagents, deionized water, changing temperature, controlled solution flow rates, and other methods are commonly used, which can accelerate the process in a certain way.

Figure 2. Rock salt specimens used in property and dissolution experiments [41].

The recent research conclusions and leaching or dissolution methods used in some literature can be seen in Table 1.

Table 1. Some leaching and dissolution methods and research conclusions.

Reinforcing Factor	Research Conclusion
High temperature and high pressure	Simulated the leaching effect of dolomite reservoirs under burial conditions and explored the dissolution laws of different types of dolomites [42].
pH and high temperature	pH affects the selective dissolution of feldspar and calcite, and the existing species of silicon and calcium in solution varies depending on the pH value [43].
CO_2 concentration	The average total dissolution rate of rock samples in an open environment is 1.43 times that of a semi-open one and 2.70 times that of a closed area. The CO_2 exchange reaction is the main influencing factor [44].

Table 1. *Cont.*

Reinforcing Factor	Research Conclusion
High temperature, high pressure, and pH	Under mixed deposition conditions, the main dissolved minerals are different at different temperature ranges [45].
Temperature and pressure	The development of buried dissolution pores are controlled by rock physical properties, acidic fluid concentration, and temperature [46].
High temperature and high pressure	The injection of acidic fluids into deep open systems caused by fracture opening or activation can lead to partial dissolution of dolomites, and the scale is related to fluid flux [47].
Deionized water	Simulations were conducted to investigate the contact leaching characteristics of three types of materials, namely cement slurry, cement fly ash, and cement slag, under the action of deionized water using relatively small water-to-solid ratios and different surface areas [48].
pH	Through uniaxial compression tests of the limestone at different angles and directions, the characteristics of limestone under dissolution were studied [49].
pH	Studied the time-varying laws of three evaluation indicators, including tensile strength of layer splitting, dissolution depth, and cumulative relative dissolution of calcium ions [50].
Deionized water	Studied the influence of mineral admixtures such as fly ash and slag on the leaching performance of cement slurry in deionized water and their dosage [51].
pH	The effect of sulfuric acid action on the leaching rate of cement mortar was studied, and the curve of the change in acid consumption of the specimen with immersion time was obtained [52].
pH	The mechanism and microscopic morphology of calcium leaching in cement-based materials [53].
Temperature, pressure, and pH	Deriving the dissolution rate of sandstone at different scales using vertical scanning interferometry [54].
pH and deionized water	Ammonium nitrate solution can effectively simulate and accelerate the leaching process of dam concrete, and the driving force for accelerating leaching is the acid-base neutralization reaction between ammonium nitrate and cement hydration products to produce highly soluble calcium nitrate [55].
pH	Under stress, the solid reaction activation energy of the solid–liquid contact surface in the reaction system increases. Stress causes deformation of the rock mass, and cracks on the surface of the rock mass continue to expand and extend under stress, increasing the solid–liquid contact area and promoting the occurrence of dissolution [56].
High temperature, high pressure, and pH	Under open system high-temperature and high-pressure conditions, organic acid dissolution of feldspar components in sandy conglomerate can effectively improve reservoir porosity and permeability [57].
Water pressure and temperature	The dissolution of carbonate rocks includes chemical and physical dissolution. As the hydrodynamic pressure increases, the ratio of simultaneous increase in chemical and physical dissolution damage tends to be the same. There is a coupling relationship between chemical dissolution and physical damage [58].
pH (controlled by the dissolution of CO_2)	Changes in porosity and permeability of carbonate rocks immersed in water containing carbon dioxide were studied [59].

Table 1. *Cont.*

Reinforcing Factor	Research Conclusion
Temperature	The direct relationship between the dissolution rate of limestone and temperature was studied [60].
Temperature, high pressure, PH	Saltwater flow and carbon dioxide in carbonate rock cores promote dissolution, leading to an increase in porosity [61].
High temperature, high pressure, and pH	Lithology, fluid type, and temperature and pressure conditions can all affect dissolution. Gypsum is more soluble than limestone and dolomite. The type of fluid has almost no effect on the dissolution of gypsum rock [62].
pH	Leaching experiments and uniaxial compressive strength tests were performed to analyze the karst development characteristics and mechanism of concrete leaching [63].
Temperature and pressure	Simulate the dissolution process of carbonate rocks [64].
pH	Study the influence of chemical and mineralogical composition on dissolution of phosphate rock samples [65].
Temperature	The diffusion coefficient of concrete is positively correlated with temperature and negatively correlated with the degree of hydration. The type of cement has an impact on the diffusion coefficient, while the water–cement ratio has no significant relationship with the diffusion coefficient [66].

2.2. Microscopic Methods

It is hard to reveal the laws and mechanisms of leaching and dissolution only using macroscopic experimental methods; so, microscopic experimental methods are also needed to obtain information such as ion distribution, microscopic morphology, and identification of the reaction products. The main technical means of microscopic experiments include scanning electron microscopy (SEM), diffraction of X-rays (XRD), 3D laser scanning, energy dispersive spectroscopy (EDS), TG-DTA, ion concentration detection, etc.

The SEM technology is to utilize a finely focused electron beam in a scanning electron microscope to bombard the specimen, getting various physical information through the interaction between electrons and the specimen. When applied to non-metallic materials, platinum or other conductive coatings are usually plated on the surface of the material. In leaching and dissolution research, SEM is often used to obtain the micro morphological characteristics of the surface of the test piece, such as the difference in the surface morphology of plastic cement-based materials before and after leaching [67], the micro morphology of C–S–H gel in the concrete test piece [53], and the development types and dissolution pore characteristics of rocks [68]. The SEM equipment is shown in Figure 3. Figure 4 shows the microscopic surface morphology of a white marble under SEM.

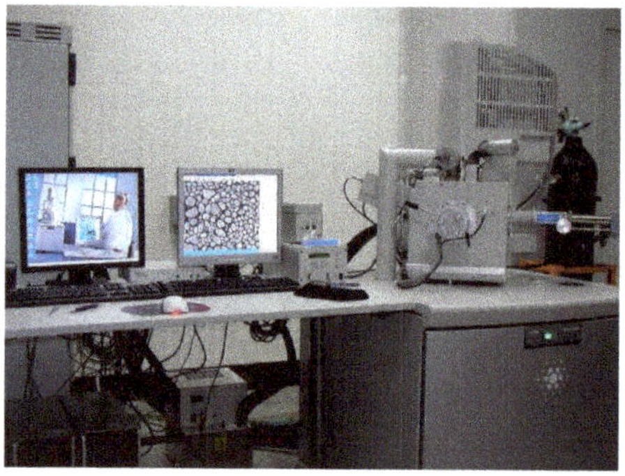

Figure 3. Scanning electron microscope.

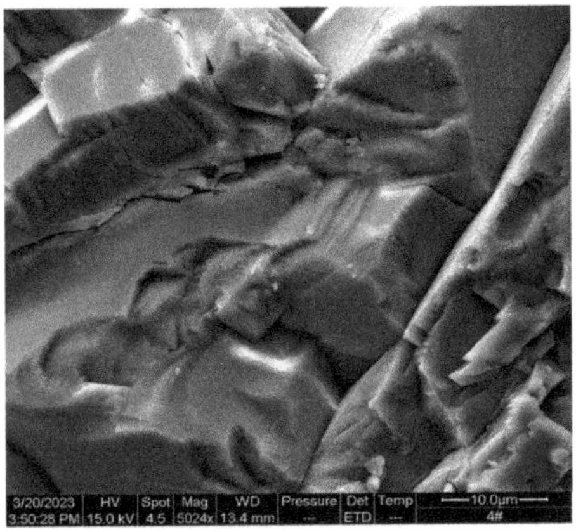

Figure 4. SEM image of white marble surface.

The XRD technology can perform phase analysis of samples, and the changes in the relative content of a certain component before and after leaching can be analyzed through X-ray diffraction patterns [69]. For example, XRD can be used to compare the differences in mineral composition of sandstone cores before and after dissolution [70], study the order degree of rocks such as dolomite [71], and judge the differences in hydration products of cement [72]. Figure 5 shows the XRD diffractograms of a marble sample after dissolution; the chemical composition can be seen from the perspective and shape of the peaks.

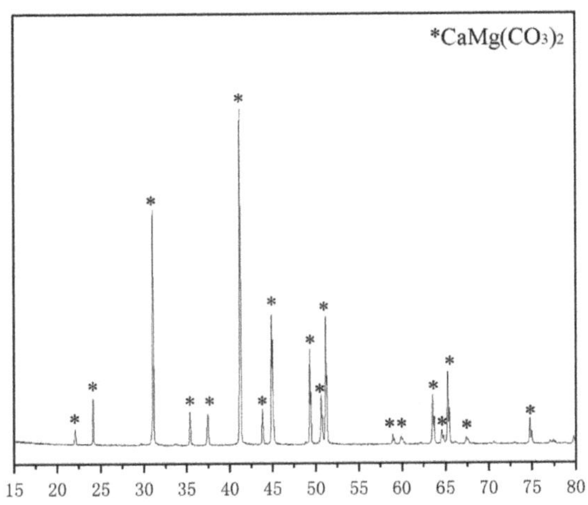

Figure 5. XRD diffractograms of a marble sample after dissolution.

Three-dimensional laser scanning uses high-speed laser and non-contact measurement to scan the target object in the area to obtain the 3D spatial point cloud dataset of the target object and convert it into a 3D spatial model in a unified coordinate state. Three-dimensional laser scanning has no contact with the specimen and is high-precision, which can accurately obtain accurate data on the surface corrosion morphology of the specimen. Three-dimensional laser scanning is widely used to provide the information about the surface morphology of exposed carbonate rocks [73–76], which can help to study the changes in porosity of rocks after dissolution [77] and can provide information about the dissolution rate and geometric evolution characteristics [78]. On this basis, 4D tomography has begun to be used to study the process of dissolution [79].

Due to the fact that various elements have their own X-ray characteristic wavelengths and the size of the characteristic wavelengths depends on the characteristic energy released during the energy level transition process, the EDS technology can utilize the different characteristic energies of X-ray photons from different elements for compositional analysis. It is often used in conjunction with other microscopic testing methods to determine the mineral composition of the sample. By combining EDS and SEM analysis methods, compared to traditional techniques such as XRD, testing objectives can be achieved at a lower cost and low mineral composition can be observed more easily [80].

The TG-DTA technology includes heating and records the quality changes of the sample during the testing process. The mass proportion of each component in the sample can be analyzed through the thermal analysis curve [81]. TG-DTA is used to record the relative content changes caused by the thermal decomposition of hydration products in concrete at a curing age of 28 days [82] and determine the hydration degree of recycled cement pastes [83].

The ion concentration detection can analyze the content of each ion in the leaching solution. For example, Mo Yunchuan [84] conducted ion composition testing on the solution leaked from various fractured limestone samples; Shutao Zhou used inductively coupled plasma mass spectrometry to determine the changes in elemental content of the solution during the leaching process [85].

The XCMT technology obtains digital images of material X-ray absorption from various angles and then obtains three-dimensional images via numerical reconstruction of a set of 2D images [86], which makes it an important technique to avoid damage to the microstructure of the sample.

In addition, nuclear magnetic resonance (NMR) is also used to observe the pore size of porous rocks [87]; digital holographic microscopy is used in the nanoscale observations of the gypsum leaching in situ [88]; electron probe microanalysis (EPMA) is used to display significant differences between dissolution and non-dissolution areas to determine the boundaries [89]; a back-scattered electron (BSE) is used to measure in situ dissolution porosity of minerals [90]; and nano indentation tests are used to quantify the embedding strength of the fracture surface [91]. Figure 6 shows the test area and obtained test results; electron microprobe analysis is used to determine the exact composition of the samples [92]. CT Scanning is used to observe the spatial distribution of pores and fractures [93].

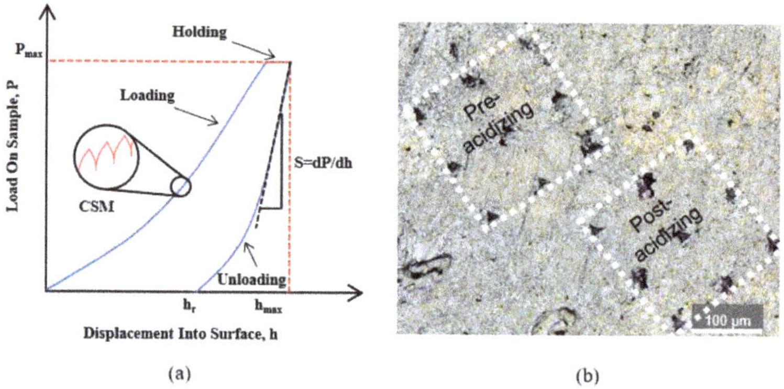

Figure 6. Nano indentation tests [91]. (**a**) Test result; (**b**) Test area

2.3. Simulation Experiment

With the improvement of numerical simulation software and computer performance, simulation experiments conducted using numerical simulation software such as COMSOL have also become more feasible and reliable. It is difficult to simulate the dissolution or leaching process using traditional in-situ or laboratory experiments due to with multiple factors or long-time spans, but simulation experiments can analyze the entire process.

Under the reasonable selection of mechanism and setting of the influencing factors, such experiments can effectively simulate the development of leaching and dissolution in real situations. An independently developed evolutionary simulation software was used to conduct simulation experiments on the relationship between crack density, precipitation, and dissolution rate in carbonate rock mass [94]; the OpenSEES was applied to establish a finite element model of concrete column components considering leaching damage and obtain the laws of the changes in compressive and flexural bearing capacity of the components [95]; COMSOL was used to conduct numerical simulation tests on the leaching of cementitious materials in various parts of a certain reservoir dam body and compare them with actual monitoring results, analyzing the areas where the dam body is more severely affected [96]; PHEEQC was used to simulate the leaching of calcium carbonate under mixed conditions such as shallow groundwater, deep circulating hot water, and soil infiltration water [97] and can also be applied for modeling the geochemical reactions in porous media [98]; and three-dimensional numerical simulations were established to conducted multiple core displacement tests on carbonate rock cores [99].

3. The Mechanism of Leaching and Dissolution

3.1. The Dissolution of Rocks

The dissolution mechanism of rocks varies depending on the differences in their main components.

The dissolution of carbonate rocks is mainly the process of carbon dioxide dissolved in water and carbonates such as $CaCO_3$ forming soluble substances such as calcium bicarbonates. Due to the fact that rocks are composed of particles of different sizes, dissolution is relatively strong in intergranular and intergranular pores, as well as in various types of fractures [100]. The dissolution of dolomite can be divided into four stages: the initial dissolution stage, secondary dissolution acceleration stage, stable dissolution rate stage, and dissolution attenuation stage; the dissolution rate of initial stage is much greater than the stable stage [101]. In addition, acidic solutions rich in hydrogen ions such as acid rain can also cause corrosion when in contact with carbonates [102,103]. The chemical reaction equation for acid-rain dissolution of carbonate rocks is as follows [104]. The meaning of the compounds can be found in Appendix A.

$$CaCO_3 + H_2SO_4 \rightarrow CO_2 + Ca^{2+} + H_2O + SO_4{}^{2-} \tag{1}$$

$$CaCO_3 + 2HNO_3 \rightarrow CO_2 + Ca^{2+} + H_2O + 2NO_3{}^{-} \tag{2}$$

Among them, Yuan Daoxian proposed a karst dynamics model to visually represent the process [105], in which the environment was divided into three phases: gas phase, liquid phase, and solid phase. Carbon dioxide enters the liquid phase and reacts with water to produce hydrogen ions. Meanwhile, calcium carbonate dissolves into calcium ions and carbonate ions, and the latter reacts with the hydrogen ions to produce soluble bicarbonate ions.

Sulfate rocks, such as gypsum and anhydrite, have different dissolution mechanisms from carbonate rocks. Research has shown that even without the participation of carbon dioxide, water still has a certain solubility in sulfate rocks [106]. Besides chemical dissolution, physical effects also cannot be ignored [102]. At the same time, impurities such as calcite or dolomite in sulfate rocks such as gypsum also affect the equilibrium of dissolution, leading to changes in solubility [107].

Silicates are often difficult to dissolve in water due to their composition, and their dissolution can be used for carbonation and storage of carbon dioxide [108].

3.2. The Leaching of Cement-Based Material

The leaching of cement-based materials mainly comes from calcium dissolution. When there is a difference between the pore solution of cement-based materials and surrounding solvents such as deionized water, the concentration gradients break the initial solid–liquid equilibrium and induce the leaching of cement-based materials [109,110]. The solid phase calcium will dissolve into the pore solution to maintain the dissolution balance of calcium ions in the pore solution and keep pH stable. However, in soft water or acidic solution, the calcium hydroxide will continue to run off under the concentration gradient of the pore solution and the substances in the external environment, which will lead to the continuous decalcification and dissolution of C–S–H gel and ettringite, leading to the gradual decrease in calcium ion concentration in the pore solution, the constant decrease in pH value, and the gradual loss of gellability of the hydration products. Ultimately, it leads to material corrosion and damage. Among them, the calcium ion concentration in the pore solution of cement-based materials in soft water is higher than that in the environment, resulting in a difference in internal and external concentration, leading to continuous outward diffusion and leaching of calcium ions in the pores, which belongs to a physical process. In acidic solutions, the hydration products of cement react chemically with hydrogen ions, causing decalcification of cement-based materials, which belongs to a chemical process [53,111,112]. A flowchart of a leaching model for Portland cement binders is shown in Figure 7.

$$Ca(OH)_2 \rightarrow Ca^{2+} + 2OH^- \tag{3}$$

$$nCaO{\cdot}2SiO_2{\cdot}nH_2O \rightarrow nCa^{2+} + H_3SiO4^- + [2n - 1]H_2O \tag{4}$$

or

$$Ca(OH)_2 + 2H^+ \rightarrow Ca^{2+} + H_2O \tag{5}$$

$$C\text{-}S\text{-}H + 2H^+ \rightarrow Ca^{2+} + H_2O + SiO_2 \tag{6}$$

$$C\text{-}A\text{-}H + 2H^+ \rightarrow Ca^{2+} + H_2O + Al_2O_3 \tag{7}$$

$$C_4AF + 2H^+ \rightarrow Ca^{2\,+} + H_2O + Al_2O_3 + Fe_2O_3 \tag{8}$$

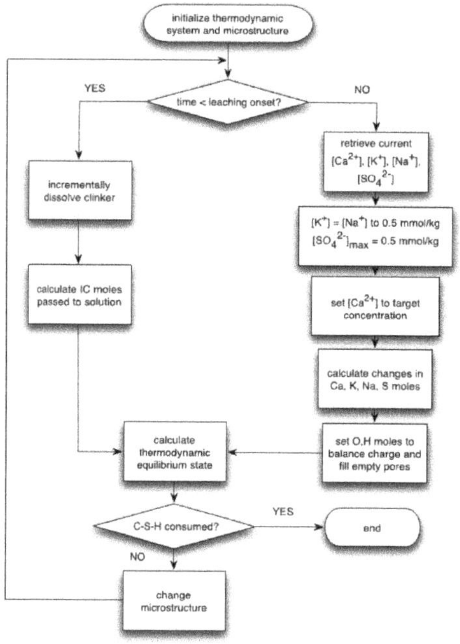

Figure 7. Flowchart of a leaching model for Portland cement binders [113].

4. The Impacts of Leaching and Dissolution on Material Properties

4.1. The Dissolution of Rocks

The impacts of dissolution on rocks are mainly reflected in the formation of the strength, dissolved pores, cracks, and surface roughness on the surface [103,114–117]. The components that can react with fluids will continuously decrease as the dissolution process progresses which leads to changes in rock mineral composition. And, the alteration in microstructure, pore characteristics, and other factors can also affect the physical and mechanical properties of rocks.

After dissolution, the porosity usually increases continuously, leading to the connection between pores in rocks [118,119], and the permeability of rocks may increase several times or even tens of times [116]. But, in some cases, the permeability of rocks may actually decrease. In a series of core flooding experiments, the calcite particles generated by dissolution blocked the pore channels in the rock core, resulting in a decrease in permeability [120]. In some cases, cracks appear in the rock. When reflected in the physical and mechanical properties of rocks, it is the increase in surface roughness and water absorption and decrease in strength [103,121]. The comparison on indicators of rocks before and after dissolution can be seen in Table 2.

Table 2. Comparison on indicators of rocks before and after dissolution.

The Type of Rocks	Engineering Property Indicators	Before Dissolution	After Dissolution	Conclusions
Dolomite Marble	Porosity	1.1%	4.7%; 7.2%; 8.2%	After dissolution, the porosity and pore size of the rock significantly increased. The development of dissolution pores and gaps occurred on the surface of the rock [122].
Dolomite Marble	Pore radius	0.05~0.32 μm	3~5 μm	
Limestone	Compressive strength	115.2 MPa	129.6 Mpa	After dynamic dissolution experiments on limestone, the compressive strength of the rock decreased by 18% and the Young's modulus reduced by 26% [123].
Limestone	Elastic modulus	26,585.5 Mpa	19,754.3 Mpa	
Mixture of dolomite, calcite, and quartz	Poisson's ratio	0.176	0.116	The rock became more easily broken under the stress after dissolution. Additionally, the Poisson's ratio also decreased by 34.09% due to salt dissolution. Therefore, salt dissolution overall reduced the rock strength of saline-lacustrine carbonate rock to some extent [116].
Mixture of dolomite, calcite, and quartz	Mineral composition	88.99% calcite, 3.17% dolomite, and 7.58% quartz	79.61% calcite, 9.68% dolomite, and 9.27% quartz	This variation in the mineral composition suggested that calcite in the carbonate samples reacted during the dissolution experiment, while the other two components had almost no reaction [117].

4.2. The Leaching of Cement-Based Materials

C–S–H gel and $Ca(OH)_2$ are important components to ensure the performance of concrete, which are closely related to the strength, density, and impermeability of cement-based material. When subjected to leaching, calcium precipitates and the microstructure changes, which leads to deterioration of the engineering properties [124,125]. The Comparison on indicators of cement-based materials before and after leaching can be seen in Table 3.

Table 3. Comparison on indicators of cement-based materials before and after leaching.

The Type of Cement-Based Materials	Engineering Property Indicators	Before Leaching	After Leaching	Conclusions
Calcium silicate hydrates	The Ca/Si atomic ratio	0.91	0.02	The leaching of cement-based materials mainly resulted in the loss of calcium, and the increase in the specific surface area of C–S–H(I) samples after leaching was mainly attributed to the modification of pore structure, which was caused by the transformation of two-dimensional structure of C–S–H(I) into C–S–H with cross-linking structure and silica gel with three-dimensional structure [53].
Calcium silicate hydrates	Specific area	<200	>275	
Fly ash concrete	Compressive strength	15 Mpa	7.5 Mpa	The compressive strength and elastic modulus of concrete with different fly ash dosages had a similar trend with leaching time, and the compressive strength continuously decreased, with the most obvious decrease in the strength of concrete without fly ash added [126].
Fly ash concrete	Elastic modulus	4500	About 3000	
Silicate concrete	Pore volume	27~30%	35~37%	There was a correlation between the dissolution of portlandite and the increase in the volume of pores larger than 200 nm [127].
Silicate concrete	Tensile strength	3.0~3.25	2.5~2.75	After leaching, all fracture property parameters rapidly decreased, including the crack initiation load,maximum load, fracture toughness, elastic modulus, tensile strength, and fracture energy [128].

5. Discussion

The main experiment methods for leaching and dissolution are mainly divided into three types: laboratory experiments, in situ field experiments, and simulation experiments. Laboratory-accelerated experiments can significantly accelerate the process by strengthening one or several factors, thereby simulating the leaching or dissolution process and studying the reaction mechanisms. In situ field experiments were mainly used for long-term exposure tests. Simulation experiments like numerical modelling tests can be used when the related parameters were obtained. Leaching and dissolution are common forms of degradation in engineering, and there has been extensive research on the mechanism and influencing factors over the past hundred years. It can be seen that the mechanism of leaching and dissolution and the conditions for the formation have not been well understood and the related knowledge is constantly improved. The methods for conducting leaching and dissolution experiments are gradually becoming diversified and standardized. However, we should also note that, currently, most experiments are only conducted under one single experimental condition, and there are few experiments that study dissolution or leaching under multiple factors. In addition, there is a lack of research on the linking relationship between the experiment results conducted in the laboratory and the natural dissolution that occurs in the field.

The strengthening factors of laboratory-accelerated experiments mainly focus on several factors such as pH, temperature, and pressure, which are also the main factors affecting the dissolution rate and are relatively easy to change in an experiment. The factors affecting dissolution and leaching can be seen in Tables 4 and 5. In addition, leaching and dissolution occurring in nature is often caused by the combined action of multiple factors. However, there are relatively few studies that involve the proportion of leaching or dissolution caused by one factor in the overall factors. If the factors selected in the experiment differ from the actual situation, it may lead to errors in the mechanism of leaching or dissolution research, which is also one of the problems that needs to be solved in the future. The influencing factors of rock dissolution and concrete leaching are slightly different, but there are many similarities.

Table 4. Factors affecting dissolution of rocks.

Type	Influence Factor
Environmental factor	Temperature
	Pressure
	Gas phase components
	Illumination
Solvent	Solvent components
	pH
Material properties	Composition
	Mineral composition
	Pore characteristics
	Surface roughness

The design of leaching and dissolution experiments can be carried out according to the following model shown in Figure 8. A leaching or dissolution experiment can be carried out using three types of methods, and the results coming from stimulation experiments can verify the results obtained from the other two kinds of experiments. Though microscopic experiment, the indicator needed by the problem can be obtained.

Table 5. Factors affecting leaching of cement-based materials.

Type	Influence Factor
Environmental factor	Temperature
	Pressure
	Gas-phase components
	Illumination
Solvent	Solvent components
	pH
Material properties	Type of cement
	Aggregate
	Water–cement ratio
	Pore characteristics
	Stress state

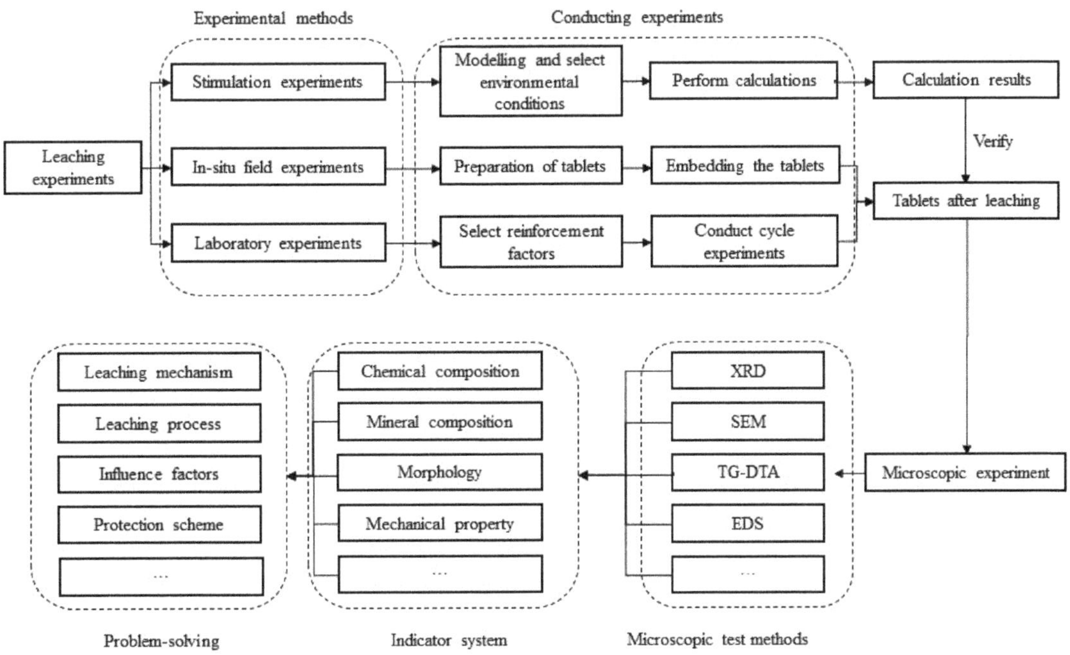

Figure 8. The flow chart of designing leaching and dissolution experiment.

There are many micro testing methods for leaching and dissolution, such as XRD, SEM, and EDS. Their purpose is focused on obtaining the chemical composition, mineral composition, surface morphology, porosity, and other purposes of the specimen. With the increasing demand for accuracy and functionality, more and more testing methods and instruments are being used in the research. At present, there are both common traditional testing methods such as XRD and SEM in research, as well as relatively new technologies put into use. In addition, the updates of portable testing instruments also provide conditions for on-site testing, which will greatly shorten testing time and facilitate the acquisition of experimental data from the test site. Overall, the new methods and traditional methods can be well combined to provide various information from multiple perspectives and with higher accuracy.

At present, the widespread and complete application of concrete and rocks in the construction field is more inclined to indoor decoration. The research on the protection of existing stone buildings is also receiving increasing attention. Therefore, the testing methods for rocks should be combined with quasi non-destructive testing or non-destructive testing, such as ultrasonic testing, 3D laser scanning, etc. For example, ultrasonic non-destructive testing is a convenient, fast, and inexpensive testing method like in Figure 9. Through two ultrasonic transmitting and receiving devices, researchers can quickly obtain information such as wave velocity and waveform to determine changes in material properties. This is a testing method worth promoting, but currently there is still room for improvement in ultrasonic testing instruments for non-metals.

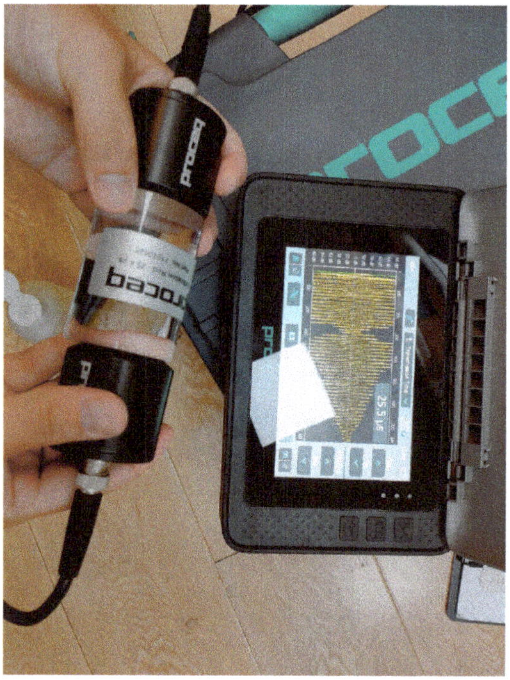

Figure 9. Ultrasonic non-destructive testing equipment.

6. Conclusions

This study investigated the factors influencing the leaching of cement-based materials and leaching of rocks, the related experiment methods were discussed and the effect of leaching and dissolution on the engineering properties were reported. The following conclusions can be drawn:

(1) There are three main types of leaching and dissolution experiments: laboratory experiments, field in situ experiments, and simulation experiments. Field in situ experiments can obtain more accurate results under real conditions but are relatively difficult to conduct; laboratory experiments are usually accelerated experiments that can fasten the entire process in a short period of time; simulation experiments can simulate long-term and dynamic processes but require careful selection of simulation environments and conditions, which are usually used as auxiliary means for other experiments.

(2) XRD, SEM, and other testing methods are widely used in the study of dissolution and leaching. With the advancement in technology, more high-precision and high-

resolution technologies have been developed, making the indicators of dissolution more adequate and accurate.

(3) There are many factors that affect leaching of cement-based materials and rock dissolution, including environmental factors, material properties, and solvent properties. These influencing factors often serve as the key for designing an accelerated experiment.

(4) The mechanism of rock dissolution varies depending on its mineral composition. Typically, rocks containing more carbonate rock components are more prone to dissolution.

(5) The leaching of cement-based materials is mainly due to the dissolution of calcium. During the process, the hydration products gradually lose their intact structure, leading to corrosion and destruction.

(6) Leaching can cause deterioration of engineering structural properties. Changing the compositions of the material can alter the process of deterioration, which is also an important direction for optimizing cement-based materials.

Author Contributions: Conceptualization, J.W. and K.L.; Methodology, L.Z., J.W. and K.L.; Software, L.Z. and J.W.; Validation, L.Z. and J.W.; Formal analysis, L.Z., M.W., S.L., L.Y. and J.W.; Investigation, L.Z., M.W., S.L., L.Y. and J.W.; Resources, J.W. and K.L.; Data curation, L.Z., M.W., S.L., L.Y. and J.W.; Writing—original draft, L.Z., M.W., S.L., L.Y., J.W. and K.L.; Writing—review & editing, L.Z., M.W., S.L., L.Y., J.W. and K.L.; Supervision, J.W. and K.L.; Project administration, J.W. and K.L.; Funding acquisition, J.W. and K.L. All authors have read and agreed to the published version of the manuscript.

Funding: This work is a part of a series of projects financially supported by the National Natural Science Foundation of China (Grant No. 52038004).

Institutional Review Board Statement: Not applicable.

Informed Consent Statement: Not applicable.

Data Availability Statement: Data are contained within the article.

Conflicts of Interest: The authors declare no conflict of interest.

Appendix A

Compound	Identification
$CaCO_3$	Calcium carbonate
H_2SO_4	Sulfuric acid
HNO_3	Nitric acid
Ca^{2+}	Calcium ion
$SO_4{}^{2-}$	Sulfate ion
$NO_3{}^{-}$	Nitrate ion
$Ca(OH)_2$	Calcium hydroxide
OH^{-}	Hydroxide ion
$nCaO{\cdot}2SiO_2{\cdot}nH_2O$	Calcium silicate
C-S-H	Hydrate calcium silicate
C-A-H	Calcium aluminium hydrate
C_4AF	Tetra calcium aluminoferrite

References

1. Zhang, Z.J. Research on the History and Compositions of Concrete. *Adv. Mater. Res.* **2014**, *988*, 207–210. [CrossRef]
2. Newell, P. An ancient battle between environment and concrete. *Nat. Rev. Chem.* **2021**, *5*, 513–514. [CrossRef] [PubMed]
3. Zhang, K.; Shen, Z.; Gan, L. Advances in cement-based materials leaching test. *Adv. Sci. Technol. Water Resour.* **2018**, *38*, 86–94. (In Chinese) [CrossRef]
4. Li, X.; Fang, K. Research on the contact dissolution of hydraulic roller compacted concrete (RCC). *Concrete* **2002**, *12*, 12–17. (In Chinese) [CrossRef]
5. Xiao, K.; Zhou, Y.; Mu, S.; Shi, L.; Gao, H. Summary of experimental methods for accelerated concrete corrosion. *Concrete* **2020**, *19*, 13–16. (In Chinese) [CrossRef]
6. Zhang, L.; Li, J.; Yu, H.; Zhan, Q. Study on the Rule of Concrete Corrosion in Harsh Environment. *IOP Conf. Ser. Earth Environ. Sci.* **2019**, *304*, 52002. [CrossRef]
7. Guo, J.; Bi, X.; Fang, R.; Li, S. Advances in the chemical dissolution methods of soluble rocks. *Hydrogeol. Eng. Geol.* **2020**, *47*, 24–34. [CrossRef]
8. Zhang, B.; Liu, J. Classification and characteristics of karst reservoirs in China and related theories. *Pet. Explor. Dev.* **2009**, *36*, 12–29. [CrossRef]
9. Su, W. Controlling Model for Rocky Desertification of Karst Mountainous Region and Its Preventing Strategy in Southwest, China. *J. Soil Water Conserv.* **2002**, *16*, 29–32. [CrossRef]
10. Cao, X.; He, Y.; Song, Z.; Li, H.; Sun, G.; Lan, J. Prediction of Depth Distribution Curve of Dissolution Rates Based on the Undulating Features of Rock Surfaces. *Teh. Vjesn. Tech. Gaz.* **2022**, *29*, 2098–2104. [CrossRef]
11. Wang, Y.; Lu, H.; Wang, J.; He, H. Effects of highly crystalized nano C-S-H particles on performances of Portland cement paste and its mechanism. *Crystals* **2020**, *10*, 816. [CrossRef]
12. Bufe, A.; Hovius, N.; Emberson, R.; Rugenstein, J.K.C.; Galy, A.; Hassenruck-Gudipati, H.J.; Chang, J. Co-variation of silicate, carbonate and sulfide weathering drives CO_2 release with erosion. *Nat. Geosci.* **2021**, *14*, 211–216. [CrossRef]
13. Gombert, P. Role of karstic dissolution in global carbon cycle. *Glob. Planet. Chang.* **2002**, *33*, 177–184. [CrossRef]
14. Liao, J.; Wu, K.; Xiong, L.; Zhao, J.; Li, X.; Zhang, C. Dissolution mechanism of a deep-buried sandstone reservoir in a deep water area: A case study from Baiyun Sag, Zhujiang River (Pearl River) Mouth Basin. *Acta Oceanol. Sin.* **2023**, *42*, 151–166. [CrossRef]
15. Akindipe, D.; Saraji, S.; Piri, M. Pore matrix dissolution in carbonates: An in-situ experimental investigation of carbonated water injection. *Appl. Geochem.* **2022**, *147*, 105483. [CrossRef]
16. Mahani, H.; Keya, A.L.; Berg, S.; Bartels, W.; Nasralla, R.; Rossen, W.R. Insights into the Mechanism of Wettability Alteration by Low-Salinity Flooding (LSF) in Carbonates. *Energy Fuels* **2015**, *29*, 1352–1367. [CrossRef]
17. Jiang, X.; Feng, T.; Zheng, Z.; Lei, M.; Zhang, W.; Ma, X.; Yi, X. A review of karst collapse mechanisms. *Carsologica Sin.* **2023**, *42*, 517–527. (In Chinese) [CrossRef]
18. Li, D.; Luo, Y. Measurement of Carbonate Rocks Distribution Area in China. *Carsologica Sin.* **1983**, 147–150.
19. Lopes, J.A.G.; Medeiros, W.E.; Oliveira, J.G.; Santana, F.L.; Araújo, R.E.B.; La Bruna, V.; Xavier, M.M.; Bezerra, F.H.R. Three-dimensional characterization of karstic dissolution zones, fracture networks, and lithostratigraphic interfaces using GPR cubes, core logs, and petrophysics: Implications for thief zones development in carbonate reservoirs. *Mar. Petrol. Geol.* **2023**, *150*, 106126. [CrossRef]
20. Zhao, Z.; Shen, Y. Rain-induced weathering dissolution of limestone and implications for the soil sinking-rock outcrops emergence mechanism at the karst surface: A case study in southwestern China. *Carbonate Evaporite* **2022**, *37*, 69. [CrossRef]
21. Laouafa, F.; Guo, J.; Quintard, M. Modelling and applications of dissolution of rocks in geoengineering. *J. Zhejiang Univ. A* **2023**, *24*, 20–36. [CrossRef]
22. Zhang, K.; Deng, X.; Gao, J.; Liu, S.; Wang, F.; Han, J. Insight into the Process and Mechanism of Water–Rock Interaction in Underground Coal Mine Reservoirs Based on Indoor Static Simulation Experiments. *ACS Omega* **2022**, *7*, 36387–36402. [CrossRef] [PubMed]
23. Barakat, Y.; Cui, Y.; Mokni, N.; Delage, P.; Bernier, F. Effects of pH and exposure time to alkaline solutions on the mineralogy of the Opalinus Clay from the lower sandy facies of Mont Terri site. *Eng. Geol.* **2022**, *306*, 106766. [CrossRef]
24. Wang, Y.; Bedrikovetsky, P.; Yin, H.; Othman, F.; Zeinijahromi, A.; Le-Hussain, F. Analytical model for fines migration due to mineral dissolution during CO_2 injection. *J. Nat. Gas Sci. Eng.* **2022**, *100*, 104472. [CrossRef]
25. Borja, R.I.; Chen, W.; Odufisan, A.R. A constitutive framework for rocks undergoing solid dissolution. *J. Mech. Phys. Solids* **2023**, *173*, 105198. [CrossRef]
26. Matsubara, H.; Yamada, T. Mathematical and numerical modelling of limestone dissolution. *Environ. Geotech.* **2022**, *9*, 274–285. [CrossRef]
27. Morse, J.W.; Arvidson, R.S. The dissolution kinetics of major sedimentary carbonate minerals. *Earth-Sci. Rev.* **2002**, *58*, 51–84. [CrossRef]
28. Johnson, K.S. Subsidence hazards due to evaporite dissolution in the United States. *Environ. Geol.* **2005**, *48*, 395–409. [CrossRef]
29. Veress, M. Karst Types and Their Karstification. *J. Earth Sci.-China* **2020**, *31*, 621–634. [CrossRef]
30. Jiang, G.; Chen, Z.; Siripornpibul, C.; Haryono, E.; Nguyen, N.X.; Oo, T.; Manzano, L.S.J.; Vongphachanh, S.; Kong, S.; Guo, F. The karst water environment in Southeast Asia: Characteristics, challenges, and approaches. *Hydrogeol. J.* **2021**, *29*, 123–135. [CrossRef]

31. Gabrovšek, F. On Concepts and Methods for the Estimation of Dissolutional Denudation Rates in Karst Areas. *Geomorphology* **2009**, *106*, 9–14. [CrossRef]
32. Krklec, K.; Domínguez-Villar, D.; Perica, D. Use of rock tablet method to measure rock weathering and landscape denudation. *Earth-Sci. Rev.* **2021**, *212*, 103449. [CrossRef]
33. Plan, L. Factors controlling carbonate dissolution rates quantified in a field test in the Austrian alps. *Geomorphology* **2005**, *68*, 201–212. [CrossRef]
34. He, H.; Wang, Y.; Wang, J. Effects of aggregate micro fines (AMF), aluminum sulfate and polypropylene fiber (PPF) on properties of machine-made sand concrete. *Appl. Sci.* **2019**, *9*, 2250. [CrossRef]
35. Krklec, K.; Marjanac, T.; Perica, D. Analysis of "Standard" (Lipica) Limestone Tablets and Their Weathering by Carbonate Staining and SEM Imaging, A Case Study on the VIS Island, Croatia. *Acta Carsologica* **2013**, *42*, 135–142. [CrossRef]
36. Li, K.; Xu, Z.; Lv, X.; Dong, Y.; Wang, H. Quantifying the Effect of Drought with Different Durations on Karst Dissolution Based on Field Control Test. *Prog. Geogr.* **2021**, *40*, 1704–1715. (In Chinese) [CrossRef]
37. Yan, W.; Zeng, C.; Xiao, S.; Lan, J.; Dai, L.; Tai, Z.; He, J.; He, C.; Di, Y. Dissolution Rate and Karst Carbon Sink of Different Land Use in Typical Dolomite Watershed with Humid Subtropical Weather. *Earth Environ.* **2021**, *49*, 529–538. (In Chinese)
38. Wu, J.; Gu, C.; Xu, S.; Zhao, X.; Huang, G. Corrosion analysis of carbonate rocks in southern Jiangsu Province. *Carsologica Sin.* **2021**, *40*, 565–571. (In Chinese)
39. Meng, J.; Chen, S.; Wang, J.; Chen, Z.; Zhang, J. Development and Application of Carbonate Dissolution Test Equipment under Thermal–Hydraulic–Chemical Coupling Condition. *Materials* **2022**, *15*, 7383. [CrossRef]
40. Rahimi, M.R.; Mohammadi, S.D.; Beydokhti, A.T. Correlation between the mineral composition and dissolution rate constant of sulfate rocks at different pressures and water flow velocities. *Bull. Eng. Geol. Environ.* **2022**, *81*, 371. [CrossRef]
41. Zhang, T.; Wang, X.; Hu, J.; Ye, J.; Wu, Y.; Wen, X. Experimental study on erosion characteristics of salt rock by water jet under low pressure. *J. Energy Storage* **2022**, *55*, 105690. [CrossRef]
42. Qiao, Z.; Lv, Y.; Chen, W.; She, M.; Shao, G.; Cao, P.; Sun, X. Origin and evolution of burial dissolved vugs in dolomite: Evidence from high-temperature and high-pressure dissolution kinetic simulation. *Marin Orig. Petrl. Geol.* **2021**, *26*, 326–334. (In Chinese)
43. Song, L.; Li, M.; Yu, H.; Sun, X.; Li, C. Selective Dissolution Process of Feldspar and Calcite Minerals under Different pH Conditons. *J. Northeast. Pet. Univ.* **2022**, *46*, 74–83. (In Chinese)
44. Lin, Y.; Ren, H.; Wu, Y.; Jia, F.; Liu, P.; Liang, J. Experimental Simulation of the Carbonate Dissolution Process under Different Occurrence Conditions. *Hydrogeol. Eng. Geol.* **2021**, *48*, 15–26. (In Chinese)
45. Cao, T.; Qu, X.; Jiang, Y.; Wang, Q.; Liu, C. Dissolution Simulation of Organic Acids and Minerals under Mixed Deposition. *Bull. Mineral. Petrol. Geochem.* **2021**, *40*, 1181–1188. (In Chinese)
46. Shen, A.; Qiao, Z.; She, M.; Meng, S.; Zhang, J.; Wang, X. Prediction of burial dissolved vugs in carbonates based on dissolution simulation: A case study of the Longwangmiao Formation dolostone reservoirs, Sichuan Basin. *Oil Gas Geol.* **2021**, *42*, 691–701. (In Chinese)
47. Liu, S.; Chen, S.; Liu, B.; Shi, K.; Liu, Y.; Zheng, H.; Luo, Q. Pore evolution of the Permian Qixia-Maokou Formations dolomite in Sichuan Basin based on in-situ dissolution simulation experiment. *Oil Gas Geol.* **2021**, *42*, 703–715. (In Chinese)
48. Huo, J.; Song, H.; Guo, Y.; Peng, P. Experimental Study on Contact Dissolution of Different Cement-based Materials under Deionized Water. *J. Hydroelectic Eng.* **2013**, *32*, 242–247. (In Chinese)
49. Li, B. Experimental Study on the Evolution Law of Mechanical Properties of Layered Limestone under Dissolution. Master's Thesis, China Three Gorges University, Yichang, China, 2021. (In Chinese).
50. Shaowei, W.; Yingli, X.; Pinghua, Z.; Hui, L.; Qifeng, L. Experimental Study on Splitting Tensil Strength of Concrete Layer Interface under Dissolution Effects. *Adv. Sci. Technol. Water Resour.* **2022**, *42*, 27–32. (In Chinese)
51. Shaoli, H. Experimental Study on the Corrosion Performance of Mineral Admixture Concrete under Soft Water Conditions. Master's Thesis, Nanjing University of Science & Technology, Nanjing, China, 2019. (In Chinese).
52. Caihong, Z.; Zhigang, S.; Min, M.; Shihua, X. Experimental Study and Numerical Simulation on Cement Mortar Corroded by Sulfuric Acid. *J. Nat. Disasters* **2021**, *30*, 112–121. (In Chinese)
53. Jin, M.; Ma, Y.; Li, W.; Huang, J.; Zeng, H.; Lu, C.; Zhang, J.; Liu, J. Degradation of C–S–H(I) at different decalcification degrees. *J. Mater. Sci.* **2022**, *57*, 19260–19279. [CrossRef]
54. Hale, S.; Trindade Pedrosa, E.; Jacob, A.; Reinhardt, M.; Enzmann, F.; Kersten, M.; Butscher, C.; Blum, P. Upscaling calcite dissolution rates in a tight reservoir sandstone. *Environ. Earth Sci.* **2022**, *81*, 303. [CrossRef]
55. Kong, X.; Chen, G.; Ji, G.; Li, S. Experim Ental Study on Accelerating Effect of Ammonium Nitrate on the Calcium Leaching of Dam Concrete. *Concrete* **2017**, *4*, 34–37. (In Chinese) [CrossRef]
56. Yan, K.; Liang, F.; Zou, H.; Li, P. Study on the Effect of Stress on Limestone Dissolution. *China Water Transp.* **2022**, *22*, 97–99. (In Chinese)
57. Shan, X.; Guo, H.; Chen, X.; Guo, X.; She, M.; Li, Y. Experimental Simulation of Burial Dissolution of Conglomerate Reservoir in Baikouquan Formation in Maxi Slope Area, Junggar Basin. *Acta Sedimentol. Sin.* **2021**, *48*, 15–26. (In Chinese) [CrossRef]
58. Liu, Q.; Lu, Y.; Zhang, F.E. Laboratory simulation experiment on dissolution of limestone under hydrodynamic pressure. *Carbonate Evaporite* **2013**, *28*, 3–11. [CrossRef]
59. Abbaszadeh, M.; Nasiri, M.; Riazi, M. Experimental investigation of the impact of rock dissolution on carbonate rock properties in the presence of carbonated water. *Environ. Earth Sci.* **2016**, *75*, 791. [CrossRef]

60. Kirstein, J.; Hellevang, H.; Haile, B.G.; Gleixner, G.; Gaupp, R. Experimental determination of natural carbonate rock dissolution rates with a focus on temperature dependency. *Geomorphology* **2016**, *261*, 30–40. [CrossRef]
61. Elkhoury, J.E.; Ameli, P.; Detwiler, R.L. Dissolution and deformation in fractured carbonates caused by flow of CO2-rich brine under reservoir conditions. *Int. J. Greenh. Gas Control.* **2013**, *16*, S203–S215. [CrossRef]
62. Hong, D.; Fan, M.; Yu, L.; Cao, J. An experimental study simulating the dissolution of gypsum rock. *Energy Explor. Exploit.* **2018**, *36*, 942–954. [CrossRef]
63. Zhang, K.; Zhu, K.; He, Y.; Zhang, Y. Experimental study on karst development characteristics of calcrete and analysis of its dissolution mechanism. *Carbonate Evaporite* **2022**, *37*, 41. [CrossRef]
64. Zhang, X.; Li, B.; Peng, J.; Qu, F.; Zhang, K.; Yang, S.; Xu, Q. Experimental analysis of dissolution reconstruction of deep dolomite reservoirs: A case study of the Cambrian dolomite reservoirs in the Tarim Basin. *Front. Earth Sci.* **2022**, *10*, 1015460. [CrossRef]
65. Tõnsuaadu, K.; Kallas, J.; Kallaste, T.; Urtson, K.; Einard, M.; Martin, R.; Kuusik, R.; Trikkel, A. Estonian Phosphate Rock Dissolution in Hydrochloric Acid: Optimization of Acid Dosage and Concentration. *Minerals* **2023**, *13*, 578. [CrossRef]
66. Samson, E.; Marchand, J. Modeling the effect of temperature on ionic transport in cementitious materials. *Cem. Concr. Res.* **2007**, *37*, 455–468. [CrossRef]
67. Gao, S.; Jiang, L.; Xu, X.; Chu, H.; Lv, Z.; Jin, W. Experimental Study on the Leaching Behavior of Plastic Cement-based Materials. *Fly Ash Compr. Util.* **2021**, *35*, 46–53. (In Chinese) [CrossRef]
68. Sun, G.; Wang, Z.; Liu, P.; Zhang, Z. Dissolved micro-pores in alkali feldspar and their contribution to improved properties of tight sandstone reservoirs: A case study from Triassic Chang-63 sub-member, Huaqing area, Ordos Basin. *Oil Gas Geol.* **2022**, *43*, 658–669. (In Chinese)
69. Chen, B.; Li, Q.; Tan, Y.; Alvi, I.H. Dissolution and Deformation Characteristics of Limestones Containing Different Calcite and Dolomite Content Induced by CO_2-Water-Rock Interaction. *Acta Geol. Sin.* **2023**, *97*, 956–971. [CrossRef]
70. Ren, L.; Yu, Z.; Zhao, J.; Lin, R.; Wu, J.; Song, Y.; Xie, M. Impact of Carbonic Acid Dissolution on Flow Characteristics of Tight Sand Stone Reservoir: Taking Chang 6 Tight Sandstone in Ordos Basin as an Example. *Pet. Geol. Oilfield Dev. Daqing* **2022**, 1–9. (In Chinese) [CrossRef]
71. Zhong, Q.; Huang, S.; Zou, M.; Tong, H.; Huang, K.; Zhang, X. Controlling factors of order degree of dolomite in carbonate rocks: A case study from Lower Paleozoic in Tahe Oilfield and Triassic in northeastern Sichuan Basin. *Lithol. Reserv.* **2009**, *21*, 50–55. (In Chinese)
72. Xu, L.; Lai, Y.; Ma, D.; Wang, J.; Li, M.; Li, L.; Gao, Z.; Liu, Y.; He, P.; Zhang, Y. Effects of Fiber and Surface Treatment on Airport Pavement Concrete against Freeze–Thawing and Salt Freezing. *Materials* **2022**, *15*, 7528. [CrossRef] [PubMed]
73. Zhong, L. Digital Analysis of Corrosion Morphology of Exposed Carbonate Rock Surface and Study of Material Accumulation Effect. Master's Thesis, Nanning Normal Univeristy, Nanning, China, 2021. (In Chinese).
74. Galindo, K.; Lins, C.; Guimarães, L.; Lima, A.; Silva, K.; Nova, A. Application of microtomography and petrography techniques for the characterization of porosity of synthetic carbonatic rock minerals before and after acidification processes. *Sci. Rep.* **2022**, *12*, 17026. [CrossRef] [PubMed]
75. Morais, S.; Lecoutre, C.; Philippot, G.; Aubert, G.; Nguyen, O.; Cario, A.; Vidal, E.; Campbell, Z.S.; Garrabos, Y.; Azaroual, M.; et al. 3D Imaging of On-Chip Porous Medium Using X-ray Laminography: Application to Reactive Flow Dissolution at Small Scales. *Processes* **2023**, *11*, 1981. [CrossRef]
76. Deng, Y.; Guo, X.; Luo, M.; Chen, X.; Kuang, Y.; Zhou, H. Study on the Microstructure and Variation Law of Carbonate Rock Dissolution Based on Scanning Electron Microscopy and CT Imaging Technology. *Carsologica Sin.* **2022**, *41*, 698–707. (In Chinese) [CrossRef]
77. Zhong, Y.; Li, Q.; Wang, R.; Yao, T. Changes of Physical and Mechanical Properties of Coral Reef Limestone under CO_2–Seawater–Rock Interaction. *Appl. Sci.* **2022**, *12*, 4105. [CrossRef]
78. Guren, M.G.; Renard, F.; Noiriel, C. Dissolution rate variability at carbonate surfaces: 4D X-ray micro-tomography and stochastic modeling investigations. *Front. Water* **2023**, *5*, 1185608. [CrossRef]
79. Cooper, M.P.; Sharma, R.P.; Magni, S.; Blach, T.P.; Radlinski, A.P.; Drabik, K.; Tengattini, A.; Szymczak, P. 4D tomography reveals a complex relationship between wormhole advancement and permeability variation in dissolving rocks. *Adv. Water Resour.* **2023**, *175*, 104407. [CrossRef]
80. Liu, P.; Wang, Z.; Zhang, Z. A New Quantitative Approach for Element-Mineral Determination Based on "EDS (Energy Dispersive Spectroscopy) Method". *Geofluids* **2021**, *2021*, 4023704. [CrossRef]
81. Neumann, C.; Faria, E.F.; Dos Santos, A.C.P. Concrete Leaching of a Hydroelectric Powerhouse Due to 40 Years of Exposure to River Water. *Constr. Build. Mater.* **2021**, *302*, 124253. [CrossRef]
82. Wang, Z. Research on the Effect and Mechanism of Nano-Particles and Fly Ash on Durability and Service-Life of Concrete in Corrosion Environment of Cold Regions. Master's Thesis, Inner Mongolia Agricultural University, Hohhot, China, 2021. (In Chinese).
83. Wang, J.; Xu, L.; Li, M.; Wang, Y.; He, H.; Xiang, D.; Li, K.; Hao, T. Investigations on Factors Influencing Physical Properties of Recycled Cement and the Related Carbon Emissions and Energy Consumptions. *J. Clean Prod.* **2023**, *414*, 137715. [CrossRef]
84. Mo, Y.; Zuo, S.; Fu, L.; Chen, S. Study on the Mechanism of AMD Seepage on Corrosion and Permeability Decline of Fractured Limestone. *Min. Saf. Environ. Prot.* **2022**, *49*, 89–94, 101. (In Chinese) [CrossRef]

85. Zhou, S.; Sun, Q.; Zhang, H.; Meng, H.; Gao, Q.; Zhou, Y. Elemental Dissolution Characteristics of Granite and Gabbro under High-temperature Water-rock Interactions. *Sci. Total Environ.* **2023**, *897*, 165455. [CrossRef] [PubMed]
86. Burlion, N.; Bernard, D.; Chen, D. X-ray microtomography: Application to microstructure analysis of a cementitious material during leaching process. *Cem. Concr. Res.* **2006**, *36*, 346–357. [CrossRef]
87. Lai, J.; Wang, G.; Fan, Z.; Chen, J.; Wang, S.; Zhou, Z.; Fan, X. Insight into the Pore Structure of Tight Sandstones Using NMR and HPMI Measurements. *Energy Fuels* **2016**, *30*, 10200–10214. [CrossRef]
88. Feng, P.; Brand, A.S.; Chen, L.; Bullard, J.W. In Situ Nanoscale Observations of Gypsum Dissolution by Digital Holographic Microscopy. *Chem. Geol.* **2017**, *460*, 25–36. [CrossRef] [PubMed]
89. Haga, K.; Sutou, S.; Hironaga, M.; Tanaka, S.; Nagasaki, S. Effects of Porosity on Leaching of Ca from Hardened Ordinary Portland Cement Paste. *Cem. Concr. Res.* **2005**, *35*, 1764–1775. [CrossRef]
90. Dorn, R.I. Digital Processing of Back-scatter Electron Imagery: Amicroscopic Approach to Quantifying Chemical Weathering. *Geol. Soc. Am. Bull.* **1995**. [CrossRef]
91. Xu, P.; Sheng, M.; Lin, T.; Liu, Q.; Wang, X.; Khan, W.A.; Xu, Q. Influences of Rock Microstructure on Acid Dissolution at a Dolomite Surface. *Geothermics* **2022**, *100*, 102324. [CrossRef]
92. Filiberto, L.H.; Putnis, C.V.; Julia, M. Factors Controlling Reaction Pathways during Fluid–rock Interactions. *Contrib. Mineral. Petrol.* **2023**, *178*, 53. [CrossRef]
93. Meng, J.; Chen, S.; Wang, J.; Chen, Z.; Zhang, J. Experimental Study on the Dissolution Characteristics and Microstructure of Carbonate Rocks under the Action of Thermal–Hydraulic–Chemical Coupling. *Materials* **2023**, *16*, 1828. [CrossRef]
94. Zhu, X.; Li, S.; Feng, X.; Kang, Z.; Fan, H.; Huang, X. Numerical Investigation on the Main Controlling Factors of the Dissolution Evolution of Three-dimensional Fracture-cavity Reservoirs. *Carsologica Sin.* **2021**, *40*, 943–951. (In Chinese) [CrossRef]
95. Yang, H. Numerical Simulation of the Degradation Law of Seismic Performance of Concrete Structure in Chloride-Leaching Environment. Master's Thesis, Nanjing University of Science & Technology, Nanjing, China, 2021. (In Chinese).
96. Zhou, D.; Shen, Z.; Ma, B.; Zhang, K. Distribution Law of Concrete Dam Mass Loss under Seepage Erosion. *Adv. Sci. Technol. Water Resour.* **2022**, *42*, 72–78. (In Chinese) [CrossRef]
97. Hou, W.; Gong, X.; Liu, F.; Li, H. Numerical Experiment of the Hydrothermal Synergistic Mixing Corrosion Mechanism for Carbonate Rocks. *Carsologica Sin.* **2022**, *42*, 775–784. (In Chinese) [CrossRef]
98. Wang, J.; Zhao, Y.; An, Z.; Shabani, A. CO_2 Storage in Carbonate Rocks: An Experimental and Geochemical Modeling Study. *J. Geochem. Explor.* **2022**, *234*, 106942. [CrossRef]
99. Hao, Y.; Smith, M.; Sholokhova, Y.; Carroll, S. CO_2-induced Dissolution of Low Permeability Carbonates. Part II: Numerical Modeling of Experiments. *Adv. Water Resour.* **2013**, *62*, 388–408. [CrossRef]
100. Zhang, L.; Zhao, Q.; Hu, X.; Han, G.; Zhao, X. Laboratory Dissolution Test on Dolomite And Its Microdissolution Mechanism. *J. Eng. Geol.* **2012**, *20*, 576–584. (In Chinese) [CrossRef]
101. Liu, W.; Liu, P.; Xu, H.; Gong, B.; Ji, F. Study on the Microstructure Evolution and Strength Damage Mechanism of Dolomite under Dissolution Condition. *Sustainability* **2022**, *14*, 11447. [CrossRef]
102. Yuan, G. Genetic Mechanism of Dissolution of Feldspars and Carbonate Minerals during Diagenesis and Its Impact on Reservoir Poroperm. Ph.D. Thesis, China University of Petroleum, Beijing, China, 2015. (In Chinese)
103. Luquot, L.; Gouze, P. Experimental determination of porosity and permeability changes induced by injection of CO_2 into carbonate rocks. *Chem. Geol.* **2009**, *265*, 148–159. [CrossRef]
104. Gibeaux, S.; Vázquez, P.; De Kock, T.; Cnudde, V.; Thomachot-Schneider, C. Weathering assessment under X-ray tomography of building stones exposed to acid atmospheres at current pollution rate. *Constr. Build. Mater.* **2018**, *168*, 187–198. [CrossRef]
105. Yuan, D. Carbon Cycle and Global Karst. *Quat. Sci.* **1993**, *13*, 1–6. (In Chinese)
106. Mohammad, R.; Seyed, D.; Alireza, T. Laboratory simulation of gypsum rock dissolution at different pressures, water-flow velocities and pH ranges. *Q. J. Eng. Geol. Hydrog.* **2023**, *56*, qjegh2021-120. [CrossRef]
107. Zaier, I.; Billiotte, J.; De Windt, L.; Charmoille, A. The impact of common impurities present in gypsum deposits on in situ dissolution kinetics. *Environ. Earth Sci.* **2023**, *82*, 31. [CrossRef]
108. Snæbjörnsdóttir, S.Ó.; Sigfússon, B.; Marieni, C.; Goldberg, D.; Gislason, S.R.; Oelkers, E.H. Carbon Dioxide Storage through Mineral Carbonation. *Earth Environ.* **2020**, *1*, 90–102. [CrossRef]
109. Mainguy, M.; Tognazzi, C.; Torrenti, J.; Adenot, F.; LCPC, P.F.; WCA. Modelling of leaching in pure cement paste and mortar. *Cem. Concr. Res.* **2000**, *30*, 83–90. [CrossRef]
110. Tang, Y.; Yin, G.; Chen, F.; Zhang, J. Solid–liquid equilibrium state and equation of cement-based materials in ammonium chloride solution. *Mater. Struct.* **2022**, *55*, 220. [CrossRef]
111. Alonso, C.; Castellote, M.; Llorente, I.; Andrade, C. Ground water leaching resistance of high and ultra high performance concretes in relation to the testing convection regime. *Cem. Concr. Res.* **2006**, *36*, 1583–1594. [CrossRef]
112. Mirsayapov, I.; Yakupov, S.; Hassoun, M. About concrete and reinforced concrete corrosion. *IOP Conf. Series. Mater. Sci. Eng.* **2020**, *890*, 12061. [CrossRef]
113. Feng, P.; Miao, C.; Bullard, J.W. A model of phase stability, microstructure and properties during leaching of portland cement binders. *Cem. Concr. Compos.* **2014**, *49*, 9–19. [CrossRef]
114. Taylor, K.C.; Nasr-El-Din, H.A.; Mehta, S. Anomalous Acid Reaction Rates in Carbonate Reservoir Rocks. *SPE J.* **2006**, *11*, 488–496. [CrossRef]

115. Liu, Z.; Yuan, D.; Wolfgang, D. Comparative study of dissolution rate-determining mechanisms of limestone and dolomite. *Environ. Geol.* **2005**, *49*, 274–279. [CrossRef]
116. Tan, Q.; Kang, Y.; You, L.; Peng, H.; Zhao, F.; Bai, J. The role of salt dissolution on the evolution of petrophysical properties in saline-lacustrine carbonate reservoirs: Pore structure, porosity–permeability, and mechanics. *J. Hydrol.* **2023**, *618*, 129257. [CrossRef]
117. Wang, H.; Zhou, Q.; Sheng, J.; Luo, Y.; Liu, J.; Liu, X. Effect of long-term infiltration on porosity-permeability evolution in carbonate rocks: An online NMR coupling penetration test. *J. Hydrol.* **2023**, *617*, 129029. [CrossRef]
118. Araújo, R.E.B.; La Bruna, V.; Rustichelli, A.; Xavier, M.M.; Agosta, F.; Bezerra, F.H.R.; Brito, M.F.P.; Barbosa, J.A.; Antonino, A.C.D. Pore network characteristics as a function of diagenesis: Implications for epigenic karstification in shallow-water carbonates. *Mar. Petrol. Geol.* **2023**, *149*, 106094. [CrossRef]
119. Ping, S.; Wang, F.; Wang, D.; Li, S.; Yuan, Y.; Feng, G.; Shang, S. Multi-scale deterioration mechanism of shear strength of gypsum-bearing mudstone induced by water-rock reactions. *Eng. Geol.* **2023**, *323*, 107224. [CrossRef]
120. Zekri, A.Y.; Shedid, S.A.; Almehaideb, R.A. Investigation of supercritical carbon dioxide, aspheltenic crude oil, and formation brine interactions in carbonate formations. *J. Pet. Sci. Eng.* **2009**, *69*, 63–70. [CrossRef]
121. Parandeh, M.; Dehkohneh, H.Z.; Soulgani, B.S. Experimental investigation of the acidizing effects on the mechanical properties of carbonated rocks. *Geoenergy Sci. Eng.* **2023**, *222*, 211447. [CrossRef]
122. Liu, C.; Li, H. Study on the Expressive Characteristics of Rock Corrosion and Its Effect on Physical and Mechanical Properties of Rock Surface Layer. *Earth Environ.* **2012**, *40*, 255–260. (In Chinese)
123. He, C.; Guo, J. Mechanism Study of Acid on Mechanical Properties of Limestone. *Chin. J. Rock Mech. Eng.* **2013**, *32*, 3016–3021. (In Chinese) [CrossRef]
124. Marinoni, N.; Pavese, A.; Voltolini, M.; Merlini, M. Long-term leaching test in concretes: An X-ray powder diffraction study. *Cem. Concr. Compos.* **2008**, *30*, 700–705. [CrossRef]
125. Müllauer, W.; Beddoe, R.E.; Heinz, D. Effect of carbonation, chloride and external sulphates on the leaching behaviour of major and trace elements from concrete. *Cem. Concr. Compos.* **2012**, *34*, 618–626. [CrossRef]
126. Chen, A. An Experimental Study on the Degradation of Mechanical Properties of Leached Fly-Ash Concrete. Master's Thesis, Zhejiang University of Technology, Hangzhou, China, 2017. (In Chinese)
127. Huang, B.; Qian, C. Experiment study of chemo-mechanical coupling behavior of leached concrete. *Constr. Build. Mater.* **2011**, *25*, 2649–2654. [CrossRef]
128. Zhang, W.; Shi, D.; Shen, Z.; Wang, X.; Gan, L.; Shao, W.; Tang, P.; Zhang, H.; Yu, S. Effect of calcium leaching on the fracture properties of concrete. *Constr. Build. Mater.* **2023**, *365*, 130018. [CrossRef]

materials

MDPI

Review

Does Current Knowledge Give a Variety of Possibilities for the Stabilization/Solidification of Soil Contaminated with Heavy Metals?—A Review

Agnieszka Lal [1] and Joanna Fronczyk [2,*]

[1] Faculty of Civil Engineering and Architecture, Lublin University of Technology, 40 Nadbystrzycka Str., 20-618 Lublin, Poland
[2] Institute of Civil Engineering, Warsaw University of Life Sciences—SGGW, 166 Nowoursynowska Str., 02-787 Warsaw, Poland
* Correspondence: joanna_fronczyk@sggw.edu.pl

Abstract: Stabilization/solidification of contaminated soil is a process that allows simultaneous strengthening of the soil structure, disposal of contamination and recycling of industrial waste, implemented as substitutes for Portland cement or additives to improve the properties of the final product obtained. Extremely intensive development of studies pertaining to the S/S process prompted the authors to systematize the binders used and the corresponding methods of binding the contamination, and to perform an analysis of the effectiveness expressed in geomechanical properties and leachability. The study pays close attention to the types of additives and binders of waste origin, as well as the ecological and economic benefits of their use. The methods of preparing and caring for the specimens were reviewed, in addition to the methods of testing the effectiveness of the S/S process, including the influence of aging factors on long-term properties. The results of the analyses carried out are presented in the form of diagrams and charts, facilitating individual evaluation of the various solutions for the stabilization/solidification of soils contaminated with heavy metals.

Keywords: binder; amendments; additives; sample preparation; unconfined compressive strength—UCS; leachability tests

1. Introduction

In-situ stabilization/solidification of soils contaminated with heavy metals is the subject of worldwide research in the fields of chemistry, geotechnics and environmental engineering. It was originally a process used under the name of stabilization, only as an economical alternative to soil replacement, in case of insufficient strength and deformation properties of the subsoil [1–10]. However, due to its effective binding of contaminants at relatively low cost [11–13] and competitive time-consumption [14,15], it has become known as a leading method of soil remediation [16–19]. It is understood both as a natural subsoil, subjected to incidental contamination, and is still used for engineering tasks (including the foundation of cubic and linear structures), as well as subsoil, which is a substructure or another layer of a landfill, in which the presence of contaminants is long-lasting and intensified. There is a number of binders available for use in the stabilization/solidification process, the choice of which govern the method of minimizing heavy metal activity in the soil [20]. The neutralization of contaminants can take place only mechanically (due to the binding of pozzolanic materials), chemically (due to the reaction of contaminants with suitable compounds, resulting in the precipitation of metals in a less harmful form), by oxidation, chemical reduction or as a combination of the above [21–23]. Systematization of the variety and multiplicity of binders that are able to implement the S/S process is required. There have been attempts to create such a database [23–26]; however, due to the intensive development of research, there are deficiencies in the reported data. Waste materials with pozzolanic properties should have a special place in this type of statement. Their importance is highlighted due to the fact that they are perfectly in line with the sustainable construction trends being promoted. Nevertheless, it is also important to define the main types of tests that allow verification of the quality of the stabilization/solidification, expressed in the mechanical properties of the subsoil and the effectiveness in immobilization of contaminants. It is also necessary to gather information on the procedures of preparing samples for testing, as well as basic parameters and methods of curing. The aforementioned elements were analyzed on the basis of the studies available worldwide, mainly in the English-language literature. The research results are intended to serve as a basis for the verification of the initial assumptions made in the authors' planned research work, while their compendium is presented in this review article.

The aim of this paper is to characterize the most popular binders used in the stabilization/solidification of heavy metal-contaminated soils and to identify the most effective and environmentally friendly solutions currently in use. The above was achieved by analyzing the results of UCS compressive strength tests and information on the leachability of contaminants from S/S products presented in the literature. It should be noted that, due to the large amount of data reported and the volume of this article, the analysis of contaminant leaching was kept to a minimum and the focus was on the mechanical properties of the treated soil. At the same time, it is pointed out that a more detailed consideration of the binding mechanism of individual metals in soil–cement mixtures and the presentation of leaching test results achieved with the analyzed binders, as an extremely important issue, should be the subject of a separate paper. In addition, this article summarizes the research methods that can be used to assess the effectiveness and sustainability of the processes for strengthening contaminated soils while reducing the mobility of contaminants in the subsoil.

2. Samples Preparation and Curing Conditions

Essential criteria for conducting optimization studies of the S/S process are the repeatability and homogeneity of the samples. The standards describing the testing of cement–soil samples [27], as well as the literature data from published scientific results, can be used in this regard. The vast majority of the articles outlining sample preparation contain a scheme of the procedure, as shown in Figure 1.

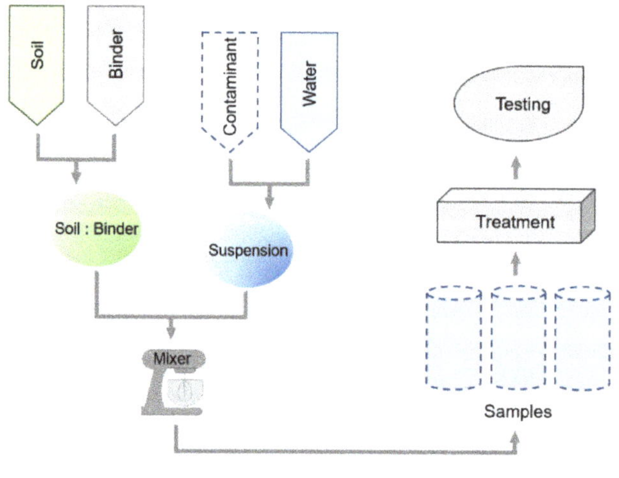

Figure 1. Scheme of the sample preparation procedure.

The soil is initially dried to a solid mass, ground and sieved through a 2.0 mm sieve, and the selected additives dried to a solid mass are blended in appropriate weight proportions until a homogeneous mixture is obtained. A specified amount of water contains the assumed concentration of the contaminant, i.e., the selected heavy metal is added. Finally, mixtures with strictly controlled moisture content, composition and degree of pollution are obtained. The prescribed procedure is modified by various researchers. In Li and Poon's study [28], contaminants in the form of lead nitrate were added directly to the dry mixture, before deionized water was fed. Goodarzi and Movahedrad introduced suspensions of the contaminant in distilled water into a dry soil sample, followed by blending the wet contaminated soil with binders [29]. However, a more correct approach appears to be to prepare the contaminated soil in advance, allowing for a thorough mixing of heavy metal ions and soil, as used in the study by Li et al. [30]. Soil with contaminants at a determined moisture content was stored under standard cure conditions for 30 days before being combined with binders at the assumed proportions. In the study by Wang et al. [31], contaminated soil taken from military sites was used; thus, an additional dosage of the contaminant was not required. In the case of the stabilization/solidification of wastes from the carbonation and hardening of steel, remediation was preceded by the initial inactivation of barium and cyanide by the addition of suitable chemicals, followed by blending with binders [32]. During the studies of Feng et al. [33], the binder was ground in order to increase its specific surface area, resulting in increased hydration. This step was carried out, despite the time-consuming nature of this method, up to a presumed degree of grinding and mixing of the ingredients, both dry parts and those with the addition of water, usually taking place mechanically at a preset time. The use of a planetary robot mixer is very common [31]. Alternatively, bench-top mixers are also used, where the mixer is inserted into a tall container filled with the mixture ingredients [34]. However, it is also common to mix the ingredients manually [32].

Filling the molds (cylindrical or cubic) with the mixture is usually performed in layers. For liquid consistency of mixtures, it is popular to use a vibrating table [35], which guarantees effective elimination of air bubbles. Alternatively, for mixtures with a plastic consistency, layers are applied with a constant compaction energy (manually [29], with the use of the Proctor apparatus [36] or other equipment with known compaction energy [37]). Few studies indicate the compaction of mixtures to the design's volumetric density (e.g., 95% ρ_{max} [33] or to the maximum density of the soil skeleton [29]). The methodology mentioned above was found in all research papers analyzed. It should be mentioned that very few studies were carried out on a macro-scale, i.e., the S/S process is carried out in-situ and samples for testing are retrieved from the ground at an appropriate treatment time [38–40]. This obviously entails a lesser stability of conditions and homogeneity of samples, but allows actual environmental factors affecting the quality of the S/S process products obtained to be reflected.

The molded samples are subjected to a constant and very rarely modified method of curing. Immediately after preparation, they are wrapped in plastic film and stored for a suitable number of days at a stably maintained temperature of 20–25 °C (±2 °C) and a humidity level of 95% (±1%). The samples are unmolded either immediately or after 24 h (eventually 48 h), that is, after the initial setting of the mixture, depending on its consistency and sensitivity. Film protection and the maintenance of constant curing conditions are emphasized in all the studies analyzed.

The required period of curing corresponds to the assumed age of the samples at the time of testing. Usually, a series of samples is prepared for testing after 28 days. Often, 7-day-old samples are additionally tested, and sometimes (especially for the mixtures containing fly ash) 90-days old samples. Less frequently, however, 1-day or 3-day-old samples are tested. For the assessment of the extended setting time, part of the samples needs to be subjected to curing for several weeks.

By analyzing the data of the moisture content of the mixtures prepared for testing purposes, it was shown that the minimum amount of water applied to the dry weight of the soil, including the binder and any additives, is 7.5%, while the maximum is 60%. It is important to emphasize the significant influence of the initial moisture content of the mixture on the final results of the S/S tests [41]. However, very few sources in the literature present studies on the optimization of moisture content to obtain the maximum compressive strength of the S/S product with the different types of binders used. Some data can be found in the study by Kogbara et al. [36,42], which shows that contaminated sandy gravel subjected to the S/S process with Pulverized Fly Ash (PFA) and Ordinary Portland Cement (OPC) at a ratio of 4:1 achieves higher compressive strength as the moisture content of the mixture increases from OMC − 2% to OMC + 5%, where OMC denotes the optimum moisture content. In contrast, the opposite trend was shown in soil without contamination (sand stabilized with PFA and OPC in a ratio of 5:1) [43], where the maximum strength was reached on the dry side of OMC. Some information can also be found in the study by Boutouil and Levacher [44]. This very important aspect was analyzed by Kogbara in his paper [45], concluding, as a general rule, that the best mechanical and binding properties for contaminants were shown by the stabilized soils at a moisture content close to the optimum value, regardless of the binder system used. Due to the scope of this study, this important issue is considered to be an element suitable for a separate case study.

Wider variability is manifested in the amount of binder incorporated into the soil. In addition to the basic comparative (zero) samples, which consist of 100% dry matter soil and 100% dry matter binder (cement or as a mixture with another binder), soil/binder mixtures are being established. Soil accounts for 10% to 95% of the dry matter. In line with ecological and economic considerations, the aim is to minimize the amount of binder used in the S/S process, while maintaining the high efficiency of this method in binding contaminants and obtaining sufficient strength characteristics. Such a compromise is achieved through the use of various activating additives, water reducers or waste materials with unique properties, presented later in this article. It is also sought to replace as much Portland cement as possible with alternative pozzolanic materials, mainly waste materials (along with possible additives). A binder-to-soil ratio of 20–35% is considered to result in the achievement of immobilization of contaminants in the soil at a level acceptable to legislation [46]. In terms of varying the amount of Portland cement in relation to other binders, the amount of alternative materials in the studies analyzed varies from 10 to as much as 100%.

3. Characteristic of Binders, Amendments and Additives

From a geotechnical point of view, the addition of the binder is expected to result in improved parameters of the subsoil for future engineering usage. However, the presence of contaminants (e.g., heavy metals) results in the need to select a binder that will additionally have an environmentally positive effect on mobility and bioavailability of contaminants [47,48]. In addition to the most commonly used binder, i.e., ordinary Portland cement, waste materials such as ground granulated blast furnace slag [34], incinerated sewage sludge ash [49], fly ash of different classes [33], phosphogypsum [38], and red mud [50] were proposed. This approach has a positive environmental impact due to the increased recovery of waste, which is in line with the trend of a closed-loop economy. Additionally, natural materials in their raw or modified state were utilized, e.g., bentonite [51], phosphate rock [52], lime [47], quicklime [38], metakaolin [53], magnesia [54] and hydrated lime [55]. Among the materials mentioned, fly ash and metakaolin can be classified as pozzolans, silicates or aluminosilicates, which, in the presence of water, take part in a reaction with lime giving the end product in the form of insoluble components [56]. The content ranges of individual phases and a graphical representation showing the grouping of materials by chemical composition are shown in Table 1 and Figures 2 and 3, respectively.

Table 1. Chemical composition of binders and additives tested (wt%).

Binder/Additive	SiO_2	Al_2O_3	SO_3	CaO	TiO_2	Fe_2O_3	MgO	K_2O		P_2O_5	Na_2O	CO_2	MnO	F	LOI	Ref.
OPC	18.99–27.4	4.41–11.5	2.25–4.52	46.6–65.72	0.17–0.51	2.31–4.03	1.02–3.71	0.24–1.31	–	0.08–0.23	0.17–0.48	–	0.05–0.06	–	0–6.19	[31,37,57–65]
GGBS	32.7–36.77	7.77–29.43	1.46–2.37	31.49–40	0.36–1.63	0.31–5.54	5.5–13.91	0.43–0.85	–	−0.01–0.04	0–0.36	–	0.28–1.02	–	−1.49–2.67	[29,31,57,60–63,66]
ISSA	27.24–31.7	13.72–17.2	2.07–3.45	6.34–10.96	0.7–5.04	17.8–27.35	2.9–3.52	2.0–2.77	–	9.23–12.28	4.4–6.52	–	–	–	nt–0.99	[28,31,67]
Class F FA	53.97	31.15	0.727	4.01	–	4.16	1.01	2.04	–		0.89	–	–	–	–	[59]
Class C FA	34.2	19.3	2.2	25.8	–	5.64	5.07	0.52	–		2.4	–	–	–	0.11	[68]
PR	1.15–6.14	0.27–1.23	–	45.93–48.4	–	nt–0.16	nt–6.96	–	–	21.9 *–25.10	–	nt–13	–	2.23–2.41	nt–13.12	[66,69]
PG	1.26–8.8	nt–0.72	39.64–55.3	37.5–47.05	–	–	nt–0.32	nt–0.32	–	nt–0.03	nt–10.03	–	–	–	–	[31,50]
RM	9.11–21.43	4.57–26.1	nt-0.67	nt–45.15	nt–3.98	9.98–59.37	nt–0.33	nt–1.56	–	nt–0.37	nt–11.51	–	0.2–6	–	nt–13.41	[12,50,59,61]
CAC	7.38	52.9	0.31	34.1	2.23	1.83	0.37	0.45	–	0.19	–	–	–	–	–	[70]
QL	1.2	–	0.012	95.4	–	–	0.85		–	–	–	–	–	–	0.55	[68]
MgO	0.9–1.1	0.12–0.41	0.05–0.28	0.5–1.39	–	0.03–0.7	95.8–89.5	0.01–3.57	–	–	–	–	nt–0.02	–	0–2.76	[29,37,54,55,57]
MK	49.55–50.30	35.22–47	0.05–0.59	0.18–0.2	–	0.52–1.05	0–0.36	0.19–0.28	–	nt–0.28	0–0.01	–	–	–	nt–4.23	[61,65]
Zeolite	69.96	13.61	–	3.61	0.02	1.38	0.51	1.79	–		1.61	–	0.03	–	7.47	[71]
CaO	–	–	–	98.9	–	–	0.2	–	–	nt	–	–	–	–	–	[72]

* available 2.53; nt: not tested; LOI: loss on ignitation; OPC: ordinary Portland cement; GGBS: ground granulated blast furnace slag; ISSA: incinerated sewage sludge ash; FA: fly ash; PR: phosphate rock; PG: phosphogypsum; RM: red mud; CAC: calcium aluminate cement; QL: quicklime; MK: metakaolin.

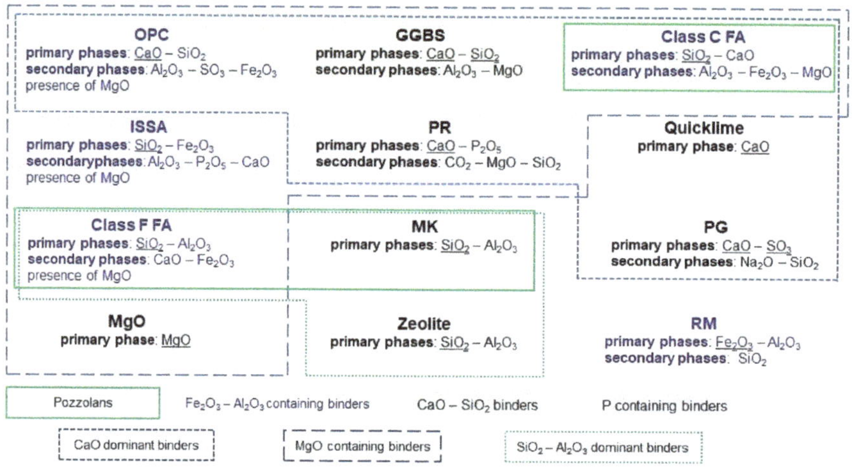

Figure 2. Classification of binders and additives for the stabilization/solidification of soil contaminated with heavy metals.

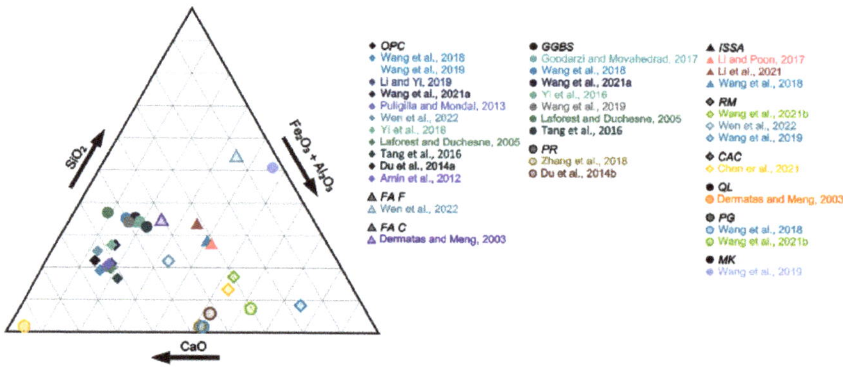

Figure 3. The ternary CaO-SiO_2-Al_2O_3 + Fe_2O_3 diagram of selected materials analyzed in the paper based on Table 1 [12,28,29,31,37,50,57–70].

The presence of individual phases affects not only the bonding process of the binder, but also the nature of the processes involved in the immobilization of contaminants. For example, the presence of phosphorus in ISSA can potentially affect the transformation of metals to insoluble forms of phosphates [67], and the presence of iron oxides (e.g., in red mud) can result in the transformation of metals to reducible forms [48]. Additionally, in order to ensure the proper course of the stabilization/solidification processes, the use of various types of activators was suggested (such as reactive MgO [29,55,57], cement [29] or hydrated lime [55] as an additive to GGBS), or mixtures of binders, additives and/or activators (such as fly ash and slag with the activator solution [58] or magnesium potassium phosphate cements mixed with FA or GGBS [73]). Zhang et al. [66] proposed a new binder produced on the basis of oxalic acid-activated phosphate rock and Du et al. [69] additionally enriched this material with monopotassium phosphate and reactive magnesia, which can be applied to Zn and Pb contaminated soils with elevated concentrations. The use of an activator (e.g., oxalic acid) allows controlling one of the most important determinants of the reaction, namely the relatively low pH that reduces the solubility of phosphate and metal compounds (e.g., Pb) [74]. Moreover, Li et al. [67] applied a similar activation method (using oxalic acid) of incinerated sewage sludge ash to immobilize high concentrations of Pb (5000 mg/kg). Considering the wide variety of approaches proposed in the literature, detailed characteristics of the most commonly selected materials for stabilization/solidification of contaminated soils, mainly in view of soil reinforcement, are presented in the following section of the paper. The materials potentially suitable for stabilization/solidification of contaminated soils, which have not found wider interest in practice, are not included in the paper. Examples of such materials include paper ash [75], high belite sulfoaluminate cement [59], municipal solid waste incineration fly ash [70], silica fumes [49,76,77], calcium carbide residue [78] and calcium sulfoaluminate cement [79]. However, this paper includes a short section dedicated to SPC binder, which, despite its lack of presence in the literature, is an environmentally friendly composite material that achieves very good strength and leachability results, and, thus, represents an interesting solution that can be analyzed in the future [72].

3.1. Ordinary Portland Cement (OPC)

The main binder used in soil stabilization/solidification is Portland cement OPC [45,80–90]. It is known to be highly effective in binding contaminants, as determined by numerous leachability tests and geochemical modelling [13]. It has been claimed [91–93] that heavy metal ions precipitate to form hydroxides and are captured by hydration products due to the alkaline environment. However, studies have provided neither direct evidence for its validity, nor sufficient reasons to refute it [94–98]. Instead, based on the results of the study, the probable adsorption of heavy metal ions on the surface of the C-S-H complex was indicated [99]. In addition, through XANES and Raman spectroscopy, it was found that the S/S process resulted in the incorporation and/or adsorption of lead onto silicate and calcium hydrates and onto ettringite [100]. Although Portland cement is still the primary binder for the stabilization/solidification of contaminated soils [101,102], the disadvantages associated with its use are well known. Some inorganic impurities, including heavy metals, interfere with the hydration of this binder, so that the engineering characteristics of S/S products are reduced [103–107]. To counteract this, an increased amount of OPC is used in relation to the soil, but this entails a significant consumption of energy to produce it and, thus, also creates a high carbon footprint [108–111]. In addition, the use of cement is fraught with uncertainty regarding the long-term effectiveness of soil binding due to its low chemical compatibility with the soil [46,112–114], its susceptibility to erosion by sulfate [56], acid rain [115], freeze-thaw cycles [116], and the predicted reduction in material characteristics over time [117]. These are the main reasons for seeking alternatives to cement among waste materials that have good pozzolanic properties. Due to the significant effect of the pH of mixtures undergoing the S/S process on the leachability of heavy metals [82], wastes that cause a slight reduction in the pH of the strongly alkaline environment created during cement hydration are particularly desirable.

3.2. Ground Granulated Blast-Furnace Slag (GGBS)

A widely used substitute for parts of Portland cement is ground granulated blast furnace slag GGBS, which is an industrial by-product of steel production [53,118]. An efficient reaction with calcium hydroxide during cement hydration forming an insoluble gel (C-S-H) is proven [119]. In addition, due to the higher specific surface area of its particles, GGBS leads to the formation of more hydration product nuclei [120]. This waste binder reduces the excessive increase in pH, thus affecting the solubility of heavy metals (most metals show the lowest solubility in solutions with pH~10), while having a positive effect on the final strength of the S/S product [31,121–123]. To mitigate the effects of heavy metals (e.g., lead) on cement hydration, phosphate- and sulfate-rich materials are often incorporated into mixtures along with GGBS for further stabilization of the pH value of the soil–cement mixture at the required level [32,124–126]. The materials of this type include the incineration sewage sludge ash ISSA or phosphogypsum. It is also known that GGBS activated with sodium compounds (NaOH, Na_2SiO_3, Na_2CO_3) with strongly alkaline properties causes an excellent increase in the compressive strength of soils [127,128], proving the beneficial effect of high pH during hydration on the final mechanical properties of the substrate, but without the presence of contaminant. The study also established that the supply of additives in the form of sulfates (e.g., salt—sodium sulfate) should be strictly controlled. Research results are available that show very good immobilization of barium and improved mechanical properties with Na_2SO_4, but also a decrease in compressive strength of the S/S product with excessive dosage of this compound [129]. GGBS is also combined with activators in the form of magnesium [130] or calcium oxides, which result in the formation of more hydrotalcite-like phases [131], which, in turn, have a beneficial effect on product strength. Magnesium and calcium oxides have many other advantages, described in a separate section of this paper. An important feature of GGBS is also its favorable grain size for the S/S process. As a fine-grained material, it fills the pores, thickening the structure and, thus, seals the subsoil, further immobilizing heavy metal pollutions [118,132]. It also decreases the porosity of cement pastes [63] and increases the density of the S/S product [133]. This type of binder is proving to be more effective than cement in Pb [46], as well as Cr immobilization, but only if the impurity is present in concentrations of no more than 2000 mg/L [62]. GBBS compounded with cement at a ratio of 1:4 also shows similar or better effects on geogenic As-containing soils than cement without any amendments [134]. Furthermore, GGBS is used in the production of geopolymer-based cement and recycled aggregate (GRAC) [135], where, with fly ash, it is a substitute for part of the cement. Nevertheless, GGBS is very frequently combined with cement, or, alternatively, lime. However, studies have indicated low compressive strengths achieved by S/S products using GGBS and lime [136].

3.3. Incinerated Sewage Sludge Ash (ISSA)

Studies of soil stabilization/solidification using ISSA and cement mixtures have indicated the high effectiveness of fly ash of this type in the adsorption, precipitation and physical immobilization of contaminants [28]. Due to the ability of the S/S product structure to buffer acids, there is an incorporation of metals into the soil structure in hydrated form, which manifests itself in a lower concentration of the harmful form of metals (lead, zinc, copper) [137]. When ISSA was used as an additive to a binder composed of OPC and GGBS, very good strength results and values competitive with other additives were obtained in leachability tests, but only at a certain molar ratio of additive to contaminant [31]. At molar ratios of 1:4 and 1:8, an excessive increase in the amount of ISSA was necessary and this resulted in an unacceptable decrease in the compressive strength of the S/S products. Stabilization/solidification studies using ISSA have also been carried out with the incorporation of an activator in the form of oxalic acid (OA), which has been successfully used in the activation of phosphate rock [67]. The aim of the addition of phosphoric acid is, in this case, for the activation of ISSA to release phosphates that react with lead to form lead phosphate hydroxides. This activator increases the effectiveness in stabilizing Pb, but its overuse can lead to the leaching of phosphate and zinc from the ISSA into the subsoil, so mixtures with these two components should be designed very responsibly.

3.4. Fly Ash (FA) and Pulverized Fly Ash (PFA)

Fly ash, as an industrial by-product, is readily used as a binder for the S/S process. For many years, it has been used as a replacement for parts of Portland cement in soil stabilization/solidification [39,42,88,117,129,138–140], especially in lead stabilization, due to its greater efficiency compared to the use of cement alone [141]. Both Class C fly ash [88] and Class F fly ash [142] are used for soil stabilization/solidification. According to ASTM C618 [143], the two classes differ primarily in the sum of SiO_2, Al_2O_3 and Fe_2O_3 contents, and calcium content. There is a lack of studies in the literature optimizing the most efficient PFA:OPC ratio, most likely due to the disproportionate amount of research effort relative to the potential benefit [46]. A ratio in the range of 1:1 to 4:1, which is usually adopted, allows satisfactory results to be obtained from the studies, verifying the effectiveness of the S/S process [144]. It should be emphasized that most PFA cannot be used as a separate binder due to their too limited free CaO content for self-hardening purposes [145]; hence, the addition of OPC is necessary, usually in an amount not less than 20% of the ash quantity. Attempts have also been made to combine PFA with lime instead of cement [146], achieving very high unconfined compressive strengths of the S/S product. [68]. In addition, it is known that high-calcium fly ash (CFA) performs very well in silicate–aluminate–phosphate geopolymerization with magnesium phosphate cement (MPC) [73]. Information concerning MPC is included in a separate section of the study. The effect of the addition of 300 g (per kg of waste) of Na_2SO_4 to the mixtures in which the waste was stabilized with fly ash with cement (25:75 ratio) and slag cement was also studied. The introduction of salt resulted in a noticeable increase in 28-day unconfined compressive strength [32]. The referenced study also showed an increased the binding efficiency of cyanide and barium contained in the stabilized waste.

3.5. Magnesium Potassium Phosphate Cement (MPC)/(MKPC)

Magnesium potassium phosphate cement is a clinker-free acid-base cement [147], which is characterized by its high strength, resulting from the reaction of MgO (dead-burnt magnesia DBM) with potassium dihydrogen phosphate, during which amorphous or crystalline k-struvite and crystallized bobierrite are formed [148,149]. In addition, chemical stabilization of lead occurs due to the reaction with residual phosphates, during which lead phosphate and pyromorphite precipitate [103]. MPC exhibits greater chemical stability than OPC, which manifests itself, among other aspects, by minimizing the adverse effects of lead on the unconfined compressive strength [150]. MPC is further used to stabilize low-activity nuclear waste (containing plutonium, neptunium, caesium, strontium, actinium, technetium and selenium) as well as cadmium, chromium, copper, nickel, lead and zinc from electro-waste [151–153]. On the basis of an analysis of the test results available in the literature, it was also found that zinc ions reduce the degree of MPC cement mineralization, although they do not affect the hydration phases of the cement [154].

3.6. Red Gypsum (RG)

Another binder used as an amendment of OPC in the S/S process is red gypsum, a by-product of the production of titanium dioxide (a white pigment widely used in the food and cosmetics industries). In combination with OPC, PFA and GGBS, and a small addition of lime, mixtures with satisfactory compressive strengths were obtained. The values turned out to be lower than those of the soil combined with cement, but higher than the assumed minimum values (0.350 MPa) [155]. In light of the test results, the gypsum-GGBS binder was considered as an environmentally friendly alternative to OPC, guaranteeing sufficient mechanical properties of the S/S product.

3.7. Phosphate Rock PR and Phosphoric Acid (PA)

Phosphate rock and phosphoric acid have the potential to convert the lead present in soil into stable forms, such as pyromorphite. For this reason, they are widely studied and even recommended for use in the stabilization/solidification of soils contaminated with this heavy metal [19,47,70,156,157]. The effectiveness of phosphorite as an additive in binding zinc is also confirmed [158,159]. When using phosphates in the S/S process, the acidic environment required for their effectiveness must be taken into consideration [69]. In addition, it is pointed out that the grade of phosphate in the ore is not consistent [66] which, for obvious reasons, affects the effectiveness of this material in the S/S process. For these reasons, researchers designing soil mixtures with stabilizing/solidifying additives often opt for the implementation of a phosphate rock activator in the form of phosphoric acid. Other additives are also used in conjunction with this raw material, including monopotassium phosphate and reactive magnesia, forming a binder indicated by the KMP symbol. Studies have confirmed the rather high early strength of KMP-stabilized products due to the hydration rate of reactive magnesia [69], as well as the possibility of increasing strength and decreasing leachability when KMP is used together with an accelerated carbonation process [160]. The aforementioned process, also investigated for reactive magnesia (in a separate part of the paper), is introduced as a factor corresponding to an accelerated version of natural weathering [161].

3.8. Phosphogypsum (PG) and Potassium Dihydrogen Phosphate (KDP)

Phosphogypsum is a by-product of phosphoric acid production, containing calcium sulfate and phosphate residues. As a phosphate-rich material, it exhibits similar properties to potassium dihydrogen phosphate KDP and ISSA, i.e., it reduces the excessive increase of pH of the soil–cement mixture. Thus, during hydration of the cement or another binder, conditions are unfavorable for the dissolution of metals [28,124,162]. Both PG and KDP are used in the S/S process mainly as additives to reduce the reactivity of contaminants. PG has been combined, among others, with Basic Oxygen Furnace Slag (BOFS) and Calcium Carbide Residue (CCR), the total composition of which is rich in dicalcium and tricalcium silicate, CaO, $CaSO_4 \cdot 2H_2O$, and $CaCl_2$, as well as phosphates, fluorides, and sulfates, to ensure that binder hydration can develop in contaminated soil [33].

3.9. Red Mud (RM)

Red mud is an alkaline by-product of the Bayer process, taking place during the production of aluminium (bauxite refining) [163]. It forms a toxic and corrosive sludge with a high pH and very high iron oxide content. It has been applied to the S/S process to stabilize heavy metals (lead, zinc and cadmium) as an additive to phosphogypsum and Portland cement [50]. A binder of this composition shows sufficient strength and binding characteristics for metals in the stabilization/solidification process at landfill sites, while also contributing to the recycling of red sludge. It has also been used successfully to reduce the bioavailability of Cd, Pb and Zn [163].

3.10. Calcium Aluminate Cement (CAC)

Calcium aluminate cement is a relatively rarely studied binder in the stabilization/ solidification of soils contaminated with heavy metals. It is more often included in the remediation of soil contaminated by organic pollutants with possible accompanying inorganic pollutants. On the basis of a study by Contessi et al., it was found that CAC exhibits a completely different mechanism in the immobilization of lead than is the case with the use of OPC [99]. Due to its high sulfate content, the reactivity is strongly shifted towards the formation of ettringite, in which case, Pb^{2+} ions are incorporated into its structure, in place of some calcium ions. The results obtained by Bougharraf et al. [164] led to the conclusion that CAC was more effective in binding organic contaminants compared to inorganic contaminants. At the same time, the binding of heavy metals was determined to be satisfactory, especially with an achieved compressive strength higher than 1 MPa. In other studies, the retention of metals (lead, copper, zinc) using CAC was determined at a level of 99.9% [165], and chromium in the form of Cr^{6+} at a level exceeding 90% [166], which was a better result compared to using OPC. However, in addition to the leachability results, the strength tests for the S/S products using calcium aluminate cement are above average. Depending on the amount of binder addition, the strength is higher than when using sulfate-resistant Portland cement from about 0.5 to even 4.0 MPa [167]. Despite the excellent strength results and the leachability of contamination, the use of CAC is very limited, due to the high cost of this raw material.

3.11. Bentonite

Bentonite is an additive to various types of binders in S/S processes. It has been found to affect the decrease in strength of the S/S product, but is excellent at improving the degree of immobilization of contaminants, both organic and inorganic [164].

3.12. Lime (CaO), Quicklime (QL), Lime Production Waste (LPW)

Lime is a material mainly used for soil stabilization. When added to soils with a high content of clay particles in the presence of water, it undergoes cation exchange and further reactions. As a result of the modification of the electrical charge density, the forces of intermolecular attraction increase, which directly results in a strengthening of the soil structure [3] and a favorable change in the Atterberg limits [168]. The long-term effectiveness of lime stabilization of loess has also been proven [169], which is extremely important for the design of permanent structures on this type of subsoil. Furthermore, hydrated lime is an activator for GGBS-type binders [170]. CaO proves to be very effective as an additive to GGBS at a ratio of 1:9, as demonstrated in the studies of contaminated soils subjected to S/S, performed after 18 months of treatment [170]. It has also been successfully used as an activator at a ratio of 15% to 85% GGBS [54] in the S/S process of soils contaminated with lead and zinc. It is also used as an additive to red gypsum (RG), GGBS and PFA to increase the alkaline pH of the mixture, resulting in better hydration and, consequently, higher strength [155]. In a study by Moon and Dermatas, it was also proven to effectively bind lead to acceptable levels using Quicklime and FA [171]. Lime production waste (LPW), obtained if the raw calcite or carbonate materials undergo burning at temperatures lower than 960 °C, is a by-product containing lime, Al_2O_3, MgO, Fe_2O_3 and other components, which is also of interests in soil stabilization, especially in mixtures with blast furnace slag and red mud [12].

3.13. Reactive Magnesia (MgO)

Magnesium oxide is usually not used as a separate binder, but as an additive to improve the properties of binder materials other than cement [172,173]. Its main advantage is to increase the effectiveness of binders in immobilizing contaminants, both inorganic and organic [16]. MgO facilitates the transport of the strongly alkaline binder to the work site, and counteracts excessive settlement of the substrate during the S/S process, thus increasing the cost-effectiveness of its application [29,55,131]. In addition, it increases the neutralizing properties for an acidic environment, and this is an extremely valuable feature when the S/S process substrate is exposed to acid rain [34]. In addition to the widespread use of MgO as a binder additive, experimental processes to stabilize/solidify contaminated soil using only reactive magnesia have also been carried out; however, the unconfined compressive strengths of the product S/S obtained were 80–90% lower than those obtained using OPC [16,70]. For this reason, an additional factor, i.e., carbonation, was incorporated into the blends under development [174,175]. Carbonation was carried out after the mixture was placed in the molds. To test the effectiveness of this solution for lead- and zinc-contaminated soils, comparative UCS and leachability tests, were conducted for the mixtures stabilized with MgO and CO_2 injection, and for the mixtures stabilized with OPC. In each case, only 5% binder dry weight and 95% soil dry weight were used. The samples containing OPC were tested after 28 days of treatment, while samples treated with MgO and saturated with carbon dioxide were tested after 72 h (lead-contaminated) and 120 h (zinc-contaminated). CO_2 was found to cause relatively high variability in the unconfined compressive strengths achieved by S/S [175,176], but at the same time, at Pb and Zn concentrations > 4000 mg/kg dry weight of soil, the strengths were significantly higher than those achieved with OPC.

3.14. Geopolymers and Geopolymeric Binder Made of NaOH-Activated Metakaolin (MK)

Great hopes were entertained in geopolymers for an environmentally friendly cement substitute, also for use in the S/S process. In addition to its high unconfined compressive strength, this innovative binder provides high resistance to acidic and high temperature environments, low permeability and extended durability [177–181]. However, despite intensive research and continuous development, this material has failed to revolutionize the binder segment. While the strength and leachability results of S/S products are competitive with those using traditional binders [182,183], the costly activators that enable the synthesis of the geopolymers (sodium or potassium hydroxides, liquid water glass) block their widespread use. The situation is partly alleviated by the fact that geopolymers based on slag [184] and fly ash [58,185] are currently being produced and studied in addition to metakaolin-based geopolymers [99,186]. These raw materials, as waste materials, increase the ecological value of the solution, which, in addition to very good characteristics, opens the possibility of its dissemination. However, the implementation of geopolymers has been quite limited so far.

3.15. Natural and Synthetic Zeolite

Both natural and synthesized zeolites from fly ash are known for their very good sorption properties, thus gaining reputation as additives that enhance the immobilization of contaminants in the S/S process. These properties are used towards the binding of petroleum substances [187] and inorganic substances in the form of heavy metals [188]. Due to the specific structure of zeolites rich in voluminous free spaces, it is possible to encapsulate large cations, such as sodium, potassium, barium, and calcium, and even molecules and catalytic groups of water, ammonia, carbonate and nitrate ions [189]. Furthermore, during the study, it was found that, due to the large reaction surface area of zeolite, resulting from its significant porosity, the S/S process using this amendment combined with OPC achieves higher unconfined compressive strength values and greater durability [190]. It was also proven that the high cation exchange capacity and high content of clinoptilolite, as well as the low concentration of potential toxic elements, cause a high potential for treating industrially contaminated soil with use of thermally treated natural zeolite [71].

3.16. SPC Binder

The reaction of single superphosphate (SSP) and calcium oxide (CaO), used simultaneously in the S/S process, leads to the formation of hydroxyapatite (HA) in the soil structure. The combination of SSP and CaO, defined as the SPC binder, thus provides benefits in the form of a densification of the structure and an increase in the unconfined compressive strength of the treated soil. In parallel, it offers an alternative to a synthetic hydroxyapatite, which, despite its many advantages, is a material that is too expensive for widespread implementation in the S/S process. Meanwhile, the environmentally friendly SPC binder, at a relatively low cost, causes the formation of calcium-phosphate crystals (including HA), which form strong cementation bonds between soil particles, and the various reaction products fill the pores (especially inter-aggregate pores) [72]. In addition, studies were performed on in situ stabilized/solidified soil with use of the SPC binder, which confirmed the good results shown during laboratory testing [40].

3.17. EnvirOceMTM

For solutions to local contamination problems, targeted mixtures are being developed for a given soil type and type of contamination, as in the case of the remediation of the former Astra military explosives Fireworks site in SE England [191]. A subsoil contaminated with high concentrations of zinc, lead and copper was subjected to an S/S process using superfine sulfate-resisting Portland cement, functioning under the name EnvirOceMTM. Satisfactory leachability results were obtained, both after 28 days and after 4 years of the S/S process. Unfortunately, the compressive strength of the S/S product was not verified.

4. Effectiveness of Mixtures and Optimization Testing Methods

The effectiveness of the S/S process is defined by the geomechanical parameters and the leachability of its products. The applicability of the solution is also strongly influenced by its durability. These properties are verified by analyzing the results of specific tests, the main types of which are summarized in this paper. Preliminary tests, which were used by the authors of the cited publications to describe the components of the designed mixtures, such as mineralogical composition, specific density, grain size, etc., were not systematized. At this point, it should be emphasized that research results indicate that no single binder can perfectly remediate all types of heavy metal-contaminated soils, and binder selection is a site-specific problem [192]. With the above in mind, a specific soil contaminated with a particular heavy metal should be tested to determine the best binder combination for stabilization/solidification purposes using data and experience described in the literature for similar cases. In contrast, valuable studies have been reported in the literature, with design charts directly indicating the proportions of water and binders in regard to the performance criterion of the S/S process of similar soils contaminated with the same heavy metals [42].

4.1. Unconfined Compression Strength

The geomechanical properties of a S/S product are mainly represented by unconfined compressive strength, and the test used and performed on mixture samples is the uniaxial compression strength test (UCS), which can be conducted according to ASTM [193]. It is conducted as a standard for both soil stabilization [194] and the S/S process. In almost all sources analyzed, the strength is tested after 28 days of treatment, often also after 7 and/or 90 days; the testing of 1-, 3- or 160-day samples is rare. The age of the test specimens is mainly determined by the mixture formulations, especially with regard to the fly ash content, which generally reaches full strength later. The results of the UCS tests are summarized for comparison purposes in Table 2.

4.2. Leaching Behavior

The Toxicity Characteristic Leaching Procedure (TCLP) is the most widely used test by which researchers determine the binding efficiency of contaminants in S/S products. Performed according to the procedures described in US EPA standards [196], it allows the comparison of effectiveness between different types of mixtures, including studies performed by different authors. Nonetheless, the TCLP test is not considered adequate by all authorities in the field of S/S testing. TCLP is mainly dedicated to the analyses of the leachability of contaminants from landfill waste, and does not reflect the conditions of the natural subsoil [167,197]. For this reason, some researchers use the Simulated Precipitation Leaching Procedure Test (SPLP) [198] instead, which allows determination of the leachability of contaminants from the ground into groundwater [28]. However, they are in a distinct minority, so the results from the more popular TCLP-type test were adopted for comparison. The data given in Table 2 were compiled from an analysis of the leachability test results presented in the papers studied. The leachability potential of contaminants from an S/S product can be estimated in two different scenarios: in the diffusion process, the most likely way corresponding to reality (semi-dynamic tank leaching [199,200]); or after grinding the product immediately before testing, which obviously maximizes the leachability of contaminants and corresponds to the worst possible scenario (Batch Leaching test [201]). In addition, contaminant leachability tests at varying pH, as well as the Acid and Base Neutralization Capacity (ANC/BNC), are used, depending on the scope of testing and the planned use of S/S products. The hydraulic conductivity test, which captures a picture of the water filtration capacity of the treated subsoil, should also be considered. In the case of many binders, the binding of contaminants also takes place by physical immobilization, so that an open filtration channel would allow the leaching of contaminants into the deeper layers of the subsoil and, consequently, into the groundwater. For this reason, hydraulic conductivity is closely related to leachability and is often tested as an indicator of process efficiency [46]. Another test that gives a picture of the leaching of contaminants with solutions of increasing aggressiveness is the BCR SEP (The Community Bureau of Reference Sequential Extraction Procedure). It allows the quantification of exchangeable fractions, reducible fractions, oxidizing fractions, and heavy metal residues in the soil. The test procedure described by Davidson et al. [202] has also been adopted in tests by other researchers. A quantitative description of the chemical stability of heavy metals in soil is possible by adopting a selected variable with an appropriate formula, such as the relative binding intensity (IR) index. S/S products are also studied in terms of pH, which is one of the factors shaping the leachability of heavy metals [203]. There are known results indicating that the lowest leachability of lead and zinc occurs when the pH of the product approximates 9.5 [204,205]. Individual countries regulate the maximum contaminant level (MCL), which corresponds to the concentration allowed in drinking water, by means of internal regulations. Due to the differences between the laws and the different ranges for individual metals, this study indicates the maximum allowed value for lead after Toxicity limitation (US EPA) standard (5 mg/L) and for zinc after Toxicity limitation (GB5085.3-2007) (100 mg/L). As mentioned in Section 4.1, a comparison of the results of individual reported studies of both UCS and leachability is not representative, and, furthermore, due to the large number of variables, impossible to present in a single chart. For this reason, Figure 4 includes data from only a few selected literature entries using different binders to show the degree of variation in the results obtained.

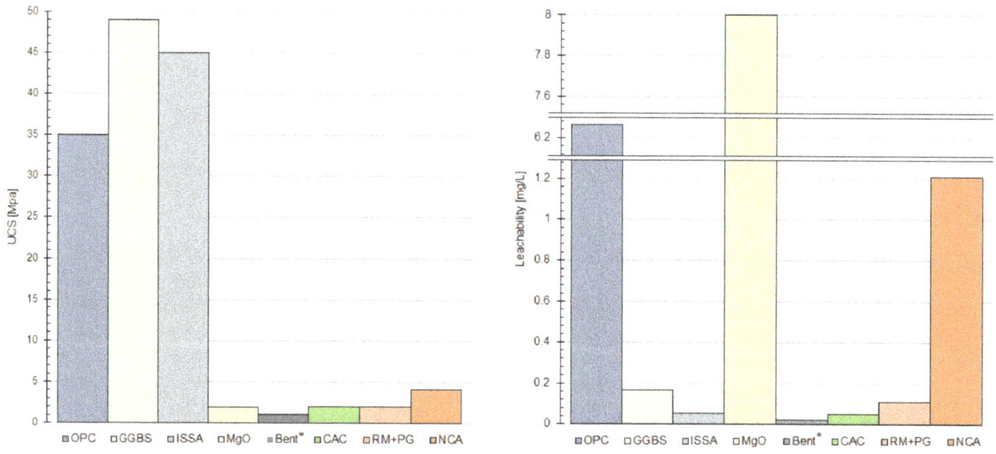

Figure 4. A few selected test results based on analyzed data: (**a**) unconfined compressive strength; (**b**) leachability. * Bent: Bentonite.

4.3. *Microstruture Investigation*

Determining the changes that have occurred in the microstructure of the soil as a result of the binder introduced requires specialized procedures and very sophisticated equipment. It is also necessary to have detailed knowledge in order to be able to analyze the results in terms of the resulting chemical products, such as phase hydration products or precipitated heavy metal compounds. A common microstructure test to determine mineralogical composition is X-ray diffraction [7] and thermogravimetric analyses [126]. Hydration heat evolution testing [103] is also implemented. In addition, scanning electron microscopy (SEM) and transmission electron microscopy (TEM) are used as confirmatory and refinement imaging tests for XRD-type studies. The TEM technique is particularly useful for detecting the presence of the cemented phase and the pozzolanic phase, i.e., by determining the C-S-A-H (calcium silicate aluminate hydrate) bonds present [3]. It is also important to determine the microporosity of the S/S-treated subsoil [33], which is made feasible by the mercury intrusion porosimetry (MIP) test [69]. Analysis of the MIP test results allows prediction of the mechanical behavior of the solidified soil and the extent to which its pores are filled by stabilization products.

Table 2. Abbreviated research results.

No	Symbol *	S/W [%]	Binder/Additive [%]			WC [%]	Heavy Metal Type	Heavy Metal [mg/kg]	UCS [MPa]	Leaching mg/L	Ref.
			PC	GGBS	ISSA					TCLP	
1	S/PC	50	50	–	–	20	Pb	1941	35	6.26	[31]
	S/BC	50	25	25	–				49	0.168	
	Pb/ISSA = 1:2	50	18.10	18.10	13.8				45	0.055	
			SC	OPC	ISSA					Leachate pH	
2	C1	80	20	–	–	24–50%	Ba	168,000	7.9	2.2	[32]
	H3	50	–	37.5	12.5				8.6	4.3	
			CAC/OPC/GGBS	TWEEN 80	BENTONIT					AFNOR NF X31-211	
3	M1	73	5/–/–	0.025	–	22	Pb	96.7	0.65	0.101	[164]
	M2	68	10/–/–	0.05	–				2.0	0.05	
	M3	68	5/–/–	0.05	5				0.45	0.197	
	M4	63	10/–/–	0.075	5				0.95	0.143	
	M5	63	5/–/–	0.075	10				0.15	0.290	
	M6	58	10/–/–	0.1	10				1.1	0.022	
	M7	73	–/3/2	0.025	–				0.4	0.290	
	M8	68	–/6/4	0.05	–				1	0.068	
	M9	68	–/3/2	0.05	5				0.22	0.300	
	M10	63	–/6/4	0.075	5				0.95	0.105	
	M11	63	–/3/2	0.075	10				0.25	0.190	
	M12	58	–/6/4	0.1	10				1	0.068	
			PC	MgO	CO_2					Leached Pb [mg/kg]	
4	Pb + PC	95	5	–	–	7.5	Pb	4000	2.0	100	[37]
	Pb + PC	95	5	–	–			16,000	0.9	1100	
	Pb + MgO + CO_2	95	–	5	used			4000	1.9	8	
	Pb + MgO + CO_2	95	–	5	used			16,000	1.95	9.8	

Table 2. *Cont.*

No	Symbol *	S/W [%]	Binder/Additive [%]			WC [%]	Heavy Metal Type	Heavy Metal [mg/kg]	UCS [MPa]	Leaching mg/L	Ref.
			$Ca(OH)_2$	MgO	GGBS		Zn/Pb			TCLP	
	CGZn0.25	75	3.75	–	21.25		0.25		4.1	0.264	
	CGZn0.5	75	3.75	–	21.25		0.5		1.9	0.220	
	CGZn1	75	3.75	–	21.25		1		0.45	0.178	
	CGPb0.25	75	3.75	–	21.25		0.25		5.9	ND	
	CGPb0.5	75	3.75	–	21.25		0.5		7.9	0.072	
5	CGPb1	75	3.75	–	21.25	40	1		7.9	0.18	[54]
	MGZn0.25	75	–	3.75	21.25		0.25		4.9	0.091	
	MGZn0.5	75	–	3.75	21.25		0.5		5.0	0.082	
	MGZn1	75	–	3.75	21.25		1		3.2	0.076	
	MGPb0.25	75	–	3.75	21.25		0.25		5.0	0.066	
	MGPb0.5	75	–	3.75	21.25		0.5		5.9	0.062	
	MGPb1	75	–	3.75	21.25		1		7.1	0.166	
			OPC	ISSA						SBET [mg/kg]	
6	H0	90	10	–		20	Pb	5000	10	36	[28]
	H0.2	90	8	2					4.6	35	
	H0.5	90	5	5					2.6	31.5	
	8	soil	OPC	GGBS						TCLP	
	O5	96	4	–					3.2	0.011	
	O4G1	96	3	1					3.7	0.015	
7	O2.5G2.5	96	2	2		20	As	170.4	6.0	0.021	[41]
	O10	92	8	–					7.5	0.06	
	O8G2	92	6	2					7.8	0.065	
	O5G5	92	4	4					9.5	0.012	
			RM	PG	OPC					TCLP	
	RPPC7.5	92.5	4.3	1.1	2.1				0.726	4.2/0.8/8	
8	RPPC10	90	5.7	1.4	2.9	nt	Zn/Pb/Cd	5000	1.1	0.8/0.3/0.9	[50]
	RPPC15	85	8.6	2.2	4.2				2.013	0.3/0.11/0.7	
	PC10	90	–	–	10				3.3	0.4/0.9/0.8	

Table 2. *Cont.*

No	Symbol *	S/W [%]	Binder/Additive [%]			WC [%]	Heavy Metal Type	[mg/kg]	UCS [MPa]	Leaching mg/L	Ref.
9			CCR	PG		16.5–17.6	Ni/Zn	6352/5352		China HJ/T 299	[33]
	10% bin/dos	90	6	3	1				0.46	0.01/0.12	
	<10%bin-dos	>90	<6	<3	<1				<0.3	>0.06/>0.7	
10			KSil	FA	KOH		Zn	15,900		TCLP	[183]
	KSil0.46KOH	21	16.8	58.8	3.4	18.8			0.75	8.67	
			OPC	FA	lime						
	OPC lime	35.8	7.1	50	7.1	22.2			2.0	<0.001	
11			NCA			13	Pb	10,000	*** 7d	*** 7d	[30]
	NCA ** 10%	90	10	–	–				4.1	1.21	
	NCA ** 20%	80	20	–	–				6.32	0.318	
	NCA ** 30%	70	30	–	–				10.12	0.075	
	NCA ** 40%	60	40	–	–				11.16	0.027	
12			SPC	–	–		Pb	9710		TCLP	[72]
		92	8	–	–	22			0.352	1.8	
		90	10	–	–	22			0.432	0.9	
13			OPC	GGBS	–	–	As	1985		TCLP	[134]
	O4G1	95	4	1	–	–			1.1	4	
	O2.5G2.5	95	2.5	2.5	–	–			1.05	5.3	
14			OPC	CaO	MgO	7.5	Pb	16,000		SBLT	[192]
	Pb + OPC	95	5	–	–				1.0	1000	
	Pb + CaO	95	–	5	–				0.18	8000	
	Pb + MgO	95	–	–	5				0.05	2	

S/W: Soil or Waste; WC: water content; TCLP: Toxicity Characteristic Leaching Procedure; SBLT: single batch leaching test; * Symbol of the sample after ref.; ** NCA: New curing agent, the composition of the binder is not specified; *** 7d: test conducted on the seventh day of treatment; GGBS: ground granulated blast furnace slag; ISSA: incinerated sewage sludge ash; PG: phosphorus gypsum; KDP: potassium dihydrogen phosphate; MKPC: magnesium potassium phosphate cement; MPP: mono-potassium phosphate; DBM: dead burnt magnesia; MPP + DBM = MPC; KMP: oxalic acid-activated phosphate rock + KH_2PO_4 + MgO [1:1:2]; SC: slag cement; CH: $Ca(OCl)_2$; SS: Na_2SO_4; OF: OPC + FA (75:25); FA: fly ash; Tween 80: additive increasing the hydrophilicity of organic parts in the soil.Careful attention should be paid to the high imperfection of the above comparison. This is because the various tests used different amounts of binders in relation to the dry weight of the soil and different moisture contents of the mixtures, and different types of subsoil were treated. Furthermore, the contaminants contained in the soils were present in different concentrations. In addition to the aforementioned factors, the UCS is also affected by the shape and size of the samples, making the comparison subject to additional distortion. Nevertheless, in the present study, the authors have quoted the values obtained, due to the often quite different order of magnitude of the compressive strength achieved. The minimum value for the compressive strength of an S/S product is commonly considered as 0.350 MPa, which is the value specified by the US EPA for S/S waste in landfills [195]. For engineering purposes, it is necessary to know the design loads and stresses arising in the soil in order to estimate the minimum compressive strength of the S/S product constituting the subsoil for the proposed development.

4.4. Electrical Resistivity

Electrical resistivity EC is a criterion that has been used over the years both in assessing the hydration process of cement pastes [206], and the mechanical and deformation properties of soils [207,208]. Its immense advantages for use in geotechnics and geoengineering are its relatively low cost and time-efficient testing and, above all, its lack of interference with the natural soil structure [209]. The work of Liu et al. showed that the electrical resistivity of soil–cement mixtures can be correlated with the strength (q_u) of UCS with a good fit [210] in an exponential function [211]. On the other hand, the study of Chen et al. gave rise to the conclusion that as the concentration of lead contamination in the soil increases, a lower magnitude of electrical resistivity is obtained, while it increases with the time of the treatment process and the binder content of the mixture [101].

4.5. Durability

Durability is understood as the degree to which the mechanical and bonding properties are retained after an adverse ageing factor, e.g., drying–wetting cycles, freeze–thaw cycles, sulfate attack or acid attack, as well as exposure of the S/S product to an environment with average erosion properties over a period of several years [212]. Due to the potential impact of acid rain, freeze–thaw, drying–wetting and other environmental factors, it is necessary to know the long-term effectiveness of the solution used in the S/S process. The types of tests listed below reflect adverse natural phenomena that can reduce the originally achieved S/S product characteristics.

4.5.1. Drying–Wetting Cycle

One of the ageing factors to which the products of the S/S process are subjected is the drying–wetting cycle. During this process, soil–cement specimens are alternately heated in an oven for 48 h and at a preset temperature (30–40 °C), followed by soaking in distilled water at a stably maintained temperature over 24h [35,131]. During testing, the weight loss after each 72-h cycle and the compressive strength are usually checked. The strength is compared with that of a control test, in which the specimens are cured under stable conditions, as described in point 2. of the paper. Data can also be found on the compressive strength and leachability of specimens subjected to this ageing factor. Depending on the type and purpose of the test, it is possible to modify any of the test conditions, e.g., soaking in water with a reduced pH, corresponding to acid rain, or being subjected to drying at a higher temperature than actually prevailing in summer (60 °C for 24 h) [50]. Basic test guidelines can be found in the ASTM Standard [213].

4.5.2. Sulfate and Acid Attack

The ageing effect of cements, and consequently of cement–soil mixtures, is also influenced by the chemical composition of the environment in which it occurs. Both sulfate and the acidic pH of liquids negatively affect the strength of cement, mainly due to the degradation of C-S-H bonds [214]. The cements based on waste materials activated with MgO show a slightly lower sensitivity [173]. In order to determine the magnitude of this effect, samples are placed in suitable solutions, e.g., 5% Na_2SO_4 or, alternatively, acid, for a period of several tens of weeks [53,179]. The impact of acid rain can also be mapped in TCLP studies using an extraction liquid with pH = 2.0 to 7.0 [34,115]. There are no general methodological guidelines for testing the effects of sulfates and acids on the strength of concrete or S/S products, and both the scarcity of research in this direction and the need for its development is emphasized by all authors addressing this challenge [215].

4.5.3. Freeze—Thaw Impact

The effect of varying temperatures is important in the near-surface layers of the subsoil, where its fluctuations have a real impact on product S/S. This is reflected in freeze-thaw cycles, during which a cement–soil sample is placed in an apparatus dedicated to this test [116]. Alternatively, the soil sample is transferred from the freeze–thaw cabinet to a water container at a stably maintained positive temperature [50]. Regardless of the equipment used, the sample is subjected to alternating temperatures of −20 °C and +25 °C, with a change in ambient temperature every 24 h [116]. Basic guidelines for the test can be found in the ASTM Standard [216]. In a long-term study by Al-Tabbaa et al., 5-year old soil–cement samples with different binder types were subjected to 12 freeze–thaw cycles, during which the freezing temperature was set at −10 °C [39]. Similar to the drying–wetting cycles, the weight loss after each freeze–thaw cycle, as well as leachability and compressive strength, are checked in this type of study.

5. Conclusions

A subsoil that requires remediation must meet certain standards, due to minimum strength parameters and a maximum allowed leachability of contaminants. The analysis of research results presented in the literature enabled the determination of effective binder and additive mixtures in the stabilization/solidification of soils contaminated with heavy metals. Based on the analysis carried out, the following conclusions were drawn:

1. The variety of proposed binders, additives and their mixtures and methods of activating the materials is very extensive in the literature, providing engineers with a wide range of options depending on the geochemical conditions of the treated site.
2. Despite its many disadvantages, the most popular binder in the S/S process is Ordinary Portland Cement.
3. Implementation of waste materials such as GGBS, FA, ISSA as amendments for part of the OPC for the stabilization/solidification process is becoming common practice, with many environmental and economic advantages.
4. Replacing part of the cement with PFA or ISSA fly ash results in a significant decrease in the strength of the S/S product, but does not increase the leachability of the contaminants.
5. The implementation of GGBS in place of part of the OPC results in an increase in strength, but significantly increases the leaching of contaminants when used in too large a quantity. The addition of an activator (e.g., MgO) significantly improves the ability of GGBS solidification.
6. Considering the frequency of undertaking S/S process studies using red gypsum, red mud, calcium aluminate cement, bentonite, zeolites and superfine sulfate-resisting Portland cement, these materials should be considered niche products, effective for use only under specific conditions.
7. In optimizing the mixture of binders and additives for the S/S process of heavy metal-contaminated soils, one of the main factors considered should remain the ecological aspect.
8. The key studies assessing the effectiveness of S/S processes of contaminated soils are UCS and leachability studies. However, the scope of the latter varies widely and often does not take into account the actual conditions in the soil medium.
9. The often-overlooked ageing tests, which take into account the effects of external factors on the mechanical and chemical stability of the resulting bonds when assessing durability, should be important in the evaluation of the S/S method.

Author Contributions: Conceptualization A.L. and J.F.; methodology A.L. and J.F.; formal analysis A.L.; resources A.L and J.F.; writing—original draft preparation A.L.; writing—review and editing J.F. All authors have read and agreed to the published version of the manuscript.

Funding: This research was supported by the "Fly ash as the precursors of functionalized materials for applications in environmental engineering, civil engineering and agriculture" no. POIR.04.04.00-00-14E6/18-00 project carried out within the TEAM-NET program of the Foundation for Polish Science, co-financed by the European Union under the European Regional Development Fund.

Institutional Review Board Statement: Not applicable.

Informed Consent Statement: Not applicable.

Data Availability Statement: Not applicable.

Conflicts of Interest: The authors declare no conflict of interest.

References

1. Reethu, B.; Kumar, M.S.; Sharath, G.; Ramanjaneyulu, B.; Manchiryal, R.K. Stabilization of clayey soil using Gypsum. *J. Stud. Res.* **2020**, *5*, 886–888. [CrossRef]
2. Consoli, N.C.; Prietto, P.D.M.; Lopes, L.d.S.; Winter, D. Control factors for the long term compressive strength of lime treated sandy clay soil. *Transp. Geotech.* **2014**, *1*, 129–136. [CrossRef]
3. Al-Mukhtar, M.; Khattab, S.; Alcover, J.F. Microstructure and geotechnical properties of lime-treated expansive clayey soil. *Eng. Geol.* **2012**, *139–140*, 17–27. [CrossRef]
4. Pakbaz, M.S.; Alipour, R. Influence of cement addition on the geotechnical properties of an Iranian clay. *Appl. Clay Sci.* **2012**, *67–68*, 1–4. [CrossRef]
5. Maghous, S.; Consoli, N.C.; Fonini, A.; Pasa Dutra, V.F. A theoretical-experimental approach to elastic and strength properties of artificially cemented sand. *Comput. Geotech.* **2014**, *62*, 40–50. [CrossRef]
6. Al-Jabban, W.; Knutsson, S.; Laue, J.; Al-Ansari, N. Stabilization of Clayey Silt Soil Using Small Amounts of Petrit, T. *Engineering* **2017**, *9*, 540–562. [CrossRef]
7. Sharma, N.K.; Swain, S.K.; Sahoo, U.C. Stabilization of a Clayey Soil with Fly Ash and Lime: A Micro Level Investigation. *Geotech. Geol. Eng.* **2012**, *30*, 1197–1205. [CrossRef]
8. Krishnakumari, M.B.; Swathika, A.; Subasree, M.; Irin Priscilla, P. Stabilization of Clayey Soil Using Various Additives. *Int. Res. J. Eng. Technol.* **2008**, *6*, 1177–1181.
9. Fayaz, M.A. Laboratory Investigation for Stabilization of Clayey Soil in Kashmir Valley. *Int. J. Res. Appl. Sci. Eng. Technol.* **2018**, *6*, 2909–2914. [CrossRef]
10. Harbottle, M.J.; Al-Tabbaa, A.; Evans, C.W. A comparison of the technical sustainability of in situ stabilisation/solidification with disposal to landfill. *J. Hazard. Mater.* **2007**, *141*, 430–440. [CrossRef] [PubMed]
11. Janiszewska, S.; Bialobrzeski, T.; Kruszyńska, E.; Ciepiela, K. Przegląd metod oczyszczania gruntów i wód gruntowych in-situ. *Prz. Geol.* **2017**, *65*, 908–915.
12. Khabidolda, S.M.; Abdiyussupov, G.G.; Omirzak, M.T. Development of Environmentally Clean Construction Materials Using Industrial Waste. *Materials* **2022**, *15*, 5726. [CrossRef]
13. Navarro, A.; Cardellach, E.; Corbella, M. Immobilization of Cu, Pb and Zn in mine-contaminated soils using reactive materials. *J. Hazard. Mater.* **2011**, *186*, 1576–1585. [CrossRef]
14. Saadeldin, R.; Siddiqua, S. Geotechnical characterization of a clay–cement mix. *Bull. Eng. Geol. Environ.* **2013**, *72*, 601–608. [CrossRef]
15. Sun, Y.; Zhang, D.; Tao, H.; Yang, Y. The Effects of Portland and Sulphoaluminate Cements Solidification/Stabilization on Semi-Dynamic Leaching of Heavy Metal from Contaminated Sediment. *Sustainability* **2022**, *14*, 5681. [CrossRef]
16. Jin, F.; Wang, F.; Al-Tabbaa, A. Three-year performance of in-situ solidified/stabilised soil using novel MgO-bearing binders. *Chemosphere* **2016**, *144*, 681–688. [CrossRef]
17. Voglar, G.E.; Leštan, D. Efficiency modeling of solidification/stabilization of multi-metal contaminated industrial soil using cement and additives. *J. Hazard. Mater.* **2011**, *192*, 753–762. [CrossRef]
18. Scanferla, P.; Ferrari, G.; Pellay, R.; Volpi Ghirardini, A.; Zanetto, G.; Libralato, G. An innovative stabilization/solidification treatment for contaminated soil remediation: Demonstration project results. *J. Soils Sediments* **2009**, *9*, 229–236. [CrossRef]
19. Palansooriya, K.N.; Shaheen, S.M.; Chen, S.S.; Tsang, D.C.W.; Hashimoto, Y.; Hou, D.; Bolan, N.S.; Rinklebe, J.; Ok, Y.S. Soil amendments for immobilization of potentially toxic elements in contaminated soils: A critical review. *Environ. Int.* **2020**, *134*, 105046. [CrossRef]
20. Wang, F.; Shen, Z.; Al-Tabbaa, A. PC-based and MgO-based binders stabilised/solidified heavy metal contaminated model soil: Strength and heavy metal speciation in early stage. *Geotechnique* **2018**, *68*, 1025–1030. [CrossRef]
21. Means, J.; Nehring, K.; Smith, L. *Application of Solidification and Stabilization to Waste Materials*; CRC Press: Boca Raton, FA, USA, 1995.
22. Bone, B.D.; Barnard, L.H.; Boardman, D.I.; Carey, P.J.; Hills, C.D.; Jones, H.M.; MacLeod, C.L.; Tyrer, M. Review of scientific literature on the use of stabilisation/solidification for the treatment of contaminated soil, solid waste and sludges (SC980003/SR2). In *The Environment Agency, Bristol*; Elsevier: Amsterdam, The Netherlands, 2004; ISBN 1 844 323 196.
23. Guo, B.; Liu, B.; Yang, J.; Zhang, S. The mechanisms of heavy metal immobilization by cementitious material treatments and thermal treatments: A review. *J. Environ. Manag.* **2017**, *193*, 410–422. [CrossRef] [PubMed]
24. Dasary, S.; Chittajallu, A.; Rekha, P.; Yerramilli, A. Stabilisation and solidification technologies for the remediation of contaminated soils and sediments: An overview. *Land Contam. Reclam.* **2005**, *13*, 23–48. [CrossRef]
25. Tajudin, S.A.A.; Azmi, M.A.M.; Nabila, A.T.A. Stabilization/Solidification Remediation Method for Contaminated Soil: A Review. *IOP Conf. Ser. Mater. Sci. Eng.* **2016**, *136*, 012043. [CrossRef]
26. Wiles, C.C. A review of solidification/stabilization technology. *J. Hazard. Mater.* **1987**, *14*, 5–21. [CrossRef]

27. *ASTM D1632-07*; Standard Practice for Making and Curing Soil-Cement Compression and Flexure Test Specimens in the Laboratory. ANSI: NewYork, NY, USA, 2020.
28. Li, J.; Poon, C.S. Innovative solidification/stabilization of lead contaminated soil using incineration sewage sludge ash. *Chemosphere* **2017**, *173*, 143–152. [CrossRef]
29. Goodarzi, A.R.; Movahedrad, M. Stabilization/solidification of zinc-contaminated kaolin clay using ground granulated blast-furnace slag and different types of activators. *Appl. Geochem.* **2017**, *81*, 155–165. [CrossRef]
30. Li, X.; Yang, R.; Li, H.; Yi, H.; Jing, H. Experimental study on solidification and stabilization of heavy-metal-contaminated soil using cementitious materials. *Materials* **2021**, *14*, 4999. [CrossRef]
31. Wang, L.; Yu, K.; Li, J.S.; Tsang, D.C.W.; Poon, C.S.; Yoo, J.C.; Baek, K.; Ding, S.; Hou, D.; Dai, J.G. Low-carbon and low-alkalinity stabilization/solidification of high-Pb contaminated soil. *Chem. Eng. J.* **2018**, *351*, 418–427. [CrossRef]
32. Vaidya, R.; Kodam, K.; Ghole, V.; Surya Mohan Rao, K.; Ghole, V.; Mohan, S. Validation of an in situ solidification/stabilization technique for hazardous barium and cyanide waste for safe disposal into a secured landfill. *J. Environ. Manag.* **2010**, *91*, 1821–1830. [CrossRef]
33. Feng, Y.S.; Du, Y.J.; Zhou, A.; Zhang, M.; Li, J.S.; Zhou, S.J.; Xia, W.Y. Geoenvironmental properties of industrially contaminated site soil solidified/stabilized with a sustainable by-product-based binder. *Sci. Total Environ.* **2021**, *765*, 142778. [CrossRef]
34. Wu, H.L.; Jin, F.; Bo, Y.L.; Du, Y.J.; Zheng, J.X. Leaching and microstructural properties of lead contaminated kaolin stabilized by GGBS-MgO in semi-dynamic leaching tests. *Constr. Build. Mater.* **2018**, *172*, 626–634. [CrossRef]
35. Li, J.S.; Xue, Q.; Wang, P.; Li, Z.Z.; Liu, L. Effect of drying-wetting cycles on leaching behavior of cement solidified lead-contaminated soil. *Chemosphere* **2014**, *117*, 10–13. [CrossRef] [PubMed]
36. Kogbara, R.B.; Yi, Y.; Al-Tabbaa, A. Process envelopes for stabilisation/solidification of contaminated soil using lime-slag blend. *Environ. Sci. Pollut. Res.* **2011**, *18*, 1286–1296. [CrossRef]
37. Li, W.; Yi, Y. Stabilization/solidification of lead- and zinc-contaminated soils using MgO and CO_2. *J. CO2 Util.* **2019**, *33*, 215–221. [CrossRef]
38. Tang, P.P.; Zhang, W.L.; Chen, Y.H.; Chen, G.; Xu, J. Stabilization/solidification and recycling of sediment from Taihu Lake in China: Engineering behavior and environmental impact. *Waste Manag.* **2020**, *116*, 1–8. [CrossRef]
39. Al-Tabbaa, A.; Boes, N. Pilot in situ auger mixing treatment of a contaminated site. Part 4. Performance at five years. *Proc. Inst. Civ. Eng. Geotech. Eng.* **2002**, *155*, 187–202. [CrossRef]
40. Xia, W.Y.; Du, Y.J.; Li, F.S.; Li, C.P.; Yan, X.L.; Arulrajah, A.; Wang, F.; Song, D.J. In-situ solidification/stabilization of heavy metals contaminated site soil using a dry jet mixing method and new hydroxyapatite based binder. *J. Hazard. Mater.* **2019**, *369*, 353–361. [CrossRef]
41. Li, J.S.; Chen, L.; Zhan, B.; Wang, L.; Poon, C.S.; Tsang, D.C.W. Sustainable stabilization/solidification of arsenic-containing soil by blast slag and cement blends. *Chemosphere* **2021**, *271*, 129868. [CrossRef]
42. Kogbara, R.B.; Al-Tabbaa, A.; Yi, Y.; Stegemann, J.A. Cement–fly ash stabilisation/solidification of contaminated soil: Performance properties and initiation of operating envelopes. *Appl. Geochem.* **2013**, *33*, 64–75. [CrossRef]
43. Arora, S.; Aydilek, A. Class F fly-ash-amended soils as highway base materials. *J. Mater. Civ. Eng.* **2005**, *17*, 640–649. [CrossRef]
44. Boutouil, M.; Levacher, D. Effect of high initial water content on cement-based sludge solidification. *Adv. Mater. Res.* **2005**, *9*, 169–174. [CrossRef]
45. Kogbara, R.B. Process Envelopes for and Biodegradation within Stabilised/Solidified Contaminated Soils. Ph.D. Thesis, Cambridge University, Cambridge, UK, 2011.
46. Kogbara, R.B. A review of the mechanical and leaching performance of stabilised/solidified contaminated soils. *Environ. Rev.* **2014**, *22*, 66–86. [CrossRef]
47. Cao, X.; Wahbi, A.; Ma, L.; Li, B.; Yang, Y. Immobilization of Zn, Cu, and Pb in contaminated soils using phosphate rock and phosphoric acid. *J. Hazard. Mater.* **2009**, *164*, 555–564. [CrossRef]
48. Komárek, M.; Vaněk, A.; Ettler, V. Chemical stabilization of metals and arsenic in contaminated soils using oxides—A review. *Environ. Pollut.* **2013**, *172*, 9–22. [CrossRef]
49. Chen, L.; Wang, L.; Cho, D.-W.W.; Tsang, D.C.W.; Tong, L.; Zhou, Y.; Yang, J.; Hu, Q.; Poon, C.S. Sustainable stabilization/solidification of municipal solid waste incinerator fly ash by incorporation of green materials. *J. Clean. Prod.* **2019**, *222*, 335–343. [CrossRef]
50. Wang, F.; Xu, J.; Yin, H.; Zhang, Y.; Pan, H.; Wang, L. Sustainable stabilization/solidification of the Pb, Zn, and Cd contaminated soil by red mud-derived binders. *Environ. Pollut.* **2021**, *284*, 117178. [CrossRef]
51. Zhang, Y.; Labianca, C.; Chen, L.; De Gisi, S.; Notarnicola, M.; Guo, B.; Sun, J.; Ding, S.; Wang, L. Sustainable ex-situ remediation of contaminated sediment: A review. *Environ. Pollut.* **2021**, *287*, 117333. [CrossRef]
52. Sun, Y.; Sun, G.; Xu, Y.; Liu, W.; Liang, X.; Wang, L. Evaluation of the effectiveness of sepiolite, bentonite, and phosphate amendments on the stabilization remediation of cadmium-contaminated soils. *J. Environ. Manag.* **2016**, *166*, 204–210. [CrossRef]
53. Özbay, E.; Erdemir, M.; Durmuş, H.I. Utilization and efficiency of ground granulated blast furnace slag on concrete properties-A review. *Constr. Build. Mater.* **2016**, *105*, 423–434. [CrossRef]
54. Jin, F.; Al-Tabbaa, A. Evaluation of novel reactive MgO activated slag binder for the immobilisation of lead and zinc. *Chemosphere* **2014**, *117*, 285–294. [CrossRef]

55. Jin, F.; Gu, K.; Al-Tabbaa, A. Strength and hydration properties of reactive MgO-activated ground granulated blastfurnace slag paste. *Cem. Concr. Compos.* **2015**, *57*, 8–16. [CrossRef]
56. Tariq, A.; Yanful, E.K. A review of binders used in cemented paste tailings for underground and surface disposal practices. *J. Environ. Manag.* **2013**, *131*, 138–149. [CrossRef] [PubMed]
57. Wang, F.; Xu, J.; Zhang, Y.; Shen, Z.; Al-Tabbaa, A. MgO-GGBS Binder–Stabilized/Solidified PAE-Contaminated Soil: Strength and Leachability in Early Stage. *J. Geotech. Geoenviron. Eng.* **2021**, *147*, 04021059. [CrossRef]
58. Puligilla, S.; Mondal, P. Role of slag in microstructural development and hardening of fly ash-slag geopolymer. *Cem. Concr. Res.* **2013**, *43*, 70–80. [CrossRef]
59. Wen, W.; Jia, L.; Xie, J.; Zhao, W.; Feng, H.; Cao, D.; Sun, F.; Han, P.; Bai, X.; He, B. Electrochemical response and effect evaluation of high belite sulphoaluminate cement combined with red mud-fly ash on solidification of Cu^{2+}-contaminated kaolin. *Case Stud. Constr. Mater.* **2022**, *17*, e01497. [CrossRef]
60. Yi, Y.; Gu, L.; Liu, S.; Jin, F. Magnesia reactivity on activating efficacy for ground granulated blastfurnace slag for soft clay stabilisation. *Appl. Clay Sci.* **2016**, *126*, 57–62. [CrossRef]
61. Wang, L.; Chen, L.; Tsang, D.C.W.; Zhou, Y.; Rinklebe, J.; Song, H.; Kwon, E.E.; Baek, K.; Ok, Y.S. Mechanistic insights into red mud, blast furnace slag, or metakaolin-assisted stabilization/solidification of arsenic-contaminated sediment. *Environ. Int.* **2019**, *133*, 105247. [CrossRef]
62. Laforest, G.; Duchesne, J. Immobilization of chromium (VI) evaluated by binding isotherms for ground granulated blast furnace slag and ordinary Portland cement. *Cem. Concr. Res.* **2005**, *35*, 2322–2332. [CrossRef]
63. Tang, Y.J.; Zuo, X.B.; He, S.L.; Ayinde, O.; Yin, G.J. Influence of slag content and water-binder ratio on leaching behavior of cement pastes. *Constr. Build. Mater.* **2016**, *129*, 61–69. [CrossRef]
64. Du, Y.J.; Jiang, N.J.; Liu, S.Y.; Jin, F.; Singh, D.N.; Puppala, A.J. Engineering properties and microstructural characteristics of cement-stabilized zinc-contaminated kaolin. *Can. Geotech. J.* **2014**, *51*, 289–302. [CrossRef]
65. Amin, M.S.; Hashem, F.S.; Mohamed, M.R. Solidification/stabilisation of Zn2+ ions in metakaolin and homra- blended cement matrices. *Adv. Cem. Res.* **2012**, *24*, 239–248. [CrossRef]
66. Zhang, Z.; Guo, G.; Wang, M.; Zhang, J.; Wang, Z.; Li, F.; Chen, H. Enhanced stabilization of Pb, Zn, and Cd in contaminated soils using oxalic acid-activated phosphate rocks. *Environ. Sci. Pollut. Res.* **2018**, *25*, 2861–2868. [CrossRef]
67. Li, J.S.; Wang, Q.; Chen, Z.; Xue, Q.; Chen, X.; Mu, Y.; Poon, C.S. Immobilization of high-Pb contaminated soil by oxalic acid activated incinerated sewage sludge ash. *Environ. Pollut.* **2021**, *284*, 117120. [CrossRef]
68. Dermatas, D.; Meng, X. Utilization of fly ash for stabilization/solidification of heavy metal contaminated soils. *Eng. Geol.* **2003**, *70*, 377–394. [CrossRef]
69. Du, Y.J.; Wei, M.L.; Reddy, K.R.; Jin, F.; Wu, H.L.; Liu, Z. Bin New phosphate-based binder for stabilization of soils contaminated with heavy metals: Leaching, strength and microstructure characterization. *J. Environ. Manag.* **2014**, *146*, 179–188. [CrossRef]
70. Chen, L.; Wang, Y.S.; Wang, L.; Zhang, Y.; Li, J.; Tong, L.; Hu, Q.; Dai, J.G.; Tsang, D.C.W. Stabilisation/solidification of municipal solid waste incineration fly ash by phosphate-enhanced calcium aluminate cement. *J. Hazard. Mater.* **2021**, *408*, 124404. [CrossRef]
71. Cadar, O.; Dinca, Z.; Senila, M.; Torok, A.I.; Todor, F.; Levei, E.A. Immobilization of Potentially Toxic Elements in Contaminated Soils Using Thermally Treated Natural Zeolite. *Materials* **2021**, *14*, 3777. [CrossRef]
72. Xia, W.Y.; Feng, Y.S.; Jin, F.; Zhang, L.M.; Du, Y.J. Stabilization and solidification of a heavy metal contaminated site soil using a hydroxyapatite based binder. *Constr. Build. Mater.* **2017**, *156*, 199–207. [CrossRef]
73. Gardner, L.J.; Bernal, S.A.; Walling, S.A.; Corkhill, C.L.; Provis, J.L.; Hyatt, N.C. Characterisation of magnesium potassium phosphate cements blended with fly ash and ground granulated blast furnace slag. *Cem. Concr. Res.* **2015**, *74*, 78–87. [CrossRef]
74. Su, X.; Zhu, J.; Fu, Q.; Zuo, J.; Liu, Y.; Hu, H. Immobilization of lead in anthropogenic contaminated soils using phosphates with/without oxalic acid. *J. Environ. Sci.* **2015**, *28*, 64–73. [CrossRef]
75. Oprčkal, P.; Mladenovič, A.; Zupančič, N.; Ščančar, J.; Milačič, R.; Serjun, V.Z. Remediation of contaminated soil by red mud and paper ash. *J. Clean. Prod.* **2020**, *256*, 120440. [CrossRef]
76. Goodarzi, A.R.; Akbari, H.R.; Salimi, M. Enhanced stabilization of highly expansive clays by mixing cement and silica fume. *Appl. Clay Sci.* **2016**, *132–133*, 675–684. [CrossRef]
77. Nedunuri, S.S.S.A.; Sertse, S.G.; Muhammad, S. Microstructural study of Portland cement partially replaced with fly ash, ground granulated blast furnace slag and silica fume as determined by pozzolanic activity. *Constr. Build. Mater.* **2020**, *238*, 117561. [CrossRef]
78. Namarak, C.; Satching, P.; Tangchirapat, W.; Jaturapitakkul, C. Improving the compressive strength of mortar from a binder of fly ash-calcium carbide residue. *Constr. Build. Mater.* **2017**, *147*, 713–719. [CrossRef]
79. Cau Dit Coumes, C.; Dhoury, M.; Champenois, J.-B.; Mercier, C.; Damidot, D. Combined effects of lithium and borate ions on the hydration of calcium sulfoaluminate cement. *Cem. Concr. Res.* **2017**, *97*, 50–60. [CrossRef]
80. Sharma, H.D.; Reddy, K.R. Geoenvironmental Engineering: Site Remediation, Waste Containment, and Emerging Waste Management Technologies. *Environ. Int.* **2004**, *35*, 50–55.
81. Batchelor, B. Overview of waste stabilization with cement. *Waste Manag.* **2006**, *26*, 689–698. [CrossRef]
82. Kogbara, R.B.; Al-Tabbaa, A.; Yi, Y.; Stegemann, J.A. PH-dependent leaching behaviour and other performance properties of cement-treated mixed contaminated soil. *J. Environ. Sci.* **2012**, *24*, 1630–1638. [CrossRef]

83. Roy, A.; Eaton, H.C.; Cartledge, F.K.; Tittlebaum, M.E. Solidification/stabilization of a heavy metal sludge by a Portland cement/fly ash binding mixture. *Waste Hazard. Mater.* **1991**, *8*, 33–41. [CrossRef]
84. Lin, S.L.; Cross, W.H.; Chian, E.S.K.; Lai, J.S.; Giabbai, M.; Hung, C.H. Stabilization and solidification of lead in contaminated soils. *J. Hazard. Mater.* **1996**, *48*, 95–110. [CrossRef]
85. Sanchez, F.; Gervais, C.; Garrabrants, A.C.; Barna, R.; Kosson, D.S. Leaching of inorganic contaminants from cement-based waste materials as a result of carbonation during intermittent wetting. *Waste Manag.* **2002**, *22*, 249–260. [CrossRef] [PubMed]
86. Yilmaz, O.; Çokça, E.; Ünlü, K. Comparison of Two Leaching Tests to Assess the Effectiveness of Cement-Based Hazardous Waste Solidification/Stabilization. *J. Eng. Environ. Sci.* **2003**, *27*, 201–212.
87. Shawabkeh, R.A. Solidification and stabilization of cadmium ions in sand–cement–clay mixture. *J. Hazard. Mater.* **2005**, *125*, 237–243. [CrossRef]
88. Moon, D.H.; Lee, J.R.; Grubb, D.G.; Park, J.H. An assessment of Portland cement, cement kiln dust and class C fly ash for the immobilization of Zn in contaminated soils. *Environ. Earth Sci.* **2010**, *61*, 1745–1750. [CrossRef]
89. Voglar, G.E.; Leštan, D. Solidification/stabilisation of metals contaminated industrial soil from former Zn smelter in Celje, Slovenia, using cement as a hydraulic binder. *J. Hazard. Mater.* **2010**, *178*, 926–933. [CrossRef]
90. Kogbara, R.B.; Al-Tabbaa, A.; Stegemann, J.A. Process Envelopes For Stabilised/Solidified Contaminated Soils: Initiation work. In Proceedings of the Fifth International Conference on Environmental Science and Technology, Bangkok, Thailand, 11–13 November 2019; American Science Press: Houston, TN, USA, 2019.
91. He, H.; Suito, H. Immobilization of hexavalent chromium in aqueous solution through the formation of $3CaO{\cdot}(Al, Fe)_2O_3$ Ca $(OH)_2{\cdot}xH_2O$ phase, ettringite and C–S–H gel. *ISIJ Int.* **2002**, *42*, 139–145. [CrossRef]
92. Gougar, M.L.D.; Scheetz, B.E.; Roy, D.M. Ettringite and C-S-H Portland Cement Phases for waste ion immobilization: A Review. *Waste Manag.* **1996**, *16*, 295–303. [CrossRef]
93. Spence, R.D.; Shi, C. *Stabilization and Solidification of Hazardous, Radioactive, and Mixed Wastes*; CRC Press: Boca Raton, FL, USA, 2004; ISBN 9781420032789.
94. Su, Y.; Yang, J.; Liu, D.; Zhen, S.; Lin, N.; Zhou, Y. Solidification/stabilization of simulated cadmium-contaminated wastes with magnesium potassium phosphate cement. *Environ. Eng. Res.* **2016**, *21*, 15–21. [CrossRef]
95. Bakhshi, N.; Sarrafi, A.; Ramezanianpour, A.A. Immobilization of hexavalent chromium in cement mortar: Leaching properties and microstructures. *Environ. Sci. Pollut. Res.* **2019**, *26*, 20829–20838. [CrossRef]
96. Chen, Q.Y.; Tyrer, M.; Hills, C.D.; Yang, X.M.; Carey, P. Immobilisation of heavy metal in cement-based solidification/stabilisation: A review. *Waste Manag.* **2009**, *29*, 390–403. [CrossRef]
97. Li, X.; He, C.; Bai, Y.; Ma, B.; Wang, G.; Tan, H. Stabilization/solidification on chromium (III) wastes by C3A and C3A hydrated matrix. *J. Hazard. Mater.* **2014**, *268*, 61–67. [CrossRef] [PubMed]
98. Li, Y.; Min, X.; Ke, Y.; Fei, J.; Liu, D.; Tang, C. Immobilization potential and immobilization mechanism of arsenic in cemented paste backfill. *Miner. Eng.* **2019**, *138*, 101–107. [CrossRef]
99. Contessi, S.; Calgaro, L.; Dalconi, M.C.; Bonetto, A.; Bellotto, M.P.; Ferrari, G.; Marcomini, A.; Artioli, G. Stabilization of lead contaminated soil with traditional and alternative binders. *J. Hazard. Mater.* **2020**, *382*, 120990. [CrossRef] [PubMed]
100. Contessi, S.; Dalconi, M.C.; Pollastri, S.; Calgaro, L.; Meneghini, C.; Ferrari, G.; Marcomini, A.; Artioli, G. Cement-stabilized contaminated soil: Understanding Pb retention with XANES and Raman spectroscopy. *Sci. Total Environ.* **2021**, *752*, 141826. [CrossRef] [PubMed]
101. Chen, L.; Du, Y.J.; Liu, S.-Y.; Jin, F. Evaluation of Cement Hydration Properties of Cement-Stabilized Lead-Contaminated Soils Using Electrical Resistivity Measurement. *J. Hazard. Toxic, Radioact. Waste* **2011**, *15*, 312–320. [CrossRef]
102. Vandaperre, L.J.; Liska, M.; Al-tabbaa, A. Hydration and Mechanical Properties of Magnesia, Pulverized Fuel Ash, and Portland Cement Blends. *J. Mater. Civ. Eng.* **2008**, *20*, 375–383. [CrossRef]
103. Wang, Y.S.; Dai, J.G.; Wang, L.; Tsang, D.C.W.; Poon, C.S. Influence of lead on stabilization/solidification by ordinary Portland cement and magnesium phosphate cement. *Chemosphere* **2018**, *190*, 90–96. [CrossRef]
104. Ling, T.C.; Poon, C.S. Use of recycled CRT funnel glass as fine aggregate in dry-mixed concrete paving blocks. *J. Clean. Prod.* **2014**, *68*, 209–215. [CrossRef]
105. Ling, T.C.; Poon, C.S. Spent fluorescent lamp glass as a substitute for fine aggregate in cement mortar. *J. Clean. Prod.* **2017**, *161*, 646–654. [CrossRef]
106. Xue, Q.; Wang, P.; Li, J.S.; Zhang, T.T.; Wang, S.Y. Investigation of the leaching behavior of lead in stabilized/solidified waste using a two-year semi-dynamic leaching test. *Chemosphere* **2017**, *166*, 1–7. [CrossRef]
107. Ouhadi, V.R.; Yong, R.N.; Deiranlou, M. Enhancement of Cement-Based Solidification/Stabilization of a lead-contaminated smectite clay. *J. Hazard. Mater.* **2021**, *403*, 123969. [CrossRef]
108. Wang, L.; Chen, S.S.; Tsang, D.C.W.; Poon, C.S.; Shih, K. Recycling contaminated wood into eco-friendly particleboard using green cement and carbon dioxide curing. *J. Clean. Prod.* **2016**, *137*, 861–870. [CrossRef]
109. Morales, L.; Garzón, E.; Romero, E.; Sánchez-Soto, P.J. Microbiological induced carbonate (CaCO3) precipitation using clay phyllites to replace chemical stabilizers (cement or lime). *Appl. Clay Sci.* **2019**, *174*, 15–28. [CrossRef]
110. Mujah, D.; Cheng, L.; Shahin, M.A. Microstructural and Geomechanical Study on Biocemented Sand for Optimization of MICP Process. *J. Mater. Civ. Eng.* **2019**, *31*, 04019025. [CrossRef]

111. Scrivener, K.L.; Kirkpatrick, R.J. Innovation in use and research on cementitious material. *Cem. Concr. Res.* **2008**, *38*, 128–136. [CrossRef]
112. Lijun, H.; Jiangshan, L.; Qiang, X.; Zhen, C.; Yaoyu, Z.; Chi, S.P. Bacterial-induced mineralization (BIM) for soil solidification and heavy metal stabilization: A critical review. *Sci. Total Environ.* **2020**, *746*, 140967. [CrossRef]
113. Li, J.S.; Beiyuan, J.; Tsang, D.C.W.; Wang, L.; Poon, C.S.; Li, X.D.; Fendorf, S. Arsenic-containing soil from geogenic source in Hong Kong: Leaching characteristics and stabilization/solidification. *Chemosphere* **2017**, *182*, 31–39. [CrossRef]
114. Shen, Z.; Jin, F.; O'Connor, D.; Hou, D. Solidification/Stabilization for Soil Remediation: An Old Technology with New Vitality. *Environ. Sci. Technol.* **2019**, *53*, 11615–11617. [CrossRef]
115. Du, Y.J.; Wei, M.L.; Reddy, K.R.; Liu, Z.P.; Jin, F. Effect of acid rain pH on leaching behavior of cement stabilized lead-contaminated soil. *J. Hazard. Mater.* **2014**, *271*, 131–140. [CrossRef]
116. Wei, M.-L.; Du, Y.J.; Reddy, K.R.; Wu, H.-L. Effects of freeze-thaw on characteristics of new KMP binder stabilized Zn- and Pb-contaminated soils. *Environ. Sci. Pollut. Res.* **2015**, *22*, 19473–19484. [CrossRef]
117. Perera, A.; Al-Tabbaa, A. *Stabilisation/Solidification Treatment and Remediation*; CRC Press: Cambridge, UK, 2005; pp. 181–191.
118. Cheah, C.B.; Chung, K.Y.; Ramli, M.; Lim, G.K. The engineering properties and microstructure development of cement mortar containing high volume of inter-grinded GGBS and PFA cured at ambient temperature. *Constr. Build. Mater.* **2016**, *122*, 683–693. [CrossRef]
119. Li, S. Microstructure and composition characterisation of three 20-year-old GGBS-OPC blended pastes. *Constr. Build. Mater.* **2016**, *123*, 226–234. [CrossRef]
120. Zhang, W.L.; Zhao, L.Y.; McCabe, B.A.; Chen, Y.H.; Morrison, L. Dredged marine sediments stabilized/solidified with cement and GGBS: Factors affecting mechanical behaviour and leachability. *Sci. Total Environ.* **2020**, *733*, 138551. [CrossRef] [PubMed]
121. Wang, L.; Chen, L.; Tsang, D.C.W.; Li, J.S.; Yeung, T.L.Y.; Ding, S.; Poon, C.S. Green remediation of contaminated sediment by stabilization/solidification with industrial by-products and CO2 utilization. *Sci. Total Environ.* **2018**, *631*, 1321–1327. [CrossRef] [PubMed]
122. Allan, M.L.; Kukacka, L.E. Blast furnace slag-modified grouts for in situ stabilization of chromium-contaminated soil. *Waste Manag.* **1995**, *15*, 193–202. [CrossRef]
123. Miturski, M.; Głuchowski, A.; Sas, W. Influence of dispersed reinforcement on mechanical properties of stabilized soil. *Materials* **2021**, *14*, 5982. [CrossRef]
124. Wang, L.; Yu, I.K.M.; Tsang, D.C.W.; Li, S.; Poon, C.S. Mixture Design and Reaction Sequence for Recycling Construction Wood Waste into Rapid-Shaping Magnesia-Phosphate Cement Particleboard. *Ind. Eng. Chem. Res.* **2017**, *56*, 6645–6654. [CrossRef]
125. Ren, P.; Ling, T.C. Roles of chlorine and sulphate in MSWIFA in GGBFS binder: Hydration, mechanical properties and stabilization considerations. *Environ. Pollut.* **2021**, *284*, 117175. [CrossRef]
126. Knapik, K. Analizy Doświadczalne i Numeryczne Wybranego Gruntu Wzmo Cnionego Popiołami Lotnymi z Procesu Fluidalnego Spalania Węgla. Ph.D Thesis, Silesian University of Technology, Gliwice, Poland, 2016.
127. Sargent, P.; Hughes, P.N.; Rouainia, M.; White, M.L. The use of alkali activated waste binders in enhancing the mechanical properties and durability of soft alluvial soils. *Eng. Geol.* **2013**, *152*, 96–108. [CrossRef]
128. Deja, J. Immobilization of Cr^{6+}, Cd^{2+}, Zn^{2+} and Pb^{2+} in alkali-activated slag binders. *Cem. Concr. Res.* **2002**, *32*, 1971–1979. [CrossRef]
129. Roy, A.; Eaton, H.C.; Cartledge, F.K.; Tittlebaum, M.E. The effect of sodium sulfate on solidification/stabilization of a synthetic electroplating sludge in cementitious binders. *J. Hazard. Mater.* **1992**, *30*, 297–316. [CrossRef]
130. Yi, Y.; Liska, M.; Jin, F.; Al-Tabbaa, A. Mechanism of reactive magnesia-ground granulated blastfurnace slag (GGBS) soil stabilization. *Can. Geotech. J.* **2015**, *53*, 773–782. [CrossRef]
131. Du, Y.J.; Bo, Y.L.; Jin, F.; Liu, C.Y. Durability of reactive magnesia-activated slag-stabilized low plasticity clay subjected to drying-wetting cycle. *Eur. J. Environ. Civ. Eng.* **2016**, *20*, 215–230. [CrossRef]
132. Limbachiya, V.; Ganjian, E.; Claisse, P. Strength, durability and leaching properties of concrete paving blocks incorporating GGBS and SF. *Constr. Build. Mater.* **2016**, *113*, 273–279. [CrossRef]
133. Pomakhina, E.; Deneele, D.; Gaillot, A.C.; Paris, M.; Ouvrard, G. 29Si solid state NMR investigation of pozzolanic reaction occurring in lime-treated Ca-bentonite. *Cem. Concr. Res.* **2012**, *42*, 626–632. [CrossRef]
134. Li, J.S.; Wang, L.; Cui, J.L.; Poon, C.S.; Beiyuan, J.; Tsang, D.C.W.; Li, X.D. Effects of low-alkalinity binders on stabilization/solidification of geogenic As-containing soils: Spectroscopic investigation and leaching tests. *Sci. Total Environ.* **2018**, *631–632*, 1486–1494. [CrossRef]
135. Xie, J.; Wang, J.; Zhang, B.; Fang, C.; Li, L. Physicochemical properties of alkali activated GGBS and fly ash geopolymeric recycled concrete. *Constr. Build. Mater.* **2019**, *204*, 384–398. [CrossRef]
136. Shah, S.J.; Shroff, A.V.; Patel, J.V.; Tiwari, K. Stabilization of fuel oil contaminated soil—A case study. *Geotech. Geol. Eng.* **2003**, *21*, 415–427. [CrossRef]
137. Li, X.D.; Poon, C.S.; Sun, H.; Lo, I.M.C.; Kirk, D.W. Heavy metal speciation and leaching behaviors in cement based solidified/stabilized waste materials. *J. Hazard. Mater.* **2001**, *82*, 215–230. [CrossRef]
138. Al-Tabbaa, A.; Evans, C.W.; Wallace, C.J. Pilot in situ auger mixing treatment of a contaminated site. Part 2. Site trial. *Proc. Inst. Civ. Eng. Geotech. Eng.* **1998**, *131*, 89–95. [CrossRef]

139. Chitambira, B. Accelerated ageing of cement stabilised/solidified contaminated soils with elevated temperatures. Ph.D. Thesis, Cambridge University, Cambridge, UK, 2004.
140. Antemir, A. Performance assessment of S/S waste-forms: Initial results from site characterisation, sampling and testing. In Proceedings of the International Conference on Stabilisation/Solidification Treatment and Remediation; CRC Press: Cambridge, UK, 2005.
141. Wang, L.; Kwok, J.S.H.; Tsang, D.C.W.; Poon, C.S. Mixture design and treatment methods for recycling contaminated sediment. *J. Hazard. Mater.* **2015**, *283*, 623–632. [CrossRef] [PubMed]
142. Vukićević, M.; Pujević, V.; Marjanowić, M.; Jocković, S. Maraš-Dragojević Fine grained soil stabilization using class F fly ash with and without cement. In Proceedings of the XVI ECSMGE Geotechnical Engineering for Infrastructure and Development, Edinburgh, UK, 13–17 September 2015; Volume 1, pp. 2671–2676. Available online: https://www.icevirtuallibrary.com/doi/abs/10.1680/ecsmge.60678.vol5.413 (accessed on 15 November 2022).
143. *ASTM C618*; Standard Specification for Coalfly Ash and Raw or Calcined Natural Pozzolan for Use as a Mineral Admixture in Concrete. ASTM Standards: West Conshohocken, PA, USA, 2003.
144. Akhter, H.; Butler, L.G.; Branz, S.; Cartledge, F.K.; Tittlebaum, M.E. Immobilization of As, Cd, Cr and PB-containing soils by using cement or pozzolanic fixing agents. *J. Hazard. Mater.* **1990**, *24*, 145–155. [CrossRef]
145. Kaniraj, S.R.; Havanagi, V.G. Compressive strength of cement stabilized fly ash-soil mixtures. *Cem. Concr. Res.* **1999**, *29*, 673–677. [CrossRef]
146. Kostarelos, K.; Reale, D.; Dermatas, D.; Rao, E.; Moon, D.H. Optimum dose of lime and fly ash for treatment of hexavalent chromium-contaminated soil. *Water Air Soil Pollut. Focus* **2006**, *6*, 171–189. [CrossRef]
147. Wang, D.L.; Chen, M.L.; Tsang, D.D.C.W. *Green Remediation by Using Low-Carbon Cement-Based Stabilization/Solidification Approaches*; Elsevier: Amsterdam, The Netherlands, 2020; ISBN 9780128179826.
148. Wang, A.J.; Zhang, J.; Li, J.M.; Ma, A.B.; Liu, L.T. Effect of liquid-to-solid ratios on the properties of magnesium phosphate chemically bonded ceramics. *Mater. Sci. Eng. C* **2013**, *33*, 2508–2512. [CrossRef]
149. Chau, C.K.; Qiao, F.; Li, Z. Microstructure of magnesium potassium phosphate cement. *Constr. Build. Mater.* **2011**, *25*, 2911–2917. [CrossRef]
150. Li, J.-S.; Xue, Q.; Wang, P.; Zhang, T.-T.; Zhao, Y. Comparison of Solidification/Stabilization of Lead Contaminated Soil between Magnesia–Phosphate Cement and Ordinary Portland Cement under the Same Dosage. *Environ. Prog. Sustain. Energy* **2016**, *35*, 88–94. [CrossRef]
151. Singh, D.; Mandalika, V.R.; Parulekar, S.J.; Wagh, A.S. Magnesium potassium phosphate ceramic for 99Tc immobilization. *J. Nucl. Mater.* **2006**, *348*, 272–282. [CrossRef]
152. Buj, I.; Torras, J.; Rovira, M.; de Pablo, J. Leaching behaviour of magnesium phosphate cements containing high quantities of heavy metals. *J. Hazard. Mater.* **2010**, *175*, 789–794. [CrossRef]
153. Buj, I.; Torras, J.; Casellas, D.; Rovira, M.; de Pablo, J. Effect of heavy metals and water content on the strength of magnesium phosphate cements. *J. Hazard. Mater.* **2009**, *170*, 345–350. [CrossRef]
154. Lai, Z.; Lai, X.; Shi, J.; Lu, Z. Effect of Zn^{2+} on the early hydration behavior of potassium phosphate based magnesium phosphate cement. *Constr. Build. Mater.* **2016**, *129*, 70–78. [CrossRef]
155. Hughes, P.N.; Glendinning, S.; Manning, D.A.C.; White, M.L. Use of red gypsum in soil mixing engineering applications. *Proc. Inst. Civ. Eng.Geotech. Eng.* **2011**, *164*, 223–234. [CrossRef]
156. Zeng, G.; Wan, J.; Huang, D.; Hu, L.; Huang, C.; Cheng, M.; Xue, W.; Gong, X.; Wang, R.; Jiang, D. Precipitation, adsorption and rhizosphere effect: The mechanisms for Phosphate-induced Pb immobilization in soils—A review. *J. Hazard. Mater.* **2017**, *339*, 354–367. [CrossRef]
157. Jiang, G.; Liu, Y.; Huang, L.; Fu, Q.; Deng, Y.; Hu, H. Mechanism of lead immobilization by oxalic acid-activated phosphate rocks. *J. Environ. Sci.* **2012**, *24*, 919–925. [CrossRef]
158. Fang, Y.; Cao, X.; Zhao, L. Effects of phosphorus amendments and plant growth on the mobility of Pb, Cu, and Zn in a multi-metal-contaminated soil. *Environ. Sci. Pollut. Res.* **2012**, *19*, 1659–1667. [CrossRef]
159. Cao, X.; Ma, L.Q.; Rhue, D.R.; Appel, C.S. Mechanisms of lead, copper, and zinc retention by phosphate rock. *Environ. Pollut.* **2004**, *131*, 435–444. [CrossRef]
160. Du, Y.J.; Wei, M.L.; Reddy, K.R.; Wu, H. liang Effect of carbonation on leachability, strength and microstructural characteristics of KMP binder stabilized Zn and Pb contaminated soils. *Chemosphere* **2016**, *144*, 1033–1042. [CrossRef]
161. Fernández Bertos, M.; Simons, S.J.R.; Hills, C.D.; Carey, P.J. A review of accelerated carbonation technology in the treatment of cement-based materials and sequestration of CO_2. *J. Hazard. Mater.* **2004**, *112*, 193–205. [CrossRef]
162. Yoo, J.C.; Beiyuan, J.; Wang, L.; Tsang, D.C.W.; Baek, K.; Bolan, N.S.; Ok, Y.S.; Li, X.D. A combination of ferric nitrate/EDDS-enhanced washing and sludge-derived biochar stabilization of metal-contaminated soils. *Sci. Total Environ.* **2018**, *616*, 572–582. [CrossRef]
163. Lee, S.H.; Lee, J.S.; Jeong Choi, Y.; Kim, J.G. In situ stabilization of cadmium-, lead-, and zinc-contaminated soil using various amendments. *Chemosphere* **2009**, *77*, 1069–1075. [CrossRef]
164. Bougharraf, N.; Louati, D.; Mosbahi, M.; Rouis, M.J.; Rigane, H. Comparison of the effectiveness of different binders in solidification/stabilization of a contaminated soil. *Arab. J. Geosci.* **2018**, *11*, 348. [CrossRef]

165. Navarro-Blasco, I.; Duran, A.; Sirera, R.; Fernández, J.M.; Alvarez, J.I. Solidification/stabilization of toxic metals in calcium aluminate cement matrices. *J. Hazard. Mater.* **2013**, *260*, 89–103. [CrossRef] [PubMed]
166. Ivanov, R.C.; Angulski da Luz, C.; Zorel, H.E.; Pereira Filho, J.I. Behavior of calcium aluminate cement (CAC) in the presence of hexavalent chromium. *Cem. Concr. Compos.* **2016**, *73*, 114–122. [CrossRef]
167. Voglar, G.E.; Leštan, D. Equilibrium leaching of toxic elements from cement stabilized soil. *J. Hazard. Mater.* **2013**, *246*, 18–25. [CrossRef]
168. Kraszewski, C.; Dreger, M.; Mitrut, M.; Przygoda, M. Stabilizacja gruntów wapnem palonym w inżynierii komunikacyjnej-teoria i praktyka. *Inżynieria Morska I Geotech.* **2020**, *1*, 82–86.
169. Ma, Y.; Chen, W. Study on the Mechanism of Stabilizing Loess with Lime: Analysis of Mineral and Microstructure Evolution. *Adv. Civ. Eng.* **2021**, *2021*, 6641496. [CrossRef]
170. Wang, F.; Wang, H.; Jin, F.; Al-Tabbaa, A. The performance of blended conventional and novel binders in the in-situ stabilisation/solidification of a contaminated site soil. *J. Hazard. Mater.* **2015**, *285*, 46–52. [CrossRef]
171. Moon, D.H.; Dermatas, D. An evaluation of lead leachability from stabilized/solidified soils under modified semi-dynamic leaching conditions. *Eng. Geol.* **2006**, *85*, 67–74. [CrossRef]
172. Vandaperre, L.J.; Liska, M.; Al-Tabbaa, A. Reactive magnesium oxide cements: Properties and applications. In Proceedings of the Sustainable Construction Materials and Technologies-International Conference on Sustainable Construction Materials and Technologies; CRC Press: Coventry, UK, 2007; pp. 397–410.
173. Al-Tabbaa, A. Reactive magnesia cement. *Eco-Efficient Concr.* **2013**, *2013*, 523–543. [CrossRef]
174. Wang, L.; Yeung, T.L.K.; Lau, A.Y.T.; Tsang, D.C.W.; Poon, C.S. Recycling contaminated sediment into eco-friendly paving blocks by a combination of binary cement and carbon dioxide curing. *J. Clean. Prod.* **2017**, *164*, 1279–1288. [CrossRef]
175. Yi, Y.; Lu, K.; Liu, S.; Al-Tabbaa, A. Properties change of reactive magnesia-stabilised soil subjected to forced carbonation. *Can. Geotech. J.* **2016**, *53*, 314–325. [CrossRef]
176. Yi, Y.; Liska, M.; Unluer, C.; Al-Tabbaa, A. Carbonating magnesia for soil stabilization. *Can. Geotech. J.* **2013**, *50*, 899–905. [CrossRef]
177. Pacheco-Torgal, F.; Abdollahnejad, Z.; Camões, A.F.; Jamshidi, M.; Ding, Y. Durability of alkali-activated binders: A clear advantage over Portland cement or an unproven issue? *Constr. Build. Mater.* **2012**, *30*, 400–405. [CrossRef]
178. Bakharev, T.; Sanjayan, J.G.; Cheng, Y.B. Resistance of alkali-activated slag concrete to acid attack. *Cem. Concr. Res.* **2003**, *33*, 1607–1611. [CrossRef]
179. Guo, B.; Pan, D.; Liu, B.; Volinsky, A.A.; Fincan, M.; Du, J.; Zhang, S. Immobilization mechanism of Pb in fly ash-based geopolymer. *Constr. Build. Mater.* **2017**, *134*, 123–130. [CrossRef]
180. Nikolić, V.; Komljenović, M.; Marjanović, N.; Baščarević, Z.; Petrović, R. Lead immobilization by geopolymers based on mechanically activated fly ash. *Ceram. Int.* **2014**, *40*, 8479–8488. [CrossRef]
181. Ji, Z.; Pei, Y. Bibliographic and visualized analysis of geopolymer research and its application in heavy metal immobilization: A review. *J. Environ. Manag.* **2019**, *231*, 256–267. [CrossRef]
182. Liu, F.; Tang, R.; Wang, B.; Yuan, X. Experimental study on solidification of PB2+ in fly ash-based geopolymers. *Sustainability* **2021**, *13*, 12621. [CrossRef]
183. Luna Galiano, Y.; Fernández Pereira, C.; Vale, J. Stabilization/solidification of a municipal solid waste incineration residue using fly ash-based geopolymers. *J. Hazard. Mater.* **2011**, *185*, 373–381. [CrossRef]
184. Mao, Y.; Muhammad, F.; Yu, L.; Xia, M.; Huang, X.; Jiao, B.; Shiau, Y.; Li, D. The Solidification of Lead-Zinc Smelting Slag through Bentonite Supported Alkali-Activated Slag Cementitious Material. *Int. J. Environ. Res. Public Heal.* **2019**, *16*, 1121. [CrossRef]
185. Nath, S.K.; Kumar, S. Influence of iron making slags on strength and microstructure of fly ash geopolymer. *Constr. Build. Mater.* **2013**, *38*, 924–930. [CrossRef]
186. Guzmán-Carrillo, H.R.; Gasca-Tirado, J.R.; López-Romero, J.M.; Apátiga-Castro Luis, M.; Rivera-Muñoz Eric, M.; Pineda-Piñón, J.; Pérez-Bueno, J.J.; Feregrino-Montes, C.; López-Naranjo, E.J.; Manzano-Ramírez, A. Encapsulation of toxic heavy metals from waste CRT using calcined kaolin base-geopolymer. *Mater. Chem. Phys.* **2021**, *257*, 123745. [CrossRef]
187. Bandura, L.; Panek, R.; Madej, J.; Franus, W. Synthesis of zeolite-carbon composites using high-carbon fly ash and their adsorption abilities towards petroleum substances. *Fuel* **2021**, *283*, 119173. [CrossRef]
188. Napia, C.; Sinsiri, T.; Jaturapitakkul, C.; Chindaprasirt, P. Leaching of heavy metals from solidified waste using Portland cement and zeolite as a binder. *Waste Manag.* **2012**, *32*, 1459–1467. [CrossRef]
189. Feng, N.; Xing, F.; Leng, F.-G. Zeolite ceramsite cellular concrete. *Mag. Concr. Res.* **2000**, *52*, 117–122. [CrossRef]
190. Allagoa, M. Solidification/Stabilisation of contaminated site soil: Field versus Laboratory results. *Appl. Sci.* **2011**, *11*, 7474.
191. Antemir, A.; Hills, C.D.; Carey, P.J.; Magnié, M.C.; Polettini, A. Investigation of 4-year-old stabilised/solidified and accelerated carbonated contaminated soil. *J. Hazard. Mater.* **2010**, *181*, 543–555. [CrossRef]
192. Li, W.; Ni, P.; Yi, Y. Comparison of reactive magnesia, quick lime, and ordinary Portland cement for stabilization/solidification of heavy metal-contaminated soils. *Sci. Total Environ.* **2019**, *671*, 741–753. [CrossRef]
193. *ASTM D4219-08*; Standard Test Method for Unconfined Compressive Strength Index of Chemical-Grouted Soils. ASTM International: Philadelphia, PA, USA, 2008.
194. Jalal, F.E.; Mulk, S.; Memon, S.A.; Jamhiri, B.; Naseem, A. Strength, Hydraulic, and Microstructural Characteristics of Expansive Soils Incorporating Marble Dust and Rice Husk Ash. *Adv. Civ. Eng.* **2021**, *2021*. [CrossRef]

195. Environmental Protection Agency (EPA). *Prohibition on the Disposal of Bulk Liquid Hazardous Waste in LandfillseStatutory Interpretive Guidance*; EPA: Washington, DC, USA, 1996.
196. US EPA, E. *Method 1311: Toxicity characteristic leaching procedure*; United States Environmental Protection Agency: Washington, DC, USA, 1992; Volume 1–35, pp. 1–35.
197. Sima, J.; Cao, X.; Zhao, L.; Luo, Q. Toxicity characteristic leaching procedure over- or underestimates leachability of lead in phosphate-amended contaminated soils. *Chemosphere* **2015**, *138*, 744–750. [CrossRef]
198. US EPA, E. *Method 1312: Synthetic Precipitation Leaching Procedure. Test Methods for Evaluating Solid Waste, Physical/Chemical Methods, SW-846*; United States Environmental Protection Agency: Washington, DC, USA, 1992.
199. *ASTM C1308-08*; Standard Test Method for Accelerated Leach Test for Diffusive Releases from Solidified Waste and a Computer Program to Model Diffusive, Fractional Leaching from Cylindrical Waste Forms. ASTM International: West Conshohocken, PA, USA, 1992.
200. US EPA (United States Environmental Protection Agency). *US EPA Method 1315. Mass Transfer Rates of Constituents in Monolithic or Compacted Granular Materials Using a Semi-dynamic Tank Leaching Procedure*; US EPA (United States Environmental Protection Agency): Washington, DC, USA, 2013.
201. *ASTM D 5233-92 (Reapproved 1999)*; Standard Test Method for Single Batch Extraction Method for Wastes. Annual Book of ASTM Standards. American Society for Testing and Materials: Philadelphia, PA, USA, 1999.
202. Davidson, C.M.; Duncan, A.L.; Littlejohn, D.; Ure, A.M.; Garden, L.M. A critical evaluation of the three-stage BCR sequential extraction procedure to assess the potential mobility and toxicity of heavy metals in industrially-contaminated land. *Anal. Chim. Acta* **1998**, *363*, 45–55. [CrossRef]
203. Leiva, C.; Arroyo, F.; Villegas, R.; Vilches, L.F. Immobilization of heavy metals (Cd, Ni or Pb) using aluminate geopolymers. *Mater. Lett.* **2018**, *227*, 184–186. [CrossRef]
204. Boardman, D.I. *Lime Stabilisation: Clay-Metal-Lime Interactions*; Electronic Theses Online Service: London, UK, 1999.
205. Wilk, C.M. *Stabilization of Heavy Metals with Portland Cement*; Portland Cement Association: Skokie, IL, USA, 1997.
206. Wei, X.; Li, Z. Early hydration process of portland cement paste by electrical measurement. *J. Mater. Civ. Eng.* **2006**, *18*, 99–105. [CrossRef]
207. Damasceno, V.M.; Fratta, D.; Bosscher, P.J. Development and validation of a low-cost electrical resistivity tomographer for soil process monitoring. *Can. Geotech. J.* **2009**, *46*, 842–854. [CrossRef]
208. Giao, P.H.; Chung, S.G.; Kim, D.Y.; Tanaka, H. Electric imaging and laboratory resistivity testing for geotechnical investigation of Pusan clay deposits. *J. Appl. Geophys.* **2003**, *52*, 157–175. [CrossRef]
209. Abedi-Koupai, J.; Mehdizadeh, H. Estimation of Osmotic Suction and Electrical Conductivity in Unsaturated Soils Using Filter Paper Method. *Geotech. Test. J.* **2008**, *31*, 142–148. [CrossRef]
210. Liu, S.Y.; Du, Y.J.; Han, L.H.; Gu, M.F. Experimental study on the electrical resistivity of soil–cement admixtures. *Environ. Geol.* **2008**, *54*, 1227–1233. [CrossRef]
211. Liu, J.; Zha, F.; Xu, L.; Kang, B.; Tan, X.; Deng, Y.; Yang, C. Mechanism of stabilized/solidified heavy metal contaminated soils with cement-fly ash based on electrical resistivity measurements. *Meas. J. Int. Meas. Confed.* **2019**, *141*, 85–94. [CrossRef]
212. Liu, J.; Xu, D.; Xiong, L.; Hills, C.; Carey, P.; Gardner, K. Comparison of properties of traditional and accelerated carbonated solidified/stabilized contaminated soils. *J. Environ. Sci.* **2008**, *20*, 593–598. [CrossRef]
213. *ASTM D4843-88*; Standard Test Method for Wetting and Drying Test of Solid Wastes. ASTM International: West Conshohocken, PA, USA, 2016.
214. Yi, Y.; Li, C.; Liu, S.; Jin, F. Magnesium sulfate attack on clays stabilised by carbide slag- and magnesia-ground granulated blast furnace slag. *Geotech. Lett.* **2015**, *5*, 306–312. [CrossRef]
215. Du, Y.-J.; Jiang, N.-J.; Shen, S.-L.; Jin, F. Experimental investigation of influence of acid rain on leaching and hydraulic characteristics of cement-based solidified/stabilized lead contaminated clay. *J. Hazard. Mater.* **2012**, *225*, 195–201. [CrossRef] [PubMed]
216. *ASTM D560-03*; Standard Test Methods for Freezing and Thawing Compacted Soil-Cement Mixtures. ASTM International: West Conshohocken, PA, USA, 2003.

materials

MDPI

Review

Cementitious Grouts for Semi-Flexible Pavement Surfaces—A Review

Muhammad Imran Khan [1,2,*], Muslich Hartadi Sutanto [2], Nur Izzi Md. Yusoff [3], Salah E. Zoorob [4], Waqas Rafiq [2], Mujahid Ali [5,*], Roman Fediuk [6,7,*] and Nikolai Ivanovich Vatin [7]

1 Department of Transportation & Geotechnical Engineering, National University of Sciences and Technology (NUST), Islamabad 44000, Pakistan
2 Department of Civil & Environmental Engineering, Universiti Teknologi PETRONAS, Seri Iskandar 32610, Malaysia
3 Department of Civil Engineering, Universiti Kebangsaan Malaysia, Bangi 43600, Malaysia
4 Construction and Building Materials Program, Kuwait Institute for Scientific Research, Safat 13109, Kuwait
5 Department of Civil Engineering, Faculty of Engineering, Universiti Malaya, Kuala Lumpur 50603, Malaysia
6 Polytechnic Institute, Far Eastern Federal University, 690922 Vladivostok, Russia
7 Peter the Great St. Petersburg Polytechnic University, 195251 St. Petersburg, Russia
* Correspondence: engineermik@gmail.com (M.I.K.); mujahid_19001704@utp.edu.my (M.A.); fedyuk.rs@dvfu.ru (R.F.)

Abstract: The hybrid type of pavement called semi-flexible or grouted macadam has gained popularity over the last few decades in various countries, as it provides significant advantages over both rigid and conventional flexible pavements. The semi-flexible pavement surface consists of an open-graded asphalt mixture with high percentage voids into which flowable cementitious slurry is allowed to penetrate due to gravitational effect. Several researchers have conducted laboratory, as well as field, experiments on evaluating the performance of semi-flexible layers using different compositions of cementitious grouts. The composition of grouts (i.e., water/cement ratio, superplasticizer, polymers, admixtures, and other supplementary materials) has a significant effect on the performance of grouts and semi-flexible mixtures. A comprehensive review of cementitious grouts and their effect on the performance of semi-flexible layers are presented and summarized in this review study. The effect of byproducts and other admixtures/additives on the mechanical properties of grouts are also discussed. Finally, recommendations on the composition of cementitious grouts have been suggested.

Keywords: cementitious grout; grouted macadam; porous asphalt mixture; semi-flexible pavement; grouting ability

1. Introduction

Pavement construction aims to sustain traffic loads and environmental stresses throughout its design life. Good riding quality, durability, fatigue, and rutting characteristics are some of the key criteria that are considered during a comprehensive pavement design process. Traditionally, there have been two main choices of road construction: flexible and rigid pavements. Flexible pavements are constructed from bituminous materials, whereas rigid pavements are constructed from concrete. Conventional flexible pavements are more susceptible to rutting and fatigue distresses when subjected to heavy traffic loading and adverse weather conditions [1]. The service life of the flexible pavement is relatively short and frequent maintenance is required to preserve the pavement condition. Furthermore, routine maintenance of flexible pavement increases its overall cost. The distresses in the flexible pavement are mainly due to choice of material selection, poor mix design, and compromise on quality during construction, as well as overloading issues. On the other hand, rigid pavements have high construction costs and are usually constructed for specialized road types. Rigid pavement construction and opening to traffic usually take a longer time due to the slow concrete setting. Furthermore, rigid pavement is also avoided due to rough riding quality and provision of joints that are provided to relieve the thermal stresses.

Therefore, a new hybrid type of pavement, which is the main subject of this review study, has gained popularity in the last few decades. It is known as semi-flexible pavement, where the surface layer incorporates a semi-flexible material. It can combine the best characteristics that flexible and rigid pavements contain separately. The semi-flexible pavement surfacing, also known as grouted macadam, is believed to have a long design life, significant resistance to rutting, high durability, and joints-free pavement with an impermeable surface [2,3]. It can provide resistance against fuel if constructed in an area exposed to fuel spillage (such as oil refineries, fuel stations, and airport aprons, etc.).

The semi-flexible pavement surfacings are comprised of an open-graded asphalt skeleton (with 25–35% voids), over which a highly flowable cementitious grout is spread and allowed to infiltrate [4,5]. Sufficient flowability of the cement grouts to fill the voids and the interconnectivity of voids are considered the critical properties of semi-flexible pavement surfacings. These surfacings can be used for heavy-loaded pavement sections, as they provide excellent resistance to rutting. They can also be used in airports, industrial areas, fuel stations, and refineries as a load-bearing and fuel-resistant surfacing. The following sections present a brief overview of research performed in the modification and improvement of binder and mixtures for flexible pavements, and then a detailed overview of cementitious grouts for semi-flexible pavement and their performance is presented.

2. Review on Modification of Asphalt Binder and Mixes for Conventional Flexible Pavements

Before moving on to semi-flexible mixtures, it is quite important to discuss the failures caused in conventional flexible pavements and the methods of improving their performance. The distresses in the pavements are the result of continuous and cumulative damage in different pavement layers caused by repeated traffic load and adverse weather conditions. Some very common distresses in flexible pavements are cracks (fatigue, longitudinal, transverse, block, reflective, and edge cracks), deformation (rutting, corrugation, showing, and depression), deterioration (potholes, raveling, stripping, polished aggregate, and pumping), mat problem (segregation and bleeding) and seal coats (rock loss, segregation, bleeding, fat spot, and delamination) [6]. Studies by various researchers reveal that the main reason behind such types of failures are due to the following: heavy traffic load, high temperatures, water infiltration, hot and humid weather, freezing temperatures, poor quality of materials (aggregate, asphalt binder, and soil), as well as poor mix design and construction practices [7–9].

Performance improvement of roads in terms of new materials is a common practice in the construction of local, as well as major, highways. Various layers of the road may need improvement, depending on the type of road, as well as the availability and characteristics of local materials. Among these layers, the sub-grade and the top hot mix asphalt (HMA) layers are the most important and critical layers. If the sub-grade deforms, the rest of the layers also exhibit deformation, and this leads to the failure of the pavement. The HMA layer, which is directly subjected to traffic and environmental stresses, needs careful treatment to improve its performance. Several techniques have been recommended by researchers to stabilize different layers of pavement around the globe. The properties of weak sub-grade soil have been significantly improved using stabilizing materials, such as: oil fuel fly ash (FFA), cement, foamed asphalt (FA) and foamed sulfur asphalt (FSA), liquid asphalt, hydrated lime, synthetic fibers, a geotextile layer, and polypropylene fibers [10–16]. Many professionals have conducted considerable research on modifications of conventional asphalt using different polymers and waste materials. A study was conducted using waste oil fly ash (OFA) as an asphalt modification. To improve the bonding with asphalt, OFA was functionalized using different chemicals, such as the carboxylic group (-COOH), the amine group (-NH_2/NH), and the C18 group ($C_{18}H_{38}O$). The treated OFA-modified binder showed improved performance in terms of reducing temperature susceptibility, increasing performance grade, increasing rutting resistance, and better resistance to low-temperature cracks, compared to the untreated OFA-modified binder [17–19].

Much literature is available related to the improvement of properties of asphalt binder and mixtures. However, very limited research is conducted on grouts and semi-flexible pavement layers. Many researchers have used crumb rubber as a bitumen modifier, and the performance in terms of fatigue life, rutting resistance, resistance to oxidation, and ageing and moisture susceptibility were evaluated [20–22]. Research was conducted on utilizing waste crumb rubber in the performance evaluation of sulfur asphalt. The rheological properties were investigated, and it was concluded that the addition of crumb rubber to sulfur asphalt increased the elastic properties, zero shear viscosity, and rutting parameter ($G^*/\sin\delta$) [23]. Crumb rubber (CR) can also be used to modify the locally available base asphalt because of its rheological properties. It was concluded that there is a considerable increase in complex modulus (G^*) and a reduction in the phase angle (δ) with an increase in CR percentage at a high temperature, which is an indication of improving the elastic properties of CR-modified binders [1,24,25]. However, it was not effective in improving the fatigue of asphalt mixes when dense aggregate gradation was used [26].

Plastics are used in a wide range of products and plays an essential role in our daily life. It is due to a high strength-to-weight ratio, low cost, and easy availability of plastics everywhere. Most of these plastics are not biodegradable by nature, which contributes to environmental concerns. Despite awareness and efforts, plastic recycling after usage is pretty low. It has been reported that about 8.8% of waste plastic is recycled, while the rest of waste plastic is simply dumped into landfill and, hence, is a serious threat to human health and the environment [27,28]. Hassani et al. [29] researched the use of Poly-Ethylene Terephthalate (PET) as a modifier in an asphalt concrete mixture as an aggregate replacement. They concluded that there is a 2.8% reduction in bulk compacted mix density with an aggregate replacement of 20% by volume with Polyethylene Terephthalate (PET) granules. It was also concluded that Marshall Stability and Marshall Quotient for the control mixtures and PET-modified mixtures (used as a partial aggregate replacement) were almost the same. Recycling of PET with asphalt meets most of the specification's requirements that make it appropriate for use in road construction, and the recycling of PET also reduces environmental pollution [30]. It was concluded that natural rubber and crumb rubber are sustainable modifiers and they have been used for a wide range of asphalt modification for a cost-effective, rutting, and pavement fatigue resistance. Results show that with an increasing percentage of CR in the mixture of stone mastic asphalt (SMA), the compaction effort is increased, and drainage of the binder is decreased to fulfil the volumetric properties of SMA [31,32]. Therefore, researchers are looking for more durable and sustainable pavement surfaces that can provide satisfactory performance and serviceability throughout their design life.

3. History of Semi-Flexible Pavement

In the early stages of developing semi-flexible pavement surfaces, researchers and agencies gave different brand names for the same construction and working principles. The first semi-flexible pavement was constructed in the 1950s in France as a fuel and abrasion-resistant surfacing, and was given the name Salviacim [33]. In the 1970s and 1980s, semi-flexible pavement construction spread throughout Europe, the South Pacific, the Far East, several countries in Africa, and North America [34]. Table 1 shows the early development of semi-flexible pavement in different countries with different brand names.

Table 1. Development and purpose of grouted macadam in the past.

Reference	Country	Brand Name	Purpose of Construction
[33,35]	France	Salviacim	To provide resistance against waste oils, fuels, and abrasion
[36]	United States	Resin-Modified Pavement (RMP)	Airport taxiways, aprons, parking lots
[37–39]	Europe	Hardicrete Heavy Duty Surfacing. Worthycim Heavy Duty Paving. Confalt	Heavy-duty surface construction
[40]	Japan	RP-Pavement (Rut Proof Pavement)	Heavy-duty surface construction
[35]	France	combi-layer	Heavy-duty surface construction

4. Semi-Flexible Pavement/Grouted Macadam

Semi-flexible pavement consists of an open-graded asphalt structure with 20 to 35% voids into which flowable cementitious slurry is allowed to penetrate due to pouring and gravitational effects, as shown in Figure 1. Conventionally, Portland cement grout has been used as a cementitious slurry by pouring into open-graded asphalt mixtures, constructing a semi-flexible layer. Numerous researchers have conducted extensive laboratory, as well as field, experiments on evaluating the performance of semi-flexible pavement, which showed significant improvement in compressive strength, stability, and durability [41,42]. Moreover, a semi-flexible surfacing combines the flexibility of asphalt with the rigidity of the cement grout. The rigid cement grout provides high compressive strength, high durability, and resistance to wear. It has been practiced on the bus lanes in urban areas [43]. Based on an extensive review by Du et al. [44], it has been proposed that a semi-flexible pavement layer could be a better solution against rutting distresses [45].

A significant component in a semi-flexible layer is the cementitious grout, which contributes to rigidity and durability, with high compressive strength. The cementitious grouts need to be highly flowable to penetrate fully into the porous asphalt skeleton. The grouts may consist of cement, water, superplasticizer, admixtures, and/or other supplementary materials. Superplasticizer and admixtures/other supplementary materials are required to achieve the required flowability without compromise on the strength properties of grouts. Various combinations of cementitious grouts, such as mild glass, Panasqueira Waste Mud, and geopolymeric grouts, were investigated to evaluate the performance of a semi-flexible pavement surface [46]. Grouts play a significant role in the performance of flexible pavements. The modified grouts showed improved results in Marshall stability, stiffness, compressive strength, and resistance against rutting [46]. Furthermore, this type of pavement surface has been developed to provide rut resistance and joints-free pavement. Initially, it was developed for heavy-loaded sections with slow-moving or static loads, particularly at intersections, for rut resistance without cracks and joints [47].

Depending on the mix design of a porous asphalt structure and grout filling, the semi-flexible pavement surface shows variability in its mechanical properties. A study was conducted to compare the fatigue performance of grouted macadam with that of a conventional blend, using the same percentage of asphalt and type of aggregate. It was revealed that the average stiffness modulus of the grouted mixture was approximately three-times higher than that of a conventional asphalt mixture [48]. A study was conducted to optimize the above-stated characteristics for cement paste (containing cement, coal ash, mineral powder, and water) and cement mortar (containing cement, coal ash, mineral powder, fine standard sand, and water) as a cementitious grout for semi-flexible pavement surfaces. Cement paste (optimum value with 0.56–0.58 water–cement ratio, 10% fly ash, and 10% mineral powder) exhibit better overall performance compared to cement mortar (optimum design with 0.61–0.63 water–cement ratio, 10% fly ash and 15% fine sand), though both grouts meet the recommended technical requirements within the recommended mixture ratio range [49]. In general, the cementitious grout for a semi-flexible pavement surface should have sufficient flexure and compressive strength, as well as better fluidity, ductility, and resistance to drying shrinkage.

Figure 1. Grouted Macadam (semi-flexible pavement surface/mixtures) [4,45,48].

Resin-Modified Pavement (RMP) was used as a grouted macadam and was suggested to be used in the rehabilitation of existing pavement, as well as in new pavement construction. RMP consists of a porous asphalt mixture with 25–35% voids and cementitious grouts (containing: cement, silica sand, fly ash, water, and resin additives). It was considered an alternative to the rigid pavement, as it has a low construction costs and can be used in any environmental conditions [36]. The layers beneath the grouted macadam layer are protected due to its impervious nature (as voids are filled with cement grout), and the stress level in the base course is significantly reduced. The speed of construction and time required to open for traffic is significantly improved compared to rigid pavements [50]. Another important property of semi-flexible pavement surfaces is the early gain in strength and the smaller amount of time required to achieve the final properties of the layer compared to rigid pavement. This type of pavement can be opened to traffic after two to three days, while a rigid pavement may need 21 to 28 days [40].

5. Factors Influencing the Performance of Grouts and Semi-Flexible Mixtures

The semi-flexible mixture/layer consists of a porous asphalt skeleton and cementitious grouts, and hence, the performance of the semi-flexible layer depends on both components. Some of the factors influencing the performance of semi-flexible layers are identified and discussed as follows.

5.1. Porous Asphalt Mixture Gradation

The selection of aggregate gradation plays a significant role in the performance of the final mix's properties. Usually, a single-sized aggregate in open-graded asphalt gradation is used for the preparation of porous asphalt mixtures. Generally, 90–95% coarse aggregate, 4–5% sand, and 2–4% filler are used in the design of open-graded asphalt mixture for semi-flexible pavement applications [51,52]. The selected gradation should provide air voids of 25 to 35% in the open-graded asphalt mixture, to allow cementitious grouts to penetrate the layer fully. As the open-graded asphalt gradation contains high percentages of course aggregate, but very low percentages of fines and fillers, a low quantity of bitumen (2 to 4% by weight aggregates) is therefore used, due to less surface area [51]. Various researchers have adopted different gradation systems in the past for the mix design of the porous asphalt skeleton, as shown in Figure 2. It can be seen that the main difference is the nominal size, between 8 to 14 mm, and most of the gradation is single-sized, with very low percentages of fine aggregates, which makes the porous asphalt skeleton. However, the common goal is to achieve the target 25–35% air voids in the mixture, in order to ease the quick infiltration of grouts due to void interconnectivity and high porosity [40,53–55]. Nevertheless, in Malaysian specifications, the nominal size is 20 mm for type-1 and 14 mm for type-2 gradation systems used in the porous asphalt skeleton for semi-flexible pavement surfaces [56]. In the Netherlands, the nominal size for the porous asphalt skeleton is 11.2 mm for type 8/11 and 16 mm for type 8/16 gradation systems [57].

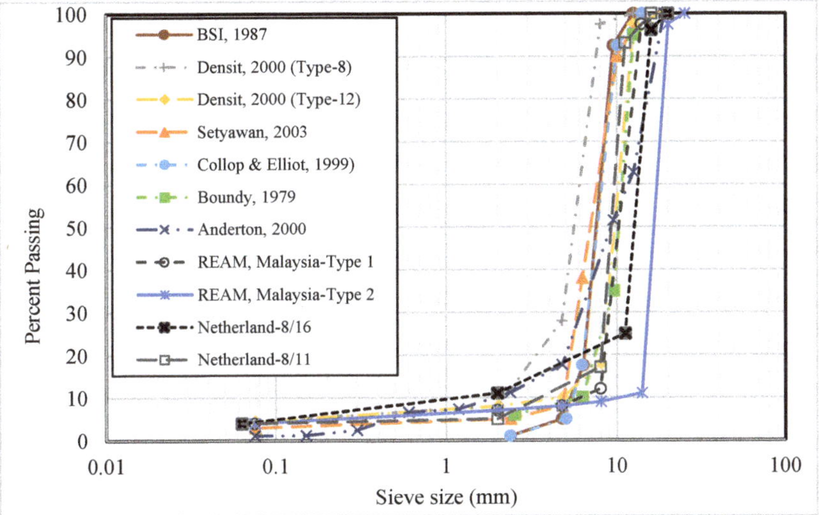

Figure 2. Porous aggregate gradation used in past studies.

A study was conducted on using REAM type-2 porous gradations with upper, lower, and midpoints of the gradation limits, as shown in Figure 3. The effect of these gradations on the volumetric and performance properties of semi-flexible mixtures was investigated. It is clear from Table 2 that different aggregate gradations of the porous asphalt skeleton have a significant effect on the resilient modulus and compressive strength properties of semi-flexible specimens. The higher the resilient modulus, the greater the material resistance against permanent deformation. Similarly, the three aggregate gradations have a different compressive strength of semi-flexible specimens using the same type of grout [58]. Furthermore, aggregate gradation (G3) with less porosity causes a delay in grout infiltration, and hence, a smaller amount of grout was utilized, due to which G3-based, semi-flexible specimens present the lowest compressive strength and least resistance to abrasion compared to G1 and G2 aggregate gradation [59]. A single-size aggregate in the range of 10–14 mm with very small percentages of fines can produce a porosity of 25 to 35% in mixtures, which allows cement slurry to penetrate easily due to interconnected voids. In other words, the aggregate gradation of a porous asphalt skeleton has a significant effect on the performance of semi-flexible pavements.

Table 2. Effect of gradation type on performance properties of semi-flexible pavement [58,59].

Gradation Type	Modulus of Elasticity (MPa)	Compressive Strength (MPa)
G1	18,000	6.09
G2	17,600	5.10
G3	17,200	4.70

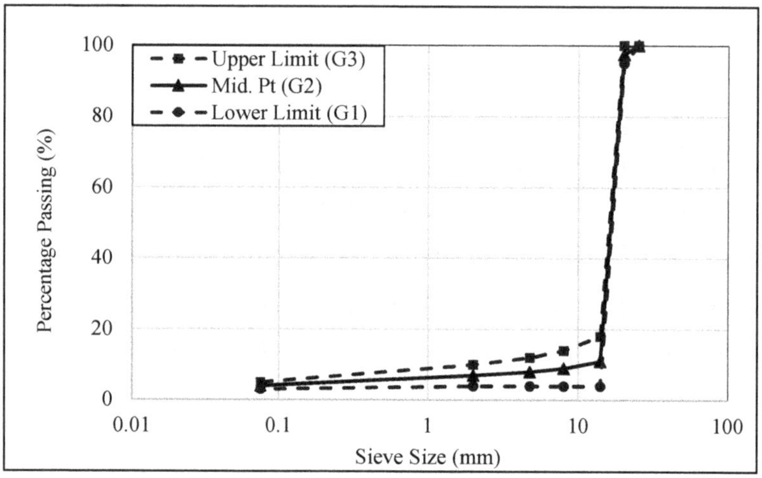

Figure 3. Aggregate gradation adopted from the Road Engineering Association Malaysia (REAM) Type-2 [56].

Similarly, a study was conducted by Saboo et al. (2019) to rank different types of porous asphalt skeleton for semi-flexible pavement, by developing a hierarchical ranking strategy [51]. The detail of seven different selected porous asphalt mixes and the evaluation criteria are presented in Tables 3 and 4, respectively. A bitumen content of 2 to 4.5% was used for all gradations (M1 to M7).

Table 3. Gradation adopted by Saboo et al. (2019) [51].

Sieve Size (mm)	Passing Percentage						
	Boundy (1979) Mix 1	**BSI (1987) Mix 2**	**Anderton (2000) Mix 3**	**Densiphalt Type 8 Mix 4**	**Densiphalt Type 12 Mix 5**	**Setyawan (2003) Mix 6**	**Hou (2017) Mix 7**
19	100	100	100	100	100	100	100
12.5	95	100	62.8	100	100	100	
9.5	35	92.5	51.5	95	95	90	44
6.3	10	17.5	35	30	12	38	
4.75	6	5	17.6	20	10	8	10
2.36	3	1	11.1	10	8	5	
1.18			7.3	5	5	3	
0.6			6.6				2
0.3			2.4				
0.15			1.2				
0.075			1.1				
Air Voids	31–32%	33–34%	27–31%	33–36%	32–35%	29–32%	30–32%
Binder Content range, % (Optimum, %)	2–3.5% (4%)	2–4% (3.5%)	2–4.5% (3.5%)	2–4.5% (4%)	2–4.5% (4%)	2–4% (4%)	2–3.5% (3.5%)
ITS (kPa)	110.46	113.35	116.52	118.83	114.42	123.32	109.85

Table 4. Acceptance/rejection criteria for the mixtures.

<table>
<tr><th>S. No</th><th>Properties of Mix</th><th>Acceptable Criteria</th><th>Initial Selection of Mixes</th><th>Final Selection</th></tr>
<tr><td>1</td><td>Draindown test</td><td><0.3%</td><td rowspan="6">Mix 1, 3.5%
Mix 1, 4.0%
Mix 2, 3.5%
Mix 3, 4.0%
Mix 3, 4.5%
Mix 4, 4.0%
Mix 4, 4.5%
Mix 5, 4.0%</td><td rowspan="6">Densiphalt type 12 (Mix 5) with 4.0% bitumen
Densiphalt type 12 (Mix 5) with 4.5% bitumen
BSI (Mix 2) with 4.0% bitumen</td></tr>
<tr><td>2</td><td>Air Voids</td><td>25% to 35%</td></tr>
<tr><td>3</td><td>Voids in coarse aggregate (VCA) ratio</td><td><1%</td></tr>
<tr><td>4</td><td>Permeability</td><td>>100 m/day</td></tr>
<tr><td>5</td><td>Cantabro loss</td><td><50%</td></tr>
<tr><td>6</td><td>Indirect Tensile Stiffness (ITS)</td><td>N/A</td></tr>
</table>

Initially, 42 mixes were prepared and evaluated for draindown, air voids, VCA, permeability, and cantabro loss. Cantabro loss is reported by measuring the percentage loss of the mass of the specimen after being subjected to 300 revolutions (at 30 rpm) in a Los Angeles abrasion machine [51]. Only eight mixes fulfilled the criteria and were further evaluated for ITS; however, finally, based on the hierarchical ranking methodology, only three mixes (Densiphalt type 12 with 4.0% bitumen, Densiphalt type 12 with 4.5% bitumen, and BSI with 4.0% bitumen) were selected as the most desirable porous asphalt mixtures that can be used for semi-flexible pavement surfaces [51].

5.2. Porosity/Voids of Asphalt Mixture

The voids ratio, or porosity, of open-graded asphalt mixture has a significant effect on the properties of the final semi-flexible mixture. Hence, with too low porosity, the voids may not be filled completely with grout, due to poor connectivity in the voids, while high porosity requires a sufficient amount of grouts, and the pavement acts as a rigid structure [60]. The study revealed that an increase in void content (from 20 to 26%) in the porous asphalt skeleton results in an improvement in the compressive, flexure, and splitting strength and resilient modulus of semi-flexible specimens, because more cementitious grout is required to fill the voids, which impart rigidity to final mixtures. However, a slight reduction in compressive strength and resilient modulus was witnessed by increasing voids beyond 26%. Moreover, the increase in void content causes a reduction in Marshall stability and an increase in the flow value of porous asphalt specimens [61]. This reduction in the stability and increase in flow value is due to a reduction in compacted density with increasing porosity.

5.3. Degree of Grout Saturation

The grouting ability of the semi-flexible mixture is measured by determining the degree of grout saturation. It indicates how many air voids (pores) of open-graded asphalt mixture are being filled with the grouts [62]. It is one of the important parameters that may influence a semi-flexible mixture's performance when subjected to heavy traffic loading and poor weather conditions [63,64]. According to the Chinese standards, the degree of grout saturation should be $\geq$92% [64]. The degree of grout saturation is not only a function of air voids in the porous asphalt mixture, but also dependent on the voids' morphological characteristics, size, and structure, which directly affect the interconnectivity of voids in porous asphalt mixtures [65].

5.4. Type of Asphalt Binder

The type of asphalt binder in a porous asphalt mixture has a substantial effect on the properties of the semi-flexible layer. Using stiffer and harder asphalt, the stiffness and resistance against the indentation compressive strength of grouted macadam increases, even at high temperatures [57,66]. The test setup of indentation compressive strength is given in Figure 4.

Figure 4. Test setup of indentation strength [57].

However, stiffer binder causes cracks in the pavement after repetition of traffic loads [60]. The effect of modified and conventional binders on the performance properties of semi-flexible specimens was analyzed. The modified binder has a notable effect on the indentation strength of semi-flexible specimens at 20 °C, and significant improvement was witnessed in the modification of conventional binder with polymers; however, similar indentation strength was noticed at 40 °C [60]. However, the selection of the type of binder is dependent on weather conditions and site location [67]. Therefore, using polymer-modified asphalt with improved higher and lower temperature performance grades (PG) is a better choice to counter the effect of hot and cold temperatures, respectively.

5.5. *Water–Cement (w/c) Ratio of Grout*

In semi-flexible pavement surfaces, the cementitious grouts are required to be highly fluid to fully penetrate the porous asphalt structure. Hence, the water/cement (w/c) ratio plays a key role in developing the strength and fluidity of cementitious grouts. Different combinations of water/cement ratios (0.45, 0.50, 0.55, and 0.60) were used to optimize the w/c ratio based on compressive strength at 7-days, 28-days, 56-days, and 90-days curing. The target strength of 15–35 MPa on 7-days curing was achieved at a 0.50 w/c ratio [42]. Khan et al. utilized a response surface methodology to optimize w/c ratio and a dose of superplasticizer to formulate cementitious grouts for semi-flexible pavement surface applications [68]. In another study, the water–cement ratio of 0.68 was achieved as the optimum ratio for Latex Resin-modified grouts based on the marsh flow cone viscosity test [69]. This amount of water produced high flowable grouts and fully penetrated the porous asphalt with a considerable durable and tough surface to resist heavy traffic load. A similar and close range of water–cement ratio (0.51 to 0.57) has been recommended based on fluidity, drying shrinkage, ductility, compressive and flexure strength, while the cement grout was modified with a polycarboxylate superplasticizer, expansion admixture, and triterpenoid saponin air-entraining agent additives (details are shown in Tables 4 and 5) [70].

The most important factor that can affect the mechanical properties of grouts and semi-flexible specimens is the fluidity (flow time) of cementitious grout to fill the voids, and hence the porosity and water–cement ratio. It was observed in a study that the porosity of the porous asphalt skeleton and flow time of grouts should be 20% and 12 s, respectively, to achieve the full penetration of grouts. However, to improve the performance properties of grouts and semi-flexible specimens, a chemical admixture (Polyacrylate Polymer) and accelerating and retarding agents were also added to grouts in different proportions [71]. So far, limited literature is available on optimizing the composition of grouts based on the physical and performance analysis of grouts and semi-flexible specimens. Based on a previous research study, a water–cement ratio in the range of 0.45 to 0.63 can be used to achieve the required fluidity and strength properties of grouts [70,72–74].

The effect of using waste tire rubber as a cement powder replacement in grouts was studied, and it was concluded that, with a w/c ratio of 0.60, ordinary Portland cement (OPC) should not be replaced by more than 20% of waste tire rubber. A small addition of tire rubber caused a reduction in compressive and flexure strength; however, the performance against low temperatures can be improved with the addition of tire rubber [75]. In short, an increase in the w/c ratio has an inverse relation with strength properties (compressive and flexural strength); the higher the w/c ratio, the less strong, and vice versa. The w/c ratio can be controlled by adding superplasticizer to improve the fluidity of grout, without compromising the strength properties [76]. The workability of grouts can be significantly improved by adding a superplasticizer [77]. Most commonly, polycarboxylate ether- superplasticizers were used in various studies [70,78–80].

5.6. Admixtures and Other Supplementary Cementitious Materials

The composition of grouts (i.e., w/c ratio, superplasticizer, polymers, and other modifying agents) has a significant effect on the performance of grouted macadam. Minimal research literature is available on different compositions and the performance of cementitious grouts for semi-flexible pavement surfaces. Controlling the viscosity of cementitious grout plays a vital role in achieving full penetration of voids in the porous asphalt skeleton and the target strength of the mixture. If the fluidity of grout is low, some voids may not be filled at the bottom of the specimen. On the other hand, if the w/c ratio of grout is high, a reduction in the overall strength of the mixture occurs. Superplasticizer can be used at a relatively low w/c ratio to achieve high fluidity [81]. Similarly, the type and concentration of superplasticizer influence the flow and strength properties of grouts. Two types of Polycarboxylic ether polymer and one type of sulphonated naphthalene formaldehyde with different dosage were used as superplasticizers to investigate flow and compressive strength. The results reveal that Polycarboxylic ether superplasticizer is superior in achieving the target flow and strength properties [82]. Malaysian specifications recommend that the target flow of 11 to 16 s (using Malaysian flow cone) and target compressive strength of 50 MPa at 1-day and 105 MPa at 28-days of cementitious grouts (grouts containing silica fume, superplasticizer, admixtures) should be used in constructing semi-flexible pavement surfaces for heavy-loaded sections [56,83].

A study was conducted by using Carboxyl latex (0 to 12%) as a partial replacement of cement in cementitious grout for a semi-flexible layer. The design parameters, water/cement ratio (0.63), mineral fillers (10%), and sand (20%) were selected based on a literature study. The optimum asphalt content (OAC) was obtained by a Marshall stability test, cantabro, and drainage test. It was concluded that cement grout containing 8% carboxyl latex showed improvement in flexure strength and better volume stability. In addition, the performance against rutting at high temperatures, low-temperature cracks, moisture damage, and fatigue were significantly improved in grout modified with carboxyl latex [72]. Different proportions of water–cement ratio, superplasticizer, and sand quantity in making grout for semi-flexible mixtures were studied. It was concluded that rupture and compressive strength were higher, but the fluidity was affected in reducing the w/c ratio from 0.3 to 0.21. However, a 0.22 w/c ratio, 0.3% fine sand, and 0.01% superplasticizer give the best combination in terms of fluidity, and compressive and rupture strength [84]. The fluidity of grouts increased with an increase in the water–cement ratio, while a reduction in compression strength and flexure strength was observed. The water–cement ratio in the range of 0.55 to 0.57 has been proposed for ordinary cement paste in a semi-flexible layer [84,85]; in addition, the acceptable ranges of using polycarboxylate superplasticizer content, expansion admixture, and air-entrained agents are 0.5–1.0%, 10%, and 0.008%, respectively, have been proposed for high-performance cement paste as a grouting material for semi-flexible pavement surfaces.

In another study, three types of grouts were used, namely, Densit grout (commercially available in the UK), Densit-modified grout, and weak grout (Densit + 25% silt). However, no significant change in the stiffness modulus and phase angle of grouted macadam was observed. However, a slight improvement in the case of Densit grout was noticed. Furthermore, grout strength and grout shrinkage are considered two characteristics used to analyze the fatigue performance of grouted macadam. The Densit grout showed the best fatigue life of macadam, while the modified grout exhibited slightly lower fatigue performance [40]. Similarly, additives and/or admixtures can be used to partially replace cement in cementitious grouts to reduce the overall cost of construction and achieve the target strength. So far, semi-flexible pavement surfaces are being designed for heavy-loaded sections, and hence, high-strength admixtures or other high-strength supplementary cement materials are necessary to be added in cementitious grouts to achieve high compressive strength (50 MPa at 1-day and 105 MPa at 28-days curing) with acceptable fluidity [56].

5.7. Industrial/Municipal Wastes and Other Polymers/Additives

The cementitious grouts contribute to making a semi-flexible layer rigid and durable. However, utilizing OPC in grouts for the construction of semi-flexible pavement is a severe threat to the environment in terms of carbon dioxide emissions. The cement industry accounts for carbon dioxide emissions, making it a critical sector for emission mitigation. The production of cement releases carbon dioxide both directly and indirectly. Furthermore, the high cost of cement requires a huge investment in constructing semi-flexible pavement surfaces using OPC grouts. Hence, researchers are investigating the use of alternative cementing materials or the partial replacement of cement with industrial/municipal wastes or byproducts to control or minimize the use of OPC in the construction industry [86,87].

Another study was conducted on utilizing fly ash and mineral powder (ground granulated blast furnace slag) in cementitious grouts for semi-flexible mixtures. Two types of grouts, namely, cement paste (cement, fly ash, mineral powder, and water) and cement mortar (cement, fly ash, mineral powder, standard sand, and water), were prepared. In this study, the optimum contents of fly ash, mineral powder (ground granulated blast furnace slag) and water–cement ratio for the cement paste were 10%, 10%, and 0.56–0.58, respectively, based on evaluating the fluidity, ductility, compressive and flexure strength (7, 28-days) and drying shrinkage properties. Similarly, for cement mortar, the optimum contents of fly ash, mineral powder, standard sand, and w/c ratio were determined as 10%, 0%, 15%, and 0.61–0.63, respectively, using the same methodology as adopted for cement paste. However, at a certain water–cement ratio, the fluidity, compressive and flexure strength of the cement paste were better than the cement mortar. The cement paste grouts with a water–cement ratio, fly ash, and mineral powder contents of 0.58, 10%, and 10%, respectively, were considered optimum compositions for semi-flexible mixtures [49].

Various combinations of cementitious grouts, such as mild glass (from recycled bottles), Panasqueira Waste Mud, and geopolymeric grouts, were investigated to evaluate the performance of semi-flexible pavement surfaces. Grouts play a significant role in the performance of flexible pavements. The modified grouts showed improved results of Marshall stability, stiffness, compressive strength, and resistance against rutting [46]. In another study, three types of cementitious grouts were used to evaluate the effect of different types of grouts on the performance of the semi-flexible layer. These grouts were standard Densit® grout (commercially available), modified grout, and weaker grout (obtained by adding 25% silt to Densit®). The water–cement ratio of 17.2% by mass of powder was used as a guideline throughout the project. The overall performance of Densit® grouts and hence the Densit® macadam were better than modified and weaker grouts in terms of compressive and fatigue strength, stiffness modulus, resistance to permanent deformation, shrinkage, and thermal cracks [40]. In another study, it was revealed that the addition of silica fume significantly improves the compressive strength of grouts, while reducing the fluidity [81]. Silica fume and fly ash substitution in cement grouts also contributes in the improvement of performance of cementitious grouts [88]. Furthermore, the target compressive strength of 90–120 MPa at 28 days was achieved when cement was replaced with pozzolanic materials such as silica fume and fly ash-based grouts, which also contribute to higher modulus of elasticity and lower porosity of grouted macadam [89].

Similarly, cross polymer resin (5 parts water + 2 parts styrene-butadiene + 1part water-reducing agent) was used to produce resin-modified pavement (RMP) mixtures. The slurry grout compositions used in this study include: cement, silica sand (18–18.5%), fly ash (18–18.5%), Latex Resin (2–3%), and water (w/c ratio 0.60–0.70). The grout compositions (such as 36.2% cement, 18.1% sand, 18.1% fly ash, 24.8% water, and 2.8% Latex Resin) were optimized based on a marsh flow cone viscosity test and were used for further field evaluation. The results from field trial sections reveal that RMP could be used as a durable and tough surfacing for military-purpose pavements, as well as for heavy vehicular loads. However, RMP does not provide significant resistance against fuel and oil damage [69]. Similarly, user guidelines were developed in 1996 by Anderton to design RMP mixtures and pavement surfaces. According to the guidelines, the composition of grouts, mix design, and performance properties, as shown in Table 5, are used for RMP mixtures [90]. Ease in construction, no joints on the surface, cost competitiveness, resistance to rutting, suited for any environmental conditions, abrasion, and wear resistance are some of the benefits of RMP.

Table 5. Grout composition and porous asphalt job mix formula for RMP [90].

Grouts Composition		**Porous Asphalt**	
Material	Weight (%)	Voids	25 to 35%
Cement (Type-1) Sand (silica Sand) Fly Ash (as Filler) Latex Resin Water	34 to 40 16 to 20 16 to 20 22 to 26 2.5 to 3.5 [w/c = 0.65–0.70, Flow time: 8 to 10 s]	Bitumen Content Penetration grade	3.5 to 4.5% by total weight 40 to 100

Various combinations of cementitious grouts were investigated from three different types of additives (Th-98 Polycarboxylate Superplasticizer, UEA Expansion Admixture, and ZY-99 Triterpenoid Saponins Air-entraining agent) and recommendations are made based on fluidity, ductility, drying shrinkage, and compressive and flexure strength tests, as described in Table 6 [70].

Table 6. The recommended range of additives and w/c ratio for grout composition [70].

Cement Paste Compositions and Additives	**Range of Variables Used**	**Recommended Values**	**Evaluated Properties of Grouts**
Water–cement ratio	0.51–0.57	0.55–0.57	
TH-928 polycarboxylate superplasticizer	0–1.0 (%)	0.5–1.0 (%)	Fluidity Ductility
UEA expansion admixture	0–12 (%)	10%	Drying shrinkage
ZY-99 Triterpenoid saponins air-entraining agent	0–0.012 (%)	0.008%	Compressive and Flexural strength

Based on the above-mentioned testing criteria, high-performance cement pasts (H1, H2, and H3) were designed for further investigation, and H2 and H3 were recommended as high early-strength cementitious grouts. The grout compositions for H1, H2, and H3 are expressed in Table 7.

Table 7. Recommended value range for High-Performance Cement Pastes (HPCP) [70].

HPCP Composition/Type	H1	H2	H3
w/c ratio	0.56	0.56	0.56
TH-928 polycarboxylate superplasticizer	0.5%	0.5%	1.0%
UEA expansion admixture	10%	10%	10%
ZY-99 Triterpenoid saponins air-entraining agent	0.008%	0.0%	0.008%

A study revealed that the target performance of semi-flexible specimens could be achieved by controlling the performance and fluidity of grouts. The addition of Polyacrylate polymer, and accelerating and retarding agents significantly improves the fluidity and performance properties of grouts and semi-flexible specimens [71]. Chemical admixtures referred to as superplasticizers or resin modifiers are usually used to make the cementitious grout highly flowable at a low amount of water [36,50,91]. However, using an excess content of superplasticizer can cause bleeding and segregation in cementitious grouts [92]. The early strength of grout can also be improved by using accelerators [50]. Different combinations of OPC, silica fume (SM), and fly ash (FA) were used by Setyawan, 2003 in his study, and no significant difference in stiffness modulus was observed at different test temperatures, as shown in Figure 5 [50].

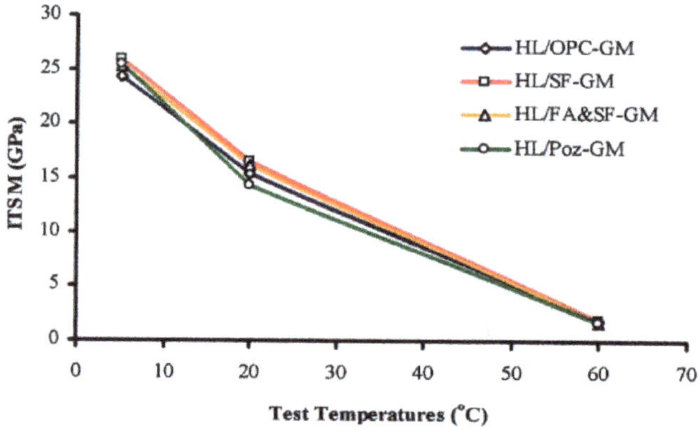

Figure 5. Effect of cementitious grout on the tensile modulus of elasticity [50].

6. Evaluation of Physical and Mechanical Properties of Cementitious Grouts

Various testing techniques have been used to evaluate the physical and mechanical properties of cementitious grouts. The most significant property of grout is to achieve adequate workability or flowability so that grout can easily and fully penetrate the porous asphalt skeleton in achieving the targeted strength. Various factors, including water–cement ratio, type and dosage of superplasticizer, type of cement, and other admixture or supplementary materials, affect the workability of grouts.

The physical and mechanical properties of cement grouts are highly dependent on the composition of the grout. The grouts can be evaluated for flowability, drying shrinkage, compressive strength (at different curing ages), and flexural strength, before use for the construction of semi-flexible pavement surfacings. Among these properties, one significant consideration is the flowability (or fluidity) of the cement grout, which is required to be fully penetrated through the depth of the porous asphalt surface layer without excessive compaction and vibration [93]. The second important consideration is the grout's compressive strength, which significantly influences the final strength and durability properties of semi-flexible mixtures.

6.1. Flowability of the Cement Grouts

Highly flowable cement grouts are required for semi-flexible pavement applications, so that they can be easily infiltrated throughout the depth of the layer. The flowability or fluidity of cement grout is dependent on cementing materials, water–cement ratio, and superplasticizer. Due to its highly flowable nature, the flowability is measured by using a flow cone apparatus. In flow cone apparatus, the required amount of grout (soon after preparing) is poured and allowed to flow out from the apparatus. The time required to empty the flow cone is measured in seconds and recorded as flow of the cement grouts. The higher the time of flow means the flowability or fluidity of grout is lower and it will be difficult to penetrate the voids in the mixture. The standard flow time, however, depends on the geometry and size of the flow cone used. Three different sizes of cones are being used and are shown in Figure 6. The Malaysian flow cone (Figure 6a) required 1000 mL of grout and the flow time is in the range of 11 s to 16 s [56,83]. Using a Malaysian flow cone, the cement grouts with a flow of 11–16 s have sufficient fluidity and are suitable for grouting SFP surfacings. On the other hand, the ASTM flow cone (Figure 6b) required 1725 mL grout and flow time is in the range of 10 to 14 s. The Marsh flow cone (as shown in Figure 6c) also requires 1000 mL cement grout; however, due to changes in geometry, flow time is in the range of 8 to 10 s [36].

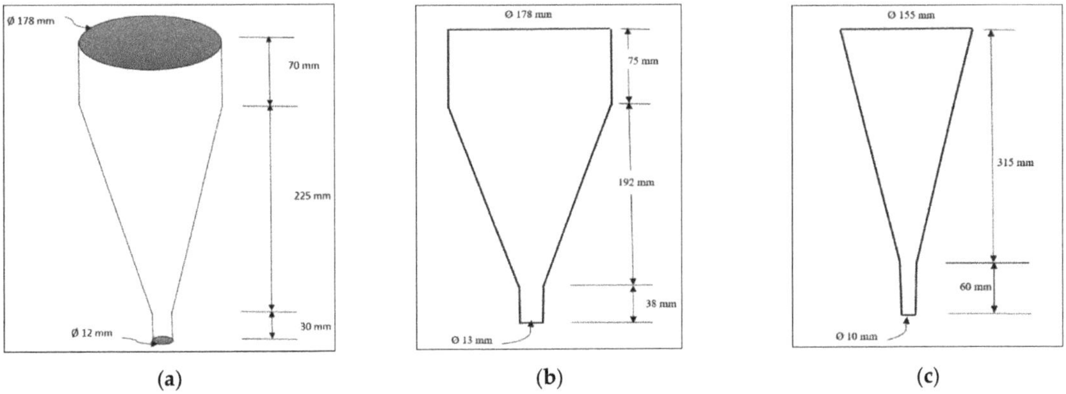

Figure 6. Flow cones with different geometries: (**a**) Malaysian flow cone (**b**) ASTM flow cone (**c**) Marsh flow cone (dimensions not-to-scale).

Flow of cement grout has been evaluated by changing the w/c ratio, supplementary cementing material (SCM), superplasticizer, and other additives. Superplasticizer can enhance the flowability at a relatively low w/c ratio, while achieving the desired strength properties [46,94]. A study reported improvement in compressive strength and flowability with the addition of 1% superplasticizer, whereas, increasing the water-to-cement ratio causes a reduction in the strength of grouts [95]. Afonso et al. (2016) reported low strength at a high w/c ratio and the incorporation of styrene–butadiene admixtures produces weak grouts [46]. However, stronger grout at a relatively low w/c ratio with enhanced compressive strength was achieved by adding milled glass and polycarboxylate superplasticizer [46].

Zhang et al. (2016) replaced cement with fly ash and mineral powder (0 to 20%), while varying the w/c ratio from 0.48 to 0.63 to formulate cement paste for semi-flexible pavement applications. The conclusion from the study indicate that increasing the water-to-cement ratio caused a significant increase in the flowability. Similarly, the increase in fly ash content causes an increase in flowability. However, the addition of mineral powder initially causes a reduction in fluidity, and then an increase was witnessed. The optimum dose of fly ash and mineral powder was recommended as 10% and at water-to-cement ratio of 0.56–0.58 for grout suitable for semi-flexible pavement applications [49]. Flow cone and flow table tests are the most common techniques used to determine the fluidity of cementitious grouts. The effect of Polycarboxylic-ether polymer superplasticizers and silica fume on fluidity and strength properties of the grout was studied, and it was concluded that sufficient workability of the grout was achieved by a 0.30 water-cement ratio with 2.0% superplasticizer and 5% silica fume by weight of OPC [96].

Therefore, based on the observations from the above-cited literature, it is important to evaluate the flowability (or flow) of cementitious grouts when designing the cementitious grouts for semi-flexible pavement surface applications. The desired flowability of grouts (while not compromising on compressive strength) plays an important role in the full penetration of voids throughout the depth of the open-graded asphalt layer.

6.2. Strength Properties of Cement Grouts

Another important property of cement grout that needs to be considered for SFP surfacing is compressive strength. It mainly depends on adding SCM and water-to-cement ratio and other additives/admixtures. It was found in the literature that different researchers have used mineral powders, gypsum, silica fume, fly ash, and ground granulated blast furnace slag (GGBS) for producing cement grouts for semi-flexible pavement applications. Zhang et al. (2016) reported a significant reduction in compressive strength while increasing the water-to-cement ratio from 0.48 to 0.63. However, flexural strength was slightly reduced [49]. Therefore, cement grout for semi-flexible pavement applications is required to be introduced with superplasticizer and SCM (such as fly ash, silica fume etc.) to enhance the strength and flowability at a relatively low water-to-cement ratio.

The combination of silica fume in addition to superplasticizer can produce grouts with sufficient fluidity and strength properties that can be recommended for semi-flexible pavement surfacings. Hence, silica fume with 5% replacement and 2.0% polycarboxylate-based superplasticizer at a 0.30 w/c ratio gives fluidity of 15 s (as determined by the Malaysian flow cone) and compressive strength at 28 days of 92.5 MPa [96]. Therefore, this grout can be recommended to produce high-strength cement grouts for semi-flexible pavement surfacings that can be used in pavements exposed to a heavy vehicular load. Moreover, the compressive strength of 57.5 MPa and 95.5 MPa at 7 and 28 days, respectively, was obtained from this grout. Furthermore, with the same grout composition, the highest flexure strength, 6.7 MPa at 7 days and 9.1 MPa at 28 days, was achieved [96]. Some commercial cement grouts are also available for the construction of semi-flexible pavement surfacing applications in the UK and Canada, etc. However, the compositions of those cement grouts are unknown. The details of some commercially available cement grouts are presented in Table 8. Furthermore, various other cement grouts that have been used in the literature for semi-flexible pavement surfacing are also presented in Table 9.

Table 8. Technical data of commercially available cement grouts.

Type of Cement Grout	Manufacturer Name/Country	Compositions	Properties	Application	Remarks
Densiphalt mortar	Densit®/UK	Composition Not Known Water = 16.2% by weight of powder	Compressive Strength 1-day = 50 MPa 7-day = 80 MPa 28-day = 110 MPa Density = 2200–2250 Kg/m^3 Setting time = 7–9 h	Warehouses, distribution centers, goods terminals, production floors, Harbors, container terminals, warehouse facilities, Bus stops and terminals, roads for heavy traffic, road crossings, Aircraft aprons, taxiways, refueling pads, taxi-holding positions	

Table 8. *Cont.*

Type of Cement Grout	Manufacturer Name/Country	Compositions	Properties	Application	Remarks
DuraTough	Lafarge/Canada	Composition Not Known	Standard grout: Compressive Strength 7-day = 35 MPa 28-day = 60 MPa High-Strength grout: Compressive Strength 7-day = 60 MPa 28-day = 100 MPa Flexural Strength 28-day = 2.5–3.0 MPA	Bus stops and transit stations, turning lanes and intersections, high-volume and heavy-load roadways, tunnels and hard-to-access areas Apron areas, taxiways, fueling/deicing areas, hangar bays, loading bays	These are the few commercially available cement grouts for semi-flexible pavement applications. Most of these grouts did not provide the compositions of cement grouts. Few grouts are provided with technical requirement. Among these grouts, nobody used waste PET and FA in the formulation of cement grouts.
Hardicrete	Miles Macadam Ltd./UK	Composition not known	No Specifications	Bus stations, waste sites and transfer stations, ports, airfields, and industrial facilities	
ULTICRETE	TARMAC A CRH company/UK	Composition not known	No Specifications	aircraft refueling areas, container ports, and distribution centers.	
CONFALT® mortar	Contec ApS/ Denmark	Portland Cement = 45–75% Silica sand = 25–45% Iron Oxide = 0–5% Amorphous Silica = 5–25%		Storage Facilities, Industrial Flooring, Cross Sections, Container Terminals, Airports, Multistorey Parking	

Table 9. Description of cement grouts from the literature.

Composition of Grouts	Strength Properties of Cement Grouts	Concluding Remarks	References
0.50 w/c ratio, 23% fly ash and 2% superplasticizer	Not determined	Semi-flexible mixtures compared to HMA • Rut resistance • Resistance to moisture damage • Lower fatigue life • Better thermal resistance	[97]
0.31 w/c ratio, pre-designed grout (JGM-301)	Compressive Strength: 34.5 MPa at 3 d curing and 42.3 MPa at 28 d curing	It was recommended that semi-flexible pavement surfacings with this grout could be used for heavily loaded pavements.	[98,99]
0.72 w/c ratio, 20% sand, 10% filler and 0–2.4% Latex	Compressive Strength: 21.6 MPa to 24.1 MPa compressive strength and 1.65 MPa to 3.54 MPa flexural strength at 7d curing	Latex in cement mortar has a positive effect on compressive and flexural strength, while negatively impacting fluidity. Semi-flexible mixtures compared to HMA • Rut resistance • Resistance to moisture damage • Better fatigue life • Better thermal resistance • Poor resistance to brittle cracking	[100]
Three interface optimizers were used: 0.25 to 0.75% Silane coupling agent, 5 to 15% Carboxyl styrene-butadiene latex, and 5 to 15% Cationic emulsified asphalt	7-days compressive Strength: (1) 42.98 MPa for OPC (2) 34.65 MPa, 36.85 MPa, 33.61 MPa for Silane coupling agent (3) 25.70 MPa, 29.70 MPa, 26.15 MPa for Carboxyl styrene-butadiene latex (4) 37.79 MPa, 33.82 MPa, 31.53 MPa for Cationic emulsified asphalt	With interface optimizers • Improvement in resistance to Drying shrinkage • Improvement in interface bonding • Improvement in resistance to low temperature cracking improve high temperature stability and moisture resistance.	[101]

Table 9. *Cont.*

Composition of Grouts	Strength Properties of Cement Grouts	Concluding Remarks	References
0.40 w/c ratio, 10% silica fume, 2% Superplasticizer, 0.04% aluminite powder (AP), 0.2% viscosity modifying agent (VMA) ≥ VMA/C, 20–60% Asphalt Emulsion (AE)/cement binder AE/C	7-days compressive Strength: CP = 26 MPa CAEP20% = 13 MPa CAEP40% = 11.3 MPa CAEP60% = 10.5 MPa 28-days compressive Strength: CP = 28.5 MPa CAEP20% = 19 MPa CAEP40% = 17.8 MPa CAEP60% = 17 MPa	• Asphalt emulsion reduces compressive strength, whereas it improves interface bonding • Rutting resistance is improved • Better moisture resistance • Poor resistance to low temperature	[102]
2–4% Naphthalene-based Superplasticizer and 0.4 to 0.6 w/c ratio and 23% Sand (different size) by weight of cement	7-days compressive Strength: For sand size < 0.6 mm gives maximum of 31.65 MPa compressive strength For sand size < 1.18 mm gives maximum compressive strength of 33.75 MPa For sand size 2.36 mm gives maximum compressive strength of 36.69 MPa	Grouts with 0.54 w/c ratio and 2% SP as optimum combination achieving the compressive strength of 24.25 MPa.	[103]
0.5 to 2% Polycarboxylate Superplasticizer, 0.4 to 0.50 water-to-cement ratio and 23% sand (with different size)	7-days compressive Strength: 13.15 MPa to 29.10 MPa (for sand size < 0.6 mm) 16.38 MPa to 30.39 MPA (for sand size < 1.18 mm) 18.45 MPa to 31.20 MPA (for sand size < 2.36 mm)	The optimal composition of grout: 0.48 w/c ratio of 0.48, and 1% SP with 23.46 MPa compressive strength of grout	[103]
0.72 water-to-cement ratio, 20% sand, 10% filler, and 1.2 to 2.4% latex powder (LP)	7-days Compressive Strength: Maximum compressive strength of 24.1 MPa at 1.2% LP and maximum flexure strength at 3.54 MPa at 2.4% LP	Fluidity decreases with increasing latex powder. 1.2% LP has highest compressive strength, whereas maximum flexural strength is achieved at 2.4% LP LP modified semi-flexible mixtures showed improved resistance to high temperature Poor resistance to fatigue and low-temperature cracks compared to HMA	[62]
0.63 w/c ratio, 10% mineral filler, 20% Sand, and 2–12% Carboxyl Latex (CL)	11.20 to 20.76 MPa of 7-days Compressive Strength and 18.56 to 31.43 MPa of 28-days Compressive Strength 4.95 to 28.74 MPa of 7-days Flexural Strength and 5.74 to 6.49 MPa of 28-days Flexural Strength	The addition of CL causes reduction in compressive strength, while improvement in flexural strength Optimum content of CL obtained was 8%. Rutting, low-temperature cracking, moisture damage and fatigue resistance of modified CL was improved compared with controlled grout	[72]
0.60 water-to-cement and cement/sand ratio of 1/0.5	10.40 MPa of 3d compressive strength and 30.7 MPa of 28d compressive strength 3.3 MPa and 7.3 MPa of flexural strength at 3d and 28d curing	Semi-flexible specimens showed better high-temperature performance, improved low-temperature cracking, moisture resistance compared with HMA.	[104]

7. Performance Evaluation of Semi-Flexible Mixtures

Several physical and performance properties have been used to investigate the suitability of semi-flexible mixtures and their comparison with the conventional flexible pavement. It includes the grouting ability, indirect tensile strength, moisture resistance, thermal expansion coefficient, flexural strength, fuel spillage resistance, stiffness modulus, fatigue life, and rutting properties of semi-flexible pavement surfacing materials [40,42,46,58,70,78,85,105]. Some of these properties that have been previously investigated and their corresponding testing methods are summarized in the following sub-sections.

7.1. Grouting Ability

The grouting ability of the semi-flexible mixture is measured by determining the degree of grout saturation. It indicates how many air voids (pores) of the open-graded asphalt mixture are being filled with the grouts [62]. It is one of the important parameters that may influence a semi-flexible mixture's performance when subjected to heavy traffic loading and poor weather conditions [63,64]. According to the Chinese standards, the degree of grout saturation should be ≥92% [64]. The degree of grout saturation is not only a function of air voids in the porous asphalt mixture, but is also dependent on the voids' morphological characteristics, size, and structure, which directly affect the interconnectivity of voids in the porous asphalt mixture [65]. Unfortunately, there is still no standardized guideline available that indicates a specific range of degree of grout saturation. However, some specific limits are defined in the literature. Studies conducted by Luo et al., 2018 and Zhong et al., 2020 suggest that the degree of grout saturation is ≥96% [62,100]. However, Shaoke et al. (2018) mention it as ≥92% in their article [64]. Similarly, some researchers suggest that the degree of grout saturation is in the range of 94 to 96% [54,63,105]. Therefore, based on the above-cited references, it could be suggested that the degree of grout saturation is in the range of 94–96%, which would justify the interconnectivity of voids and uniform grout penetration through the surface layer.

7.2. Indirect Tensile Strength and Moisture Resistance

The indirect tensile strength (ITS) test has been used to determine the tensile strength of semi-flexible mixtures, as well as moisture resistance when conditioned in water. The ITS of semi-flexible surfacing/grouted macadam also depends on the type of aggregate gradations used in the mix design. In a study, the ITS of grouted macadam was determined for six different aggregate gradations (AG) based on different nominal maximum aggregate sizes (NMAS) [106], and the results are shown in Table 10.

Table 10. Six types of gradations and the corresponding ITS values [106].

Designations	Description of Gradation	ITS @ 35 °C
AG1	NMAS of 19 mm and Mid-point gradation of surface dressing	1.41 MPa
AG2	NMAS of 13.2 mm and Mid-point gradation of surface dressing	1.71 MPa
AG3	NMAS of 13.2 mm and Mid-point gradation of OGPC	1.63 MPa
AG4	NMAS of 13.2 mm and Modified Mid-point gradation of surface dressing	1.57 MPa
AG5	NMAS of 9.5 mm and Mid-point gradation used by Oliveira	1.29 MPa
AG6	NMAS of 19 mm and Mid-point gradation of Bituminous Macadam	0.81 MPa

Similarly, three different open-graded asphalt gradations were adopted from ASTM, New Jersey, and New Zealand specifications, and indirect tensile strength was determined. The ASTM gradations depict the highest ITS value of about 1.6 MPa at 4.5 bitumen content, followed by New Jersey with an ITS value of about 1.4 MPa. The lowest ITS value was obtained for New Zealand, with about 1.35 MPa [107]. However, no data are available related to air voids obtained for these selected gradations.

Furthermore, the ITS test can also be used to evaluate the moisture susceptibility of semi-flexible mixtures. The ITS is also determined for the samples conditioned in a water bath at 60 °C for 24 h before testing. The samples are then brought to a test temperature of 25 °C. The ITS obtained at these conditions can be represented as ITS_{wet}. The ratio of ITS_{wet} and ITS_{dry} expressed as tensile strength ratio (TSR) represents the material resistance to moisture damage. In a study, a TSR of 0.97 was obtained for grouted macadam, whereas TSR for HMA mixtures was 0.89. The TSR results show that grouted macadam has superior performance against moisture damage than HMA [97]. Another study also concluded that semi-flexible mixtures using a cement–asphalt emulsion paste as a grout have higher ITS values and showed better resistance to moisture damage in terms of higher TSR values [102]. These studies indicated that grout phase materials in semi-flexible mixtures have an encouraging effect and are the appropriate coating in the asphalt mixture mastic [63]. The semi-flexible mixtures (grouted macadam) gain the ITS of 2.33 MPa, whereas bituminous concrete with only 0.99 MPa [108]. Furthermore, semi-flexible mixtures exhibit enhanced resistance to moisture and freeze-thaw damage compared with bituminous concrete in terms of TSR values [62,108]. The indirect tensile strength of semi-flexible mixtures also increases with an increase in curing age, due to the involvement of the cement grout's hydration process.

7.3. Compressive and Flexural Strength

It is usually required to determine the compressive and flexural strength of semi-flexible pavement surfacing after a certain curing period; however, it is not required in conventional flexible pavement material design. The semi-flexible pavement surfacing material behaves between flexible and rigid pavement materials, depending on the testing method and test conditions (most notably, temperature). Therefore, due to the inclusion of cement grout, semi-flexible materials have a flexural behavior under the application of traffic load.

A study was conducted using waste tire rubber powder (3, 4, 5%) as an additive and activated natural zeolite (0 to 25%) as a cement replacement in cementitious grouts for semi-flexible pavement surfacing [2,109]. The increasing percentage of zeolite mineral from 0 to 15% replacement contributes to the significant increase in compressive strength, while a slight decrease was witnessed beyond 15%. The reaction of calcium hydroxide (produced due to the hydration of cement) with silica (present in zeolite mineral) results in calcium silicate hydrate formation. This contributes to an increase in the cement grouts' density and, hence, causes an increase in the compressive strength of semi-flexible mixtures. The increase in waste tire rubber powder also causes improvement in compressive strength [109]. The behavior of improvement in flexural strength due to waste tire rubber and zeolite mineral was observed as quite similar to the compressive strength. The flexural strength of semi-flexible mixtures increased with increasing zeolite content up to 15% [109].

Gradation type, bitumen content, and curing age also significantly influence the compressive and flexural strength of semi-flexible mixtures. Sunil, 2020 [107] uses ASTM, New Jersey, and New Zealand-based porous asphalt gradation to produce semi-flexible mixtures. Bitumen content of 4%, 4.5%, and 5% by weight of aggregates was used to prepare porous asphalt mixtures for all three gradations. Optimum compositions of cement grouts with a cement:sand ratio of 1:1, w/c ratio of 0.45, and superplasticizer dose of 0.4% were used for all mixture combinations. It was reported that maximum compressive and flexural strength was at 4.5% bitumen content. Similarly, the gradations adopted from ASTM showed higher compressive and flexural strength at all bitumen contents and curing ages, followed by the New Jersey gradation. Moreover, all the mixtures' compressive and flexural strengths significantly increased with increasing curing age, from 1 day to 28 days [107].

7.4. Stiffness

The stiffness of the asphalt material and grouted macadam is considered one of the important input parameters for the structural design of pavement layers. The stiffness property indicates how the traffic load is distributed effectively to underneath layers. The stiffness property is used in ranking different asphalt mixtures. The stiffness modulus is also used to evaluate the structural behavior of pavement [110].

Stiffness properties of semi-flexible materials or grouted macadam depend on the type of binder, mix design of the asphalt mixture, type of grouting materials, and testing conditions (i.e., temperature, loading cycle, frequency). It is a well-known fact that HMA behaves as a viscoelastic material, and the properties are changed with a temperature change. However, semi-flexible behaves differently depending on test type and conditions. It is evident from the literature that the stiffness of semi-flexible pavement surfacing materials also changes with temperature changes. The stiffness decreases with an increase in test temperature, and vice versa [36,40]. Therefore, semi-flexible pavement surfacing materials also behave visco-elastically in nature, from the perspective of the stiffness modulus.

7.5. Fatigue Life

The fatigue life of semi-flexible pavement surfacing depends on many factors, such as gradation type, properties of the binder used, the combination of cement grouts and types, and input conditions of fatigue testing. In a study, ASTM, New Jersey, and New Zealand-based porous asphalt gradation were used to produce semi-flexible mixtures. Bitumen contents of 4%, 4.5%, and 5% by weight of aggregates were used to prepare porous asphalt mixtures for all three gradations. Optimum compositions of cement grouts with a cement:sand ratio of 1:1, w/c ratio of 0.45, and superplasticizer dose of 0.4% were used for all mixture combinations. The semi-flexible mixtures with ASTM gradations showed the highest fatigue life at different stress ratios (10, 20, and 30%) compared with New Jersey and New Zealand gradations [107].

However, the application of semi-flexible pavement surfacings (grouted macadam) is quite different from conventional HMA. In order to compare the fatigue life of semi-flexible mixtures with conventional HMA based on strain, a study has shown that semi-flexible pavements have a lower fatigue life compared with HMA [97]. This is due to the brittle behavior of the semi-flexible material because of the cement grouts, which cause fatigue failure at low strain compared with conventional flexible pavement materials. Semi-flexible mixtures can withstand higher maximum stress compared to HMA. However, a sudden and abrupt fall in stress happened in semi-flexible mixtures, as shown in Figure 7 [97]. However, a study conducted by Oliveira et al. (2008) based on the strain-controlled test method, concluded that semi-flexible pavement surfacing has a higher fatigue life compared with HMA [111].

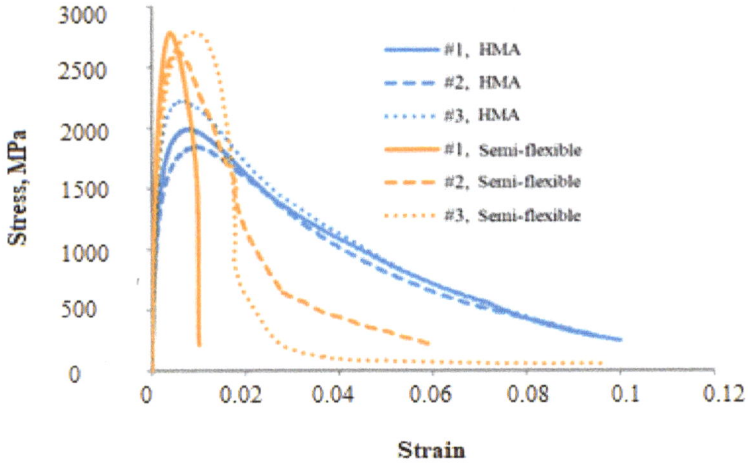

Figure 7. Stress–strain relationship of fatigue test results [97].

The resistance to fatigue failure of semi-flexible mixtures may be lower than conventional asphalt mixtures due to higher stiffness and brittleness. However, the overall fatigue performance of semi-flexible pavement surfacing materials should be investigated based on their application. It also depends on what the surface application will be used for. What are the loading scenario and design life? The tensile strain response of semi-flexible mixtures in the layer can significantly be reduced due to a high modulus and can compensate for its fatigue performance [97]. Consequently, a comprehensive pavement layer of semi-flexible mixtures is structurally evaluated considering field conditions of traffic loading, material characteristics, and environmental conditions.

Several types of laboratory tests are available to investigate the fatigue resistance of asphalt mixtures. These include a two-point bending test on trapezoidal specimens or prismatic-shaped specimens, a three-point bending test on prismatic-shaped specimens, a four-point bending test on prismatic-shaped specimens, and an indirect tensile test on cylindrical-shaped specimens. Moreover, based on the loading mode, fatigue tests are divided into two categories: control stress mode and control strain mode. In control stress mode, constant stress is applied, and fatigue life is defined as a complete failure of the specimen. However, in control strain mode, the complete failure of specimens is unlikely to occur. Therefore, fatigue life is defined as when the stiffness dropped to half of the initial value [112,113].

7.6. Rutting

Semi-flexible pavement surfacing materials showed superior performance against rutting, and the deformation could simply be negligible. Few studies evaluated the rutting (permanent deformation) in semi-flexible pavement surfacing material and the results were very satisfactory. Due to a higher stiffness modulus, semi-flexible pavements experienced no permanent deformation throughout their design life.

A study concluded that a semi-flexible mixture with New Zealand and ASTM gradation at 4.5% bitumen content showed quite similar deformation results and completed 20,000 cycles without failure and negligible deformation was witnessed [107]. Similarly, semi-flexible mixtures produced from different grouts exhibit the smallest rut depth and the highest dynamic stability of more than 13,000 times/mm at 60 °C temperature [101]. Semi-flexible mixtures depict high-temperature stability compared with HMA mixtures [101].

The dynamic stability and, hence, the rutting resistance of grouted macadam was recorded as 2.5-times higher than dense-graded asphalt mixtures. The dynamic stability of grouted macadam experienced more than 11,000 passes/mm, while asphalt concrete less than 5000 passes/mm [100]. Different laboratory tests are available to determine the extent of rutting in asphalt mixtures. These include the Asphalt Pavement Analyzer (APA), Hamburg Wheel-track Test, Asphalt Mixture Performance Tester (AMPT), and High-Temperature Indirect Tensile Test [114]. APA is used to measure rutting directly by producing deformation under an applied load. In this test, the traffic load is simulated by rolling a loaded wheel at a high temperature over an inflated (specified pressure) rubber hose [114,115]. The Hamburg Wheel-track Test is also used to measure rutting directly under the application of wheel loads. In this test, the traffic load is simulated by moving a loaded steel wheel directly on the mixture's surface. The Hamburg Wheel-track test can also be used to assess the moisture sensitivity of asphalt mixtures [114,116].

8. Compatibility of Cement Grout with Asphalt Mixture (Microstructure Characterization)

In conventional asphalt mixtures, the aggregate acts as a skeleton that transfers the load, while the asphalt acts as a binding/glue medium that binds aggregates together. However, in semi-flexible mixtures, the situation is a little bit complex, because it is a composite material with at least three phases (aggregate, bitumen matrix, and cement grout). Cai et al. (2019) evaluated the reinforcement mechanism of a semi-flexible pavement mixture and come to the conclusion that some parts of the cement grouts have an additional interlocking effect with the aggregate skeleton [117].

Furthermore, the interlocking effect of the skeleton could be improved by increasing the porosity of open-graded asphalt mixtures and, as a result, produces a positive effect on the reinforcement of semi-flexible mixtures [98,117]. The interlocking effect of grouting cement to produce an additional skeleton is much related to the interconnected voids within the open-graded asphalt mixtures [117]. A study was conducted to observe the interface of asphalt and cement grout in semi-flexible mixtures, which concluded that increasing porosity and gradation type improves the contact of grout and asphalt concrete mixtures [99]. Zhong et al. (2020) utilized latex-modified cement mortar for grouting open-graded asphalt concrete to produce semi-flexible pavement surfacing. A pull-out test was used to investigate the grout and asphalt concrete interface. The results indicate that latex-modified mortars proved to have good adhesion with asphalt concrete when compared to adhesion between the control cement grout and asphalt concrete [100].

Using interface modifiers in cement grouts can also result in improved adhesion or compatibility of cement grout with asphalt concrete. Xu et al. (2020) [101] investigated three interface modifiers (namely, silane coupling agent, carboxylated styrene–butadiene latex, and cationic emulsified asphalt) in cement mortar and investigated the interface bonding along with other performance properties, as shown in Figure 8. They concluded that the cationic emulsified asphalt interface modifier (Figure 8c) produces a good adhesion bond with asphalt and no obvious cracks were found during the microscopic study. While the other interface modifiers showed some cracks at the joint of grout and asphalt, as shown in Figure 8a,b [101].

(a) Silane coupling agent
(b) Carboxyl Styrene-Butadiene Latex
(c) Cationic Emulsified Asphalt

Figure 8. Effect of interface modifiers on grout–asphalt adhesion [101].

However, despite using modified grouts or interface modifiers, the interface of cement grout and asphalt concrete is still the weak spot and may lead to cracking after the application of an external load or environmental stresses. The semi-flexible pavement surfacing (grouted macadam) is a composite structure, having at least three phases (such as aggregates, bitumen, and cement grout). The cement grout is injected in the open-graded asphalt mixture (when it is cooled) and, therefore, there is a possibility of compatibility (adhesion) issues between the cement grout and asphalt mixture. To investigate the compatibility and adhesion between cementitious grout and asphalt mixture, SEM analysis was performed. Two types of the mixture were investigated: semi-flexible with a control grout (SFP-CN) and semi-flexible with an irradiated PET grout (SFP-IrP). The micrographs from the SEM test of the control grout-based semi-flexible specimen is shown in Figure 9.

Figure 9. SEM micrograph of semi-flexible with control cement grout.

It can be observed in Figure 9 that there are some micro-cracks at the interface of the asphalt mixture. However, no obvious cracks are found in semi-flexible specimens grouted with irradiated PET (Figure 10). It indicates that the adhesion bonding between the cement grout and asphalt mixture has been improved. The cracks in SFP-CN may be attributed due to the shrinkage of cementitious grout. Moreover, the uneven morphology and the microcracks may be induced during the cutting and polishing of specimen preparation for SEM analysis [118]. The improved adhesion between the cement grout and asphalt mixture in SFP-IrP is attributed to its better compatibility compared to SFP-CN.

Figure 10. SEM micrograph of semi-flexible with Irradiated PET-based grout.

9. Conclusions

According to the reviewed articles regarding semi-flexible pavement surfaces, it can be summarized that this type of pavement layer can produce a high-performance pavement surface. Following are the main conclusions drawn from this review:

- It can be concluded that the durability and performance properties of semi-flexible pavement surfaces largely depend on the mix design of the porous asphalt skeleton and composition of cementitious grouts. The selection of aggregate gradation and type of bitumen in porous asphalt mixtures and their effect on final semi-flexible specimens have been presented. Similarly, the effect of different compositions of cementitious grout and other additives/admixtures on the performance of semi-flexible mixtures have been explained.
- Single-size aggregate is used to achieve the required porosity of porous asphalt mixtures. A single aggregate size in the range of 8–12 mm nominal size can be used in porous mixtures to achieve the void ratio of 25–35%. Furthermore, polymer-modified binder can be used in the porous asphalt skeleton to improve its resistance against high and low-temperature performance. Municipal and/or industrial wastes/byproducts could be a better choice in binder modification towards performance improvement and environmental sustainability.
- The grouts are required to be highly flowable to penetrate through the porous asphalt skeleton. The flow value of 11 to 16 s is recommended while using the flow cone of the size used in ASTM C939. Depending on the composition of grouts, a w/c ratio of 0.45–0.65 could be used to achieve the required fluidity of cementitious grouts. Moreover, superplasticizer (0 to 2%) is used to achieve high fluidity at a relatively low w/c ratio (0.30 to 0.40). Other pozzolanic and cementitious products (such as fly ash, silica fume, fine limestone, and ground granulated blast furnace slag, etc.) could be used to achieve medium-to-high-strength grouts and, hence, medium-to-high-strength semi-flexible mixtures, and achieve the sustainability goals. Fly ash, silica fume and ground granulated blast furnace slag up to 10%, 5%, and 10%, respectively, can be used to replace cement, while achieving the required strength properties.
- The performance of cementitious grout is significantly related to the performance of semi-flexible mixtures. Therefore, the grouts are carefully designed to produce high-performance semi-flexible pavement surfaces.
- The fatigue life of semi-flexible mixtures can be evaluated using two-point or four-point bending fatigue tests. However, a four-point bending fatigue test could give better and more reliable results compared to the two-point fatigue test, because the failure of beams occurs in the uniform bending moment area.
- It is recommended to use interface modifiers to enhance the bonding between the asphalt mixture and cement grouts that help in reducing cracks in the pavement.
- Further study could be conducted on utilizing industrial/municipal wastes and other byproducts as a cement replacement in cementitious grouts for semi-flexible layers. It is a known fact that the production of cement releases carbon dioxide both directly and indirectly. Hence, there is a need for the development of durable, sustainable, and cost-effective cementitious grouts for semi-flexible pavement surfaces by utilizing a large percentage of municipal and industrial wastes/byproducts.

Funding: The research is partially funded by the Ministry of Science and Higher Education of the Russian Federation under the strategic academic leadership program 'Priority 2030' (Agreement 075-15-2022-311 dated 20 April 2022).

Institutional Review Board Statement: Not applicable.

Informed Consent Statement: Not applicable.

Data Availability Statement: Not applicable.

Conflicts of Interest: The authors declare no conflict of interest.

References

1. Amin, M.; Khan, M.; Saleem, M. Performance Evaluation of Asphalt Modified with Municipal Wastes for Sustainable Pavement Construction. *Sustainability* **2016**, *8*, 949. [CrossRef]
2. Koting, S.; Karim, M.R.; Mahmud, H.B.; Hamid, A.; Anjang, N. Mechanical Properties of Cement-Bitumen Composites for Semi-Flexible Pavement Surfacing. *Balt. J. Road Bridge Eng.* **2014**, *9*, 191–199. [CrossRef]
3. Zoorob, S.; Hassan, K.; Setyawan, A. Cold mix, cold laid semi-flexible grouted macadams, mix design and properties. In Proceedings of the 4th European Symposium on Performance of Bituminous and Hydraulic Materials in Pavements, Nottingham, UK, 11–12 April 2002; pp. 105–112.
4. Oliveira, J.R.; Pais, J.C.; Thom, N.H.; Zoorob, S.E. A Study of the Fatigue Properties of Grouted Macadams. *Int. J. Pavements* **2007**, *6*, 112–123.
5. Pratelli, C.; Betti, G.; Giuffrè, T.; Marradi, A. Preliminary In-Situ Evaluation of an Innovative, Semi-Flexible Pavement Wearing Course Mixture Using Fast Falling Weight Deflectometer. *Materials* **2018**, *11*, 611. [CrossRef]
6. Muslim, N.H.; Mohamed, M.I.; Amin, Z.M.; Shafaghat, A.; Ismail, M.; Keyvanfar, A. Pavement Structural Assessment Using Automated Tools: A Comparative Study. *Malays. J. Civ. Eng.* **2017**, *29*. [CrossRef]
7. Abo-Qudais, S.; Mulqi, M.W. New chemical antistripping additives for bituminous mixtures. *J. ASTM Int.* **2005**, *2*, 1–11. [CrossRef]
8. Hafeez, I.; Kamal, M.; Mirza, M.; Bilal, S. Laboratory fatigue performance evaluation of different field laid asphalt mixtures. *Constr. Build. Mater.* **2013**, *44*, 792–797. [CrossRef]
9. Tayfur, S.; Ozen, H.; Aksoy, A. Investigation of rutting performance of asphalt mixtures containing polymer modifiers. *Constr. Build. Mater.* **2007**, *21*, 328–337. [CrossRef]
10. Al-Malack, M.H.; Abdullah, G.M.; Al-Amoudi, O.S.B.; Bukhari, A.A. Stabilization of indigenous Saudi Arabian soils using fuel oil flyash. *J. King Saud Univ. Eng. Sci.* **2016**, *28*, 165–173. [CrossRef]
11. Asi, I.M. Stabilization of Sebkha Soil using Foamed Asphalt. *J. Mater. Civ. Eng.* **2001**, *13*, 325–331. [CrossRef]
12. Wahhab, H.I.A.-A.; Baig, M.G.; Mahmoud, I.A.; Kattan, H.M. Study of road bases construction in Saudi Arabia using foam asphalt. *Constr. Build. Mater.* **2012**, *26*, 113–121. [CrossRef]
13. Abdullah, G.M.; Wahhab, H.I.A.-A. Evaluation of foamed sulfur asphalt stabilized soils for road applications. *Constr. Build. Mater.* **2015**, *88*, 149–158. [CrossRef]
14. Aiban, S.A.; Siddiqi, Z.U.; Al-Amoudi, B.; Al-Ahmadi, H.M.; Asi, I.M. Effect of geotextiles on the load-carrying capacity and deformation characteristics of sabkha soil. *Geosynth. Int.* **2006**, *13*, 98–110. [CrossRef]
15. Aiban, S.; Al-Ahmadi, H.; Asi, I.; Siddique, Z.; Al-Amoudi, O. Effect of geotextile and cement on the performance of sabkha subgrade. *Build. Environ.* **2006**, *41*, 807–820. [CrossRef]
16. Al-Amoudi; Baghabra, O.S.; Khaqan, K.; Al-Kahtani; Saban, N. Stabilization of a Saudi calcareous marl soil. *Constr. Build. Mater.* **2010**, *24*, 1848–1854. [CrossRef]
17. Shawabkeh, R.; Khan, M.J.; Al-Juhani, A.A.; Wahhab, H.I.A.-A.; Hussein, I.A. Enhancement of surface properties of oil fly ash by chemical treatment. *Appl. Surf. Sci.* **2011**, *258*, 1643–1650. [CrossRef]
18. Parvez, M.A.; Wahhab, H.I.A.-A.; Shawabkeh, R.A.; Hussein, I.A. Asphalt modification using acid treated waste oil fly ash. *Constr. Build. Mater.* **2014**, *70*, 201–209. [CrossRef]
19. Wahhab, H.I.A.-A.; Hussein, I.A.; Parvez, M.; Shawabkeh, R.A. Use of modified oil fly ash to enhance asphalt concrete performance. *Mater. Struct.* **2015**, *48*, 3231–3240. [CrossRef]
20. Bertollo, S.; Bernucci, L.B.; Fernandes, J.L.; Leite, L.M. Mechanical properties of asphalt mixtures using recycled tire rubber produced in Brazil–A laboratory evaluation. In Proceedings of the Washington: S.n. In: 83rd TRB Annual Meeting, Washington, DC, USA, 11–12 January 2004.
21. Potgieter, C. Bitumen Rubber Asphalt in South Africa: Conventional Techniques. In Proceedings of the 3rd Eurasphalt and Eurobitume Congress, Vienna, Austria, 12–14 May 2004.
22. Lo Presti, D. Recycled Tyre Rubber Modified Bitumens for road asphalt mixtures: A literature review. *Constr. Build. Mater.* **2013**, *49*, 863–881. [CrossRef]
23. Anwar Parvez, M.; Al-Mehthel, M.; Al-Abdul Wahhab, H.I.; Hussein, I.A. Utilization of sulfur and crumb rubber in asphalt modification. *J. Appl. Polym. Sci.* **2014**, *131*, 40046. [CrossRef]
24. Khan, I.M.; Kabir, S.; Alhussain, M.A.; Almansoor, F.F. Asphalt design using recycled plastic and crumb-rubber waste for sustainable pavement construction. *Procedia Eng.* **2016**, *145*, 1557–1564. [CrossRef]
25. Sutanto, M.; Bala, N.; Al Zaro, K.; Sunarjono, S. Properties of Crumb Rubber and Latex Modified Asphalt Binders using Superpave Tests. *MATEC Web Conf.* **2018**, *203*, 05007. [CrossRef]
26. Al-Abdul Wahhab, H.I. Effect of Modifiers and Additives on Fatigue Behavior of Asphalt Concrete Mixes in the Gulf. *Int. J. Pavement Res. Technol.* **2012**, *5*, 326–332. Available online: http://www.ijprt.org.tw/reader/pdf.php?id=255 (accessed on 10 April 2021).
27. Gu, L.; Ozbakkaloglu, T. Use of recycled plastics in concrete: A critical review. *Waste Manag.* **2016**, *51*, 19–42. [CrossRef]
28. Singh, N.; Hui, D.; Singh, R.; Ahuja, I.; Feo, L.; Fraternali, F. Recycling of plastic solid waste: A state of art review and future applications. *Compos. Part B Eng.* **2017**, *115*, 409–422. [CrossRef]

29. Hassani, A.; Ganjidoust, H.; Maghanaki, A.A. Use of plastic waste (poly-ethylene terephthalate) in asphalt concrete mixture as aggregate replacement. *Waste Manag. Res.* **2005**, *23*, 322–327. [CrossRef] [PubMed]
30. Ahmadinia, E.; Zargar, M.; Karim, M.R.; Abdelaziz, M.; Shafigh, P. Using waste plastic bottles as additive for stone mastic asphalt. *Mater. Des.* **2011**, *32*, 4844–4849. [CrossRef]
31. Al-Sabaeei, A.; Yussof, N.I.M.; Napiah, M.; Sutanto, M. A Review of Using Natural Rubber in the Modification Of Bitumen and Asphalt Mixtures Used for Road Construction. *J. Teknol.* **2019**, *81*, 81–88. [CrossRef]
32. Al-Sabaeei, A.; Muniandy, R.; Hassim, S.; Napiah, M. Effect of Tire Crumb Rubber (TCR) on the Gyratory Compaction of Stone Mastic Asphalt. *Int. J. Civ. Eng. Technol.* **2018**, *9*, 1138–1150.
33. Roffe, J.C. *Salviacim-Introducing the Pavment*; Jean Lefebvre Enterprise: Paris, France, 1989.
34. Ahlrich, R.C.; Anderton, G.L. Resin Modified Pavement in Airfield Applications. In Proceedings of the Airport Pavement Innovations. Theory to Practice: Proceedings of Conference, Vicksburg, MS, USA, 8–20 September 1993.
35. Van de Ven, M.; Molenaar, A. Mechanical Characterization of Combi-layer. *Asph. Paving Technol.* **2004**, *73*, 1–22.
36. Anderton, G.L. *Engineering Properties of Resin Modified Pavement (RMP) for Mechanistic Design*; Engineer Research and Development Center Vicksburg MS Geotechnical Lab: Vicksburg, MS, USA, 2000.
37. British Board of Agrement (BBA). *Hardicrete Heavy Duty Surfacing, Agrement Certificate No 88/1969*; British Board of Agrement (BBA): Watford, UK, 1994.
38. British Board of Agrement (BBA). *Worthycim Heavy Duty Paving, Agrement Certificate No 87/1900*; British Board of Agrement (BBA): Watford, UK, 1996.
39. ApS, C. A Jointless Wearing Course with High Strength for High Loads. Available online: http://www.confalt.com/confaltr.html (accessed on 15 December 2018).
40. Oliveira, J.R. Grouted Macadam: Material Characterisation for Pavement Design. Ph.D. Thesis, University of Nottingham, Nottingham, UK, 2006.
41. Setyawan, A. Asessing the compressive strength properties of semi-flexible pavements. *Proc. Eng.* **2013**, *54*, 863–874. [CrossRef]
42. Karami, M. Application of the cementitious grouts on stability and durability of semi flexible bituminous mixtures. *AIP Conf. Proc.* **2017**, *1903*, 020012. [CrossRef]
43. Public Works Department of Malaysia (JKR). *Bituminous Road Surfacing in High Stress Areas: Malaysian Experience*; Public Works Department of Malaysia: Kuala Lumpur, Malaysia, 2002.
44. ASTM-C293. *Standard Test Method for Flexural Strength of Concrete (Using Simple Beam with Center-Point Loading)*; ASTM: Philadelphia, PA, USA, 2001.
45. Du, Y.; Chen, J.; Han, Z.; Liu, W. A review on solutions for improving rutting resistance of asphalt pavement and test methods. *Constr. Build. Mater.* **2018**, *168*, 893–905. [CrossRef]
46. Afonso, M.L.; Dinis-Almeida, M.; Pereira-de-Oliveira, L.A.; Castro-Gomes, J.; Zoorob, S.E. Development of a semi-flexible heavy duty pavement surfacing incorporating recycled and waste aggregates—Preliminary study. *Constr. Build. Mater.* **2016**, *102*, 155–161. [CrossRef]
47. Sundahl, J.; Hede, J. Semi-flexible Materials for Heavy-Duty Pavements. In Proceedings of Proceedings of the 6th International Conference on the Bearing Capacity of Roads and Airfields, Lisbon, Portugal, 24–26 June 2002.
48. Corradini, A.; Cerni, G.; D'Alessandro, A.; Ubertini, F. Improved understanding of grouted mixture fatigue behavior under indirect tensile test configuration. *Constr. Build. Mater.* **2017**, *155*, 910–918. [CrossRef]
49. Zhang, J.; Cai, J.; Pei, J.; Li, R.; Chen, X. Formulation and performance comparison of grouting materials for semi-flexible pavement. *Constr. Build. Mater.* **2016**, *115*, 582–592. [CrossRef]
50. Setyawan, A. Development of Semi-Flexible Heavy-Duty Pavements. Ph.D. Thesis, University of Leeds, Leeds, UK, 2003.
51. Saboo, N.; Ranjeesh, R.; Gupta, A.; Suresh, M. Development of hierarchical ranking strategy for the asphalt skeleton in semi-flexible pavement. *Constr. Build. Mater.* **2019**, *201*, 149–158. [CrossRef]
52. Khan, M.I.; Sutanto, M.H.; Napiah, M.B.; Zoorob, S.E.; Al-Sabaeei, A.M.; Rafiq, W.; Ali, M.; Memon, A.M. Investigating the mechanical properties and fuel spillage resistance of semi-flexible pavement surfacing containing irradiated waste PET based grouts. *Constr. Build. Mater.* **2021**, *304*, 124641. [CrossRef]
53. Ricci, R. *Laboratory Study of Grouted Macadams Impregnated with Mine Waste Geopolymeric Binder*; University Di Bologna: Bologna, Italy, 2012.
54. Hou, S.; Xu, T.; Huang, K. Aggregate Gradation Influence on Grouting Results and Mix Design of Asphalt Mixture Skeleton for Semi-Flexible Pavement. *J. Test. Eval.* **2017**, *45*, 591–600. [CrossRef]
55. Setiawan, A. Design and properties of hot mixture porous asphalt for semi-flexible pavement applications. *Media Tek. Sipil* **2009**, *5*, 41–46.
56. REAM. *Road Engineering Association of Malaysia, "Speciication of Semi-Rigid Wearing Course"*; Road Engineering Association of Malaysia (REAM): Shah Alam, Malaysia, 2007.
57. Fediuk, R.S.; Smoliakov, A.K.; Timokhin, R.A.; Batarshin, V.O.; Yevdokimova, Y.G. Using thermal power plants waste for building materials. *IOP Conf. Ser. Earth Environ. Sci.* **2018**, *87*, 092010. [CrossRef]
58. Husain, N.M.; Mahmud, H.; Karim, M.R.; Hamid, N. Effects of aggregate gradations on properties of grouted Macadam composite pavement. In Proceedings of the 2010 2nd International Conference on Chemical, Biological and Environmental Engineering, Cairo, Egypt, 2–4 November 2010; IEEE: Piscataway Township, NJ, USA; pp. 128–131.

59. Husain, N.M.; Karim, M.R.; Mahmud, H.B.; Koting, S. Effects of Aggregate Gradation on the Physical Properties of Semiflexible Pavement. *Adv. Mater. Sci Eng.* **2014**, *2014*, 529305. [CrossRef]
60. Plug, C.P.; de Bondt, A.H.; van der Woerd, B.J.; Steensma, G. *Improved Performance Grouted Macadams—High Performance Applications of Grouted Macadam*; Research & Development Wegbouwkundige Werkdagen: Doorwerth, The Netherlands, 2006.
61. Yang, B.; Weng, X. The influence on the durability of semi-flexible airport pavement materials to cyclic wheel load test. *Constr. Build. Mater.* **2015**, *98*, 171–175. [CrossRef]
62. Luo, S.; Yang, X.; Zhong, K.; Yin, J. Open-graded asphalt concrete grouted by latex modified cement mortar. *Road Mater. Pavement Des.* **2018**, *21*, 1–17. [CrossRef]
63. Hassani, A.; Taghipoor, M.; Karimi, M.M. A state of the art of semi-flexible pavements: Introduction, design, and performance. *Constr. Build. Mater.* **2020**, *253*, 119196. [CrossRef]
64. An, S.; Ai, C.; Ren, D.; Rahman, A.; Qiu, Y. Laboratory and Field Evaluation of a Novel Cement Grout Asphalt Composite. *J. Mater. Civil. Eng.* **2018**, *30*, 04018179. [CrossRef]
65. Vavrik, W.R.; Pine, W.J.; Huber, G.; Carpenter, S.H.; Bailey, R. The bailey method of gradation evaluation: The influence of aggregate gradation and packing characteristics on voids in the mineral aggregate (with discussion). *J. Assoc. Asph. Paving Technol.* **2001**, *70*.
66. Menninga, H. *Grouted Macadams, Material Research (Combination Layers, An Exploratory Material Research on combination Layers), TU Delft*; Traffic Engineering Section: Delft, The Netherlands, 2002.
67. Densit-a/s. *Densiphalt Handbook*; Densit-a/s: Aalborg, Denmark, 2000.
68. Makul, N.; Fediuk, R.; Amran, M.; Zeyad, A.M.; Murali, G.; Vatin, N.; Klyuev, S.; Ozbakkaloglu, T.; Vasilev, Y. Use of Recycled Concrete Aggregates in Production of Green Cement-Based Concrete Composites: A Review. *Crystals* **2021**, *11*, 232. [CrossRef]
69. Ahlrich, R.C.; Anderton, G.L. Construction and Evaluation of Resin Modified Pavement; Technical Report: GL-91-13; Army Engineer Waterways Experiment Station Vicksburg MS Geotechnical Lab. 1991. Available online: https://erdc-library.erdc.dren.mil/jspui/bitstream/11681/12765/1/TR-GL-91-13.pdf (accessed on 30 December 2018).
70. Pei, J.; Cai, J.; Zou, D.; Zhang, J.; Li, R.; Chen, X.; Jin, L. Design and performance validation of high-performance cement paste as a grouting material for semi-flexible pavement. *Constr. Build. Mater.* **2016**, *126*, 206–217. [CrossRef]
71. Bang, J.W.; Lee, B.J.; Kim, Y.Y. Development of a Semirigid Pavement Incorporating Ultrarapid Hardening Cement and Chemical Admixtures for Cement Grouts. *Adv. Mater. Sci Eng.* **2017**, *2017*, 9. [CrossRef]
72. Wang, D.; Liang, X.; Jiang, C.; Pan, Y. Impact analysis of Carboxyl Latex on the performance of semi-flexible pavement using warm-mix technology. *Constr. Build. Mater.* **2018**, *179*, 566–575. [CrossRef]
73. Martínez-Barrera, G.; Menchaca-Campos, C.; Hernández-López, S.; Vigueras-Santiago, E.; Brostow, W. Concrete Reinforced with Irradiated Nylon Fibers. *J. Mater. Res.* **2006**, *21*, 484–491. [CrossRef]
74. Ahmedzade, P.; Fainleib, A.; Günay, T.; Starostenko, O.; Kovalinska, T. Effect of gamma-irradiated recycled low-density polyethylene on the high-and low-temperature properties of Bitumen. *Int. J. Polym. Sci.* **2013**, *2013*, 9. [CrossRef]
75. Huang, C.; Hong, J.X.; Lin, J.T.; Deng, C.; Li, L. Utilization of waste rubber powder in semi-flexible pavement. *Key Eng. Mater.* **2014**, *599*, 361–367. [CrossRef]
76. Khan, M.I.; Huat, H.Y.; Dun, M.H.b.M.; Sutanto, M.H.; Jarghouyeh, E.N.; Zoorob, S.E. Effect of Irradiated and Non-Irradiated Waste PET Based Cementitious Grouts on Flexural Strength of Semi-Flexible Pavement. *Materials* **2019**, *12*, 4133. [CrossRef]
77. Vijaya, S.S.A.; Karim, M.R.; Mahmud, H.; Ishak, N.F. Influence of Superplasticiser on Properties of Cementitious Grout Used in Semi-Flexible Pavement. In Proceedings of the Eastern Asia Society for Transportation Studies Volume 7, The 8th International Conference of Eastern Asia Society for Transportation Studies, Surabaya, Indonesia, 1 January 2009; p. 296.
78. Hu, S.; Huang, S.; Ding, Q.; Zhang, R. Study on the Cementitious Mortar for Semi-Flexible Pavement. In Proceedings of the 1st International Conference on Microstructure Related Durability of Cementitious Composites, Nanjing, China, 13–15 October 2008.
79. Fang, B.; Xu, T.; Shi, S. Laboratory study on cement slurry formulation and its strength mechanism for semi-flexible pavement. *J. Test. Eval.* **2016**, *44*, 907–913. [CrossRef]
80. Ding, Q.; Zhao, M.; Shen, F.; Zhang, X. Mechanical behavior and failure mechanism of recycled semi-flexible pavement material. *J. Wuhan Univ. Technol. -Mater. Sci.* **2015**, *30*, 981–988. [CrossRef]
81. Husain, N.M.; Ismail, N.N.; Ismail, K.M.; Syamsunur, D.; Taib, M.M. Development of high workability grout on semi rigid wearing course. *Conf. Ser. Mater. Sci. Eng.* **2019**, *512*, 012033. [CrossRef]
82. Koting, S.; Mahmud, H.; Karim, M.R. Influence of superplasticizer type and dosage on the workability and strength of cementitious grout for semi-flexible pavement application. *J. East. Asia Soc. Transp. Stud.* **2007**, *7*, 2156–2167.
83. Public Works Department. *Description of Workmanship and Materials (Semi-Rigid Pavement), Section E*; City Hall Kuala Lumpur: Kuala Lumpur, Malaysia, 2003.
84. Wang, Y.J.; Guo, C.Y.; Tian, Y.F.; Wang, J.J. Design of Mix Proportion of Cement Mortar with High-Performance Composite Semi-Flexible Pavement. *Adv. Mater. Res.* **2013**, *641–642*, 342–345. [CrossRef]
85. Cai, J.; Pei, J.; Luo, Q.; Zhang, J.; Li, R.; Chen, X. Comprehensive service properties evaluation of composite grouting materials with high-performance cement paste for semi-flexible pavement. *Constr. Build. Mater.* **2017**, *153*, 544–556. [CrossRef]
86. Khan, M.I.; Sutanto, M.H.; Napiah, M.B.; Khan, K.; Rafiq, W. Design optimization and statistical modeling of cementitious grout containing irradiated plastic waste and silica fume using response surface methodology. *Constr. Build. Mater.* **2021**, *271*, 121504. [CrossRef]

87. Khan, M.I.; Sutanto, M.H.; Napiah, M.B.; Zoorob, S.E.; Yusoff, N.I.M.; Usman, A.; Memon, A.M. Irradiated polyethylene terephthalate and fly ash based grouts for semi-flexible pavement: Design and optimisation using response surface methodology. *Int. J. Pavement Eng.* **2020**, 1–16. [CrossRef]
88. Khan, M.I.; Sutanto, M.H.; Napiah, M.B.; Zoorob, S.E. Cementitious Grouts Containing Irradiated Waste Polyethylene Terephthalate. In *Cement Industry*; Saleh, H.M., Ed.; IntechOpen: London, UK, 2020.
89. Hassan, K.; Setyawan, A.; Zoorob, S. Effect of cementitious grouts on the properties of semi-flexible bituminous pavements. In Proceedings of the 4th European Symposium on Performance of Bituminous and Hydraulic Materials In Pavements, Bitmat 4, Nottingham, UK, 11–12 April 2002; pp. 113–120.
90. Anderton, G.L. User's Guide: Resin Modified Pavement; No. WES/MP/GL-96-7; Army Engineer Waterways Experiment Station Vicksburg MS Geotechnical Lab. 1996. Available online: https://apps.dtic.mil/sti/pdfs/ADA307526.pdf (accessed on 20 December 2018).
91. Boundy, R. Development of a Resin/Cement grouted coated Macadam Surfacing Material. Master's Thesis, University of Nottingham, Nottingham, UK, 1979.
92. Alsadey, S. Effects of super plasticizing and retarding admixtures on properties of concrete. In Proceedings of the International Conference on Innovations in Engineering and Technology, Bangkok, Thailand, 25–26 December 2013; pp. 271–274.
93. Khan, M.I.; Sutanto, M.H.; Khan, K.; Iqbal, M.; Napiah, M.B.; Zoorob, S.E.; Klemeš, J.J.; Bokhari, A.; Rafiq, W. Effective use of recycled waste PET in cementitious grouts for developing sustainable semi-flexible pavement surfacing using artificial neural network (ANN). *J. Clean. Prod.* **2022**, *340*, 130840. [CrossRef]
94. Anagnostopoulos, C.A. Effect of different superplasticisers on the physical and mechanical properties of cement grouts. *Constr. Build. Mater.* **2014**, *50*, 162–168. [CrossRef]
95. Imran, K.M.; Sutanto, M.H.; Napiah, M.B.; Zahid, M.; Usman, A. Optimization of Cementitious Grouts for Semi-Flexible Pavement Surfaces Using Response Surface Methodology. *IOP Conf. Ser. Earth Environ. Sci.* **2020**, *498*, 012004. [CrossRef]
96. Koting, S.; Karim, M.R.; Mahmud, H.; Mashaan, N.S.; Ibrahim, M.R.; Katman, H.; Husain, N.M. Effects of using silica fume and polycarboxylate-type superplasticizer on physical properties of cementitious grout mixtures for semiflexible pavement surfacing. *Sci. World J.* **2014**, *2014*, 7. [CrossRef]
97. Zhang, W.; Shen, S.; Goodwin, R.D.; Wang, D.; Zhong, J. Performance Characterization of Semi-Flexible Composite Mixture. *Materials* **2020**, *13*, 342. [CrossRef]
98. Cai, X.; Yang, J.; Chen, X.; Zhang, J.; Zhang, H. Interlocking property evaluation of dual skeleton in semi-flexible pavement material by micromechanical model and X-ray computed tomography. *Constr. Build. Mater.* **2020**, *254*, 118934. [CrossRef]
99. Cai, X.; Zhang, J.; Zhang, H.; Yao, Z.; Chen, X.; Yang, J. Identification of microstructural characteristics in semi-flexible pavement material using micromechanics and nano-techniques. *Constr. Build. Mater.* **2020**, *246*, 118426. [CrossRef]
100. Zhong, K.; Sun, M.; Zhang, M.; Qin, Y.; Li, Y. Interfacial and mechanical performance of grouted open-graded asphalt concrete with latex modified cement mortar. *Constr. Build. Mater.* **2020**, *234*, 117394. [CrossRef]
101. Xu, Y.; Jiang, Y.; Xue, J.; Tong, X.; Cheng, Y. High-Performance Semi-Flexible Pavement Coating Material with the Microscopic Interface Optimization. *Coatings* **2020**, *10*, 268. [CrossRef]
102. Zarei, S.; Ouyang, J.; Yang, W.; Zhao, Y. Experimental analysis of semi-flexible pavement by using an appropriate cement asphalt emulsion paste. *Constr. Build. Mater.* **2020**, *230*, 116994. [CrossRef]
103. Saboo, N.; Khalpada, V.; Sahu, P.K.; Radhakrishnan, R.; Gupta, A. Optimal proportioning of grout constituents using mathematical programming for semi flexible pavement. *Int. J. Pavement Res. Technol.* **2019**, *12*, 297–306. [CrossRef]
104. Zhang, H.; Liang, S.; Ma, Y.; Fu, X. Study on the mechanical performance and application of the composite cement–asphalt mixture. *Int. J. Pavement Eng.* **2019**, *20*, 44–52. [CrossRef]
105. Al-Qadi, I.L.; Gouru, H.; Weyers, R. Asphalt Portland cement concrete composite: Laboratory evaluation. *J. Transp. Eng.* **1994**, *120*, 94–108. [CrossRef]
106. Shukla, M.; Gottumukkala, B.; Nagabhushana, M.N.; Chandra, S.; Shaw, A.; Das, S. Design and evaluation of mechanical properties of cement grouted bituminous mixes (CGBM). *Constr. Build. Mater.* **2021**, *269*, 121805. [CrossRef]
107. Sunil, S.; Varuna, M.; Nagakumar, M.S. Performance evaluation of semi rigid pavement mix. *Mater. Today: Proc.* **2020**. [CrossRef]
108. Bharath, G.; Shukla, M.; Nagabushana, M.; Chandra, S.; Shaw, A. Laboratory and field evaluation of cement grouted bituminous mixes. *Road Mater. Pavement Des.* **2019**, 1–19. [CrossRef]
109. Hamzani; Munirwansyah; Hasan, M.; Sugiarto, S. Determining the properties of semi-flexible pavement using waste tire rubber powder and natural zeolite. *Constr. Build. Mater.* **2021**, *266*, 121199. [CrossRef]
110. *BS-EN-12697-26*; Bituminous Mixtures–Test Methods for Hot Mix Asphalt–Part 26: Stiffness. The British Standards Institution: London, UK, 2004.
111. Oliveira, J.; Sangiorgi, C.; Fattorini, G.; Zoorob, S. *Investigating the Fatigue Performance of Grouted Macadams*; Institution of Civil Engineers-Transport: London, UK; pp. 115–123.
112. Di Benedetto, H.; de La Roche, C.; Baaj, H.; Pronk, A.; Lundström, R. Fatigue of bituminous mixtures. *Mater. Struct.* **2004**, *37*, 202–216. [CrossRef]
113. Tayebali, A.A.; Deacon, J.A.; Coplantz, J.S.; Monismith, C.L. Modeling fatigue response of asphalt-aggregate mixtures. *Asph. Paving Technol.* **1993**, *62*, 385.

114. *NCHRP-REPORT-673*; A Manual for Design of Hot Mix Asphalt with Commentary. Transportation Research Board: Washington, DC, USA, 2011; Volume 673.
115. AASHTO. *Standard Method of Test for Determining Rutting Susceptibility of Hot Mix. Asphalt (HMA) Using the Asphalt Pavement Analyzer (APA)*; AASHTO: Washington, DC, USA, 2006.
116. *A.T. 324*; Standard Method of Test for Hamburg Wheel-Track Testing of Compacted Hot Mix Asphalt (HMA). Aashto: Washington, DC, USA, 2011.
117. Cai, X.; Zhang, H.; Zhang, J.; Chen, X.; Yang, J.; Hong, J. Investigation on reinforcing mechanisms of semi-flexible pavement material through micromechanical model. *Constr. Build. Mater.* **2019**, *198*, 732–741. [CrossRef]
118. Tan, H.; Xiong, Z.; Gong, M.; Chen, J.; Hong, J. Investigation on the Influences of Curing Time on the Cracking Resistance of Semiflexible Pavement Mixture. *Adv. Mater. Sci. Eng.* **2021**, *2021*, 8855565. [CrossRef]

materials

MDPI

Article

Relationship Between the Carbonation Depth and Microstructure of Concrete Under Freeze–Thaw Conditions

Shuhua Zhang [1,2], Guangrong Tan [1], Zhiqiang Qi [1], Bin Tian [3], Jijun Cao [1] and Bofu Chen [3,*]

1 The Seventh Geological Brigade of Hubei Geological Bureau, Yichang 443000, China; shuhuazhang2022@163.com (S.Z.)
2 Key Experiment of Geological Resources and Geological Engineering in Yichang, Yichang 443000, China
3 College of Hydraulic and Environmental Engineering, Three Gorges University, Yichang 443000, China
* Correspondence: chenbofu@ctgu.edu.cn

Abstract: Concrete structures in cold regions are affected by freeze–thaw cycles (FTCs) and carbonation, which lead to the premature failure of concrete structures. The carbonation depth, relative dynamic elastic modulus (RDEM), compressive strength, porosity, and pore size distribution of concrete under FTC conditions were tested through an accelerated carbonation experiment to study the carbonation performance evolution. The freeze–thaw effect mechanism on concrete carbonation was further analyzed via the obtained relationship between carbonation depth and pore structure. The results showed that the FTC, as a powerful source of concrete damage, accelerates the carbonation reaction. Carbonization products fill some microcracks caused by the freeze–thaw process, improve the compressive strength and dynamic elastic modulus, and alleviate the damage to concrete caused by the FTC. After carbonization under freeze–thaw damage conditions, the content of macropores with d > 1000 nm decreases, while the content of transition pores with $d \leq 10$ nm increases, which is the direct reason for the decrease in porosity and the improvement in strength. Therefore, the carbonation durability of concrete under freeze–thaw conditions can be improved by controlling the content of macropores with d > 1000 nm and increasing the content of transition pores with a pore size of $10 \text{ nm} \leq d < 100$ nm. In addition, the relationship between carbonation depth and pore structure under freeze–thaw conditions was established, and the research results can provide a reference for the study of the carbonation performance of concrete under freeze–thaw conditions.

Keywords: concrete; freeze–thaw; carbonization; microstructure; carbonation depth

1. Introduction

CO_2 is everywhere in the atmospheric environment, and concrete structures are inevitably affected [1–3]. In fact, the carbonization of concrete is a precondition of steel corrosion and, thus, is one of the key problems that affects the durability of reinforced concrete structures. In this scenario, CO_2 in the atmosphere intrudes into the concrete and reacts with the hydration product $Ca(OH)_2$, which leads to a decrease in the alkalinity of the concrete around the steel bar and makes the steel bar in the concrete lose its passivation and protection function [4]. When the carbonation depth exceeds the thickness of the protective layer, the concrete no longer protects the steel bar from the action of air and water, the alkaline environment around the steel bar disappears, the passive film is destroyed, the steel bar rusts, and the concrete cracks [5,6]. Therefore, the carbonation depth of concrete is crucial to ensuring the safety and quality of construction projects [7,8] and must be predicted when studying the durability of reinforced concrete structures [9,10]. Furthermore, given that it is mainly environmental factors that affect concrete carbonation, concrete carbonation models should also consider the influence of environmental conditions [11]. The highest and lowest temperature records of 303 weather stations in China from 1961 to 2003 show that the extreme average low temperature in northeast China reached −30 °C from December to February of the following year [12]. According to the meteorological bureau of Xinjiang Uygur Autonomous Region in China, the lowest temperature in winter in Xinjiang is −43.3 °C [13]. Concrete structures in cold regions of China inevitably suffer from freeze–thaw cycle (FTC) damage during their service [14–17], and FTC damage is one of the main factors leading to the deterioration of reinforced concrete structures in cold areas, seriously affecting their service life [18,19]. Moreover, the FTC and carbonation are not simple superpositions of two separate actions but one complex coupling effect. Concrete structures in cold areas are subjected to the alternating action of FTC and carbonation, and so the carbonation depth of concrete is bound to be affected by the FTC.

At present, three main types of models can predict the carbonation depth of concrete: theoretical models [20,21], empirical models [22,23], and semitheoretical/semiempirical models [10,24,25]. Tan et al. [26] used X-ray diffraction and scanning electron microscopy (SEM) to analyze the influence of the microstructure of polypropylene-fiber-reinforced machine-made sand concrete on carbonation resistance, and they established a prediction model of carbonation depth. Geng et al. [27] established a prediction model of the carbonation depth of recycled concrete through regression analysis of experimental data, which can be used to predict the carbonation depth of recycled concrete structures in atmospheric environments. Through a rapid carbonation test, Zhang et al. studied the influence of the carbonation time and fly ash content on the carbonation performance of concrete and obtained a prediction model of the carbonation depth of fly ash concrete through curve fitting of the test data [28]. In accordance with the practical carbonation prediction model and nonlinear regression technology based on the least square method, a multifactor calculation model of concrete carbonation depth in a standard carbonation environment was established [29]. The carbonation microstructure and porosity of concrete change in FTC and carbonation environments, affecting the carbonation rate and carbonation depth [30,31]. Substantial research has been conducted on the durability of concrete under the action of single factors, such as FTC and carbonation [32,33]. For instance, some scholars performed a systematic study on a prediction model of the carbonation depth of concrete under single carbonation [2,3], and many conclusions and empirical formulas have reached consensus in academic circles and have been widely used and verified in practical projects. However, most concrete structures in practical engineering are affected by various environmental factors at the same time, and the failure of such concrete structures in actual service is complex and serious. Therefore, the research conclusions and empirical formulas obtained under single environmental conditions have certain limitations. In an actual service environment, the existence of the FTC and carbonation makes the durability evolution of concrete more complicated, particularly since the FTC changes the pore structure, which has a more complicated and unpredictable effect on carbonation.

In view of the importance of the carbonation depth for evaluating the carbonation resistance of concrete and predicting the corrosion initiation time of steel bars, models of the relationship between carbonation depth and micropore structure must be studied under FTC damage conditions. In this study, a carbonation experiment is conducted on concrete in an FTC environment. Using the carbonation depth, RDEM, compressive strength, porosity, and pore size distribution (PSD) as durability indicators, the influence of the FTC on the carbonation depth and carbonation performance of concrete is studied, and the evolution law of the carbonation resistance of concrete in an FTC environment is analyzed from macro and micro perspectives.

2. Experimental Program

The experimental program includes an FTC and accelerated carbonization. The macroscopic parameters tested include the carbonation depth, RDEM, and compressive strength, while the microscopic parameters include porosity and PSD. The test procedure is shown in Figure 1.

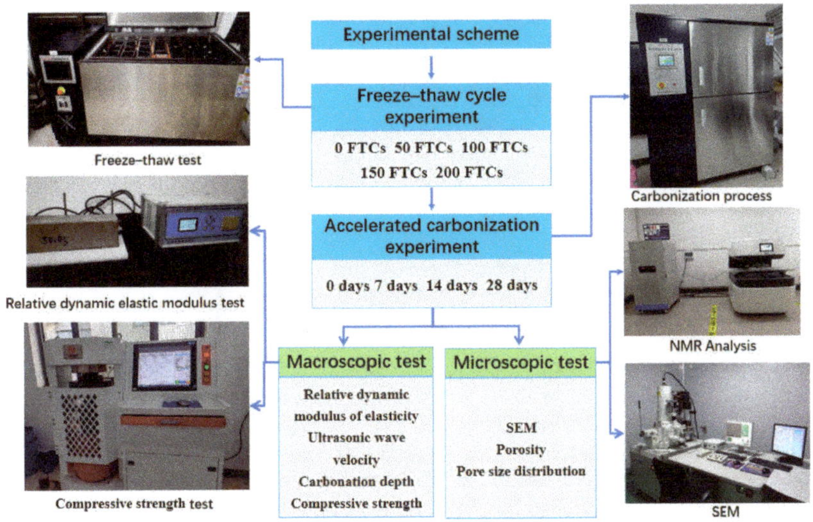

Figure 1. Flow diagram of the experiment.

2.1. Materials

The cement used was ordinary Portland cement P.O. 42.5, which is produced by Gezhouba Cement Factory in China, with a density of 3150 kg/m^3 and a specific surface area of 342 m^2/kg. Grade II fly ash produced by Yichang Chengkun Building Materials Co., Ltd., Yichang, China. is adopted, with fineness of 8.7% and loss on ignition of 7%. Sand with a fineness modulus of 2.78 was used as fine aggregate, produced in Yichang, China, with a silt content of 0.81%. Two-graded aggregate is selected for concrete, which is divided into two grades: 5~20 mm and 20~40 mm, of which small stone and medium stone are in a 1:1 ratio of artificial crushed stone aggregate, requirements of the composite fuller grading curve [34]. Additives, such as a highly efficient naphthalene water-reducing agent and an air-entraining agent, were included. The select MH-JS-102 type water-reducing agent is produced by Hunan Minghuang Technology Development Co., Ltd., Changsha, China, and the air-entraining agent is Q8122AE high-efficiency air-entraining agent produced by China Shaanxi Qinfen Building Materials Co., Ltd., Xi'an, China. The mixing water is taken from ordinary tap water in Yichang City, Hubei Province. After testing, the air content of fresh concrete is 5%. The mix proportion of the concrete specimens is shown in Table 1.

Table 1. Design of mix proportion of the concrete specimens (in kilograms per cubic meter).

Water–Cement Ratio	Cement	Fly Ash	Sand	Stone	Water	Water-Reducing Agent	Air-Entraining Agent
0.35	295	74	711	1202	129	1%	0.2%
0.40	258	65	728	1240	129	1%	0.2%
0.45	229	57	742	1264	129	1%	0.2%

2.2. Specimen Preparation

In a research experiment on the carbonation durability of concrete in rockfill dams under FTC damage conditions, a rapid carbonation process was conducted after the FTC. During the FTC, the concrete specimens were basically in a water-saturated state; given that the solubility of CO_2 in water is low and the moisture content of the specimen was high, this meant that CO_2 could not invade the interior of the concrete to undergo a carbonation reaction. Meanwhile, considering the particularity of the actual service situation of concrete in extremely cold and cold regions, concrete in water level fluctuation zones generally undergoes an FTC in cold winter and spring. In hot summer and autumn, when the reservoir water level drops, the water content of the original FTC site decreases under solar radiation, and carbonation reactions occur due to the influence of CO_2 in the atmosphere. Thus, an FTC followed by carbonation could simulate the effect of the FTC and carbonation on the durability of concrete within a year more accurately. In addition, before carbonation, concrete must be dried in a drying oven for 48 h at 60 °C to exclude the effect of internal water on carbonation. The carbonation process was performed with an ambient temperature of 20 °C. The carbonization test employed prismatic specimens with side lengths of 100 mm × 100 mm × 400 mm to test the RDEM and 100 mm × 100 mm × 100 mm cubic specimens to test the compressive strength and carbonization depth.

2.3. Freeze–Thaw Process and Accelerated Carbonization Process

FTC and carbonation tests were conducted in accordance with the operating procedures in the "Standard of Test Methods for Long-Term Performance and Durability of Ordinary Concrete" (GB/T50082-2009) [35]. The poured concrete specimens were cured in a standard curing room for 24 days, then soaked in water for 4 days until they were saturated with water and then taken out. Then, the specimens were dried with a rag, and the initial state (compressive strength, dynamic elastic modulus, pore structure) was tested before the freeze–thaw process. In the process of the FTC test, the parameters of the FTC tester are set: the center temperature of the specimen is (-17 ± 2) °C and (5 ± 2) °C, the FTC lasts no more than 4 h at a time, and the melting time is not less than 1/4 of the whole FTC. The real-time curve during the FTC is shown in Figure 2. After 50, 100, 150, and 200 FTCs, the concrete specimen was taken out to be tested, the surface moisture was dried, and the specimen was placed in a drying oven. The drying oven temperature was set to 60 °C, and carbonization commenced after 48 h of drying.

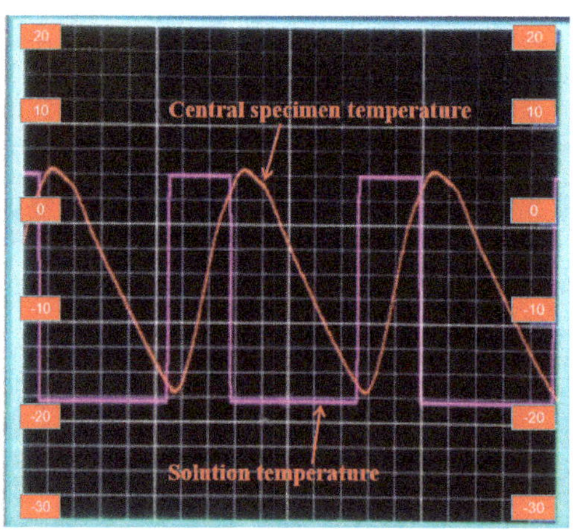

Figure 2. Real-time curve display of FTC.

Concrete carbonation was conducted using an HTX-12X microcomputer concrete carbonation test box produced by Suzhou Donghua Test Instrument Co., Ltd., Suzhou, China. The dried specimen was placed on the bracket of the concrete carbonization test box, where the distance between the specimens should be greater than 50 mm. The sealing of the carbonization test box was checked before the switch was turned on. Carbonation was started when the CO_2 concentration inside the carbonization box was steady at $(20 \pm 1)\%$, the temperature between the carbonation boxes was (20 ± 2) °C, and the relative humidity was controlled to $(70 \pm 2)\%$. After the carbonization test began, the parameters on the display of the carbonization box were checked every 4 h and adjusted in time to ensure that the temperature, humidity, and CO_2 concentration in the carbonization box were within the set ranges. On the 7th, 14th, and 28th days, concrete specimens were taken out for measurement of the ultrasonic wave velocity, compressive strength, and carbonation depth.

2.4. Macro and Micro Test Program

The RDEM and compressive strength were tested in accordance with the "Standard for Test Methods of Long-Term Performance and Durability of Ordinary Concrete" [35]. After the compressive strength test, samples with a thickness, length, and width of 1, 3, and 3 mm were cut from the cross-section of the specimen using a diamond saw and then dried and treated via high-pressure gold spraying. Subsequently, the samples were observed using a JSM-7500F SEM produced by Japan Electronics (Tokyo, Japan) with the best resolution of 1 nm (15 KV), and the micromorphological changes in the concrete after carbonation under freeze–thaw damage conditions were observed through scanning electron microscope experiments. To comprehensively understand the micropore structure information of concrete after the FTC and carbonation, samples that underwent 0, 100, and 200 FTCs and carbonation for 14 and 28 days were selected for nuclear magnetic resonance (NMR) analysis, and their porosity and PSD were tested. The test sample was processed into a cylinder with a diameter of 50 mm and a height of 50 mm using a drilling machine and a cutting machine to eliminate the influence of uneven freeze–thaw damage on the final experimental results relating to micropore structure. The porosity and PSD were tested using the MesoMR23-060H-I NMR analyzer produced by Suzhou Newman Analytical Instrument Co., Ltd., Suzhou, China. Before NMR testing, samples should be vacuumed and saturated. Thus, the specimen was placed into a vacuum water-retaining testing machine, the pumping time was set for 4 h, and the specimen was finally soaked in clean water for 24 h. To eliminate the influence of water evaporation on the test results, when the sample was taken out of the water, the water on the surface was wiped off, and the sample was wrapped in plastic and then placed on the NMR analyzer for measurement. Then, the porosity and PSD results could be obtained.

3. Experimental Results

3.1. Carbonization Depth

The essence of carbonation reaction is acid–base neutralization reaction, which is a process in which the alkalinity of pore solution in concrete gradually decreases. The pH value of pore solution in uncarbonized concrete is generally about 13, and the greater the degree of carbonation, the lower the pH value. At present, the indicators for testing the carbonation depth of concrete mainly include phenolphthalein alcohol indicators and rainbow indicators [36,37]. In this study, the phenolphthalein indicator method was used to measure the carbonation depth of concrete. After the splitting test was completed, the specimen cross-section was cleaned with a soft brush to remove the residual powder, and a 1% concentration of phenolphthalein solution was sprayed on the splitting surface. In this test, the areas that have not undergone carbonization appear pink due to their alkalinity, while the areas that have undergone carbonization do not change color due to their neutrality. Then, after 30 s, a vernier caliper was used to measure the distance from the splitting surface to the junction of the discoloration area to obtain the depth of carbonization. The measurement of carbonization depth is shown in Figure 3. By measuring the depth of carbonation, the degree of concrete carbonation can be simply evaluated, and its compressive strength and durability can be determined. The carbonization depth of each point was measured on both sides (without discoloration) at marked intervals of 10 mm, and the average of multiple measurements was taken as the final carbonization depth.

The average carbonation depth of concrete at each test age can be calculated as follows:

$$\overline{d_i} = \frac{1}{n}\sum_{i=1}^{n} d_i, \quad (1)$$

where $\overline{d_i}$ is the average carbonization depth, *mm*; d_i is the carbonization depth at measuring point *i*, *mm*; *n* is the total number of measurement points.

Figure 3. Measurement of carbonation depth of concrete (**a**) pre-carbonization specimen, (**b**) carbonized specimen.

3.1.1. Influence of FTCs on the Depth of Carbonization

The relationship between the carbonation depth and the number of FTCs for concrete specimens with 0.35, 0.40, and 0.45 water–binder ratios at 7, 14, and 28 days under FTC conditions was obtained on the basis of the color reaction of phenolphthalein as shown in Figure 4.

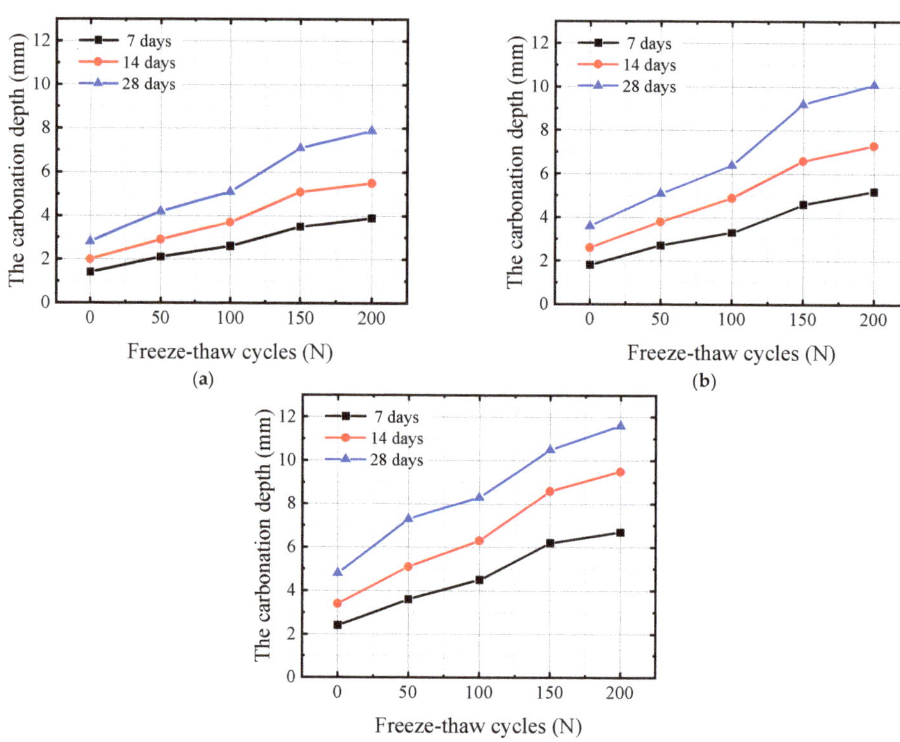

Figure 4. Relationship between the number of FTCs and the depth of carbonation for (**a**) 0.35 water–binder ratio, (**b**) 0.40 water–binder ratio, (**c**) 0.45 water–binder ratio.

As shown in Figure 4, as the number of FTCs increases, the carbonization depth of all three groups of specimens increases. After 150 FTCs, the growth rate of the carbonation depth of the concrete decreases. During the FTC process, the pore water inside the concrete experiences volume expansion when it freezes; thus, the pores inside the concrete experience a certain degree of expansion and extension. After multiple FTCs, the porosity inside the concrete increases to a certain extent. The increases in porosity, microcracks, and connected pores all provide channels for CO_2, which is conducive to its diffusion into the interior of concrete and promotes the carbonation reaction. Therefore, the carbonation depth increases with an increase in the number of FTCs.

After 150 FTCs, the carbonization reaction product $CaCO_3$ adhering to the pore walls simultaneously envelops some $Ca(OH)_2$, affecting the progress of the reaction, and blocks some pores in the concrete, affecting the diffusion rate of CO_2 and thus affecting the progress of the carbonization reaction. Therefore, the growth rate of carbonization depth decreases in the later stage of FTCs.

3.1.2. Relationship Between Carbonization Age and Carbonization Depth

As shown in Figure 5, the relationship between carbonization age and carbonization depth was plotted to investigate the influence of carbonization age on carbonization depth under freeze–thaw conditions.

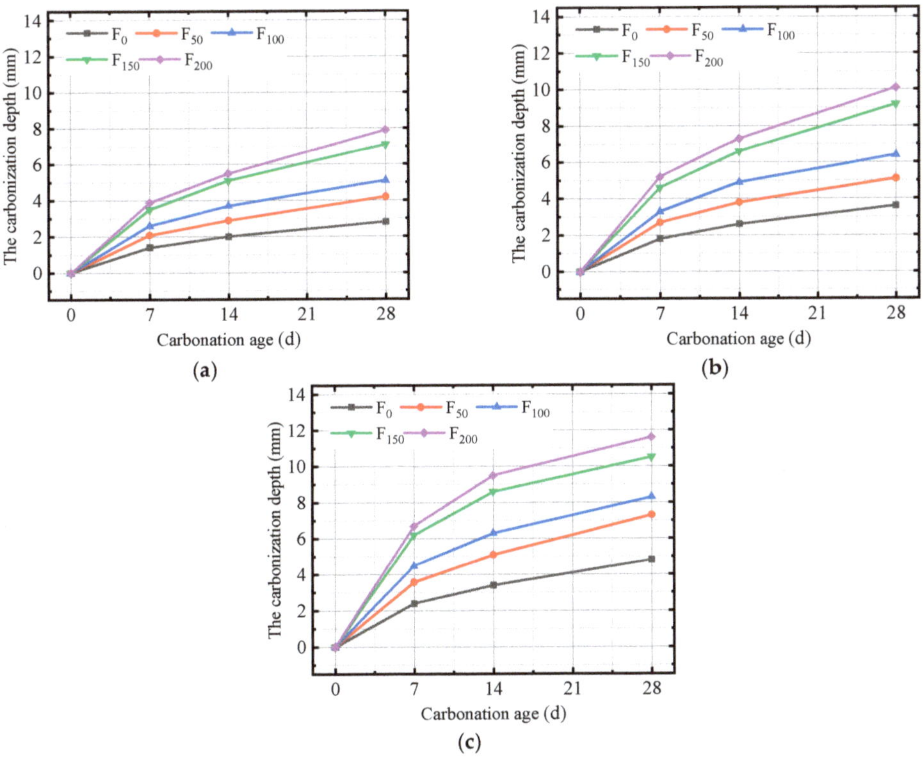

Figure 5. Relationship between carbonation age and the depth of carbonation for (**a**) 0.35 water–binder ratio, (**b**) 0.40 water–binder ratio, (**c**) 0.45 water–binder ratio.

Figure 5 shows that the carbonization depth of concrete increases with an increase in the number of FTCs and the carbonization age, and the carbonization depth of concrete after carbonization under the FTC condition is greater than that under the action of the carbonization test alone, which also indicates that the FTC can promote the carbonization reaction of concrete. With the extension of the carbonization age, the carbonization rate gradually decreases. The carbonation rate of the specimens with the three different water–binder ratios is faster than that of the concrete when carbonated for 7 days, and the carbonation rate gradually slows down after 14 days of carbonation.

In the early stage of carbonization, the FTC leads to an increase in porosity, and a relatively large CO_2 concentration difference exists between the concrete interior and the surface layer, which provides more channels for the diffusion of CO_2. In addition, the $Ca(OH)_2$ content involved in the carbonization reaction is sufficient in the concrete, resulting in the acceleration of carbonization.

With the continuous progress of the reaction, the content of $Ca(OH)_2$ in the solid phase is reduced, affecting the speed of the carbonization reaction. At the same time, the carbonization product $CaCO_3$ continues to accumulate and fill the pores inside the concrete, increasing the compactness of the concrete and slowing down the diffusion rate of CO_2 to the inside of the concrete, which contributes to the slowing down of the carbonization rate in the later stage.

3.1.3. Effect of Water–Binder Ratio on Carbonization Depth

The water–binder ratio has a great influence on the porosity of concrete, and the porosity connectivity directly affects the diffusion rate of CO_2 in concrete. The relation between the water–binder ratio and carbonization depth of concrete at different ages under the action of FTCs is shown in Figure 6.

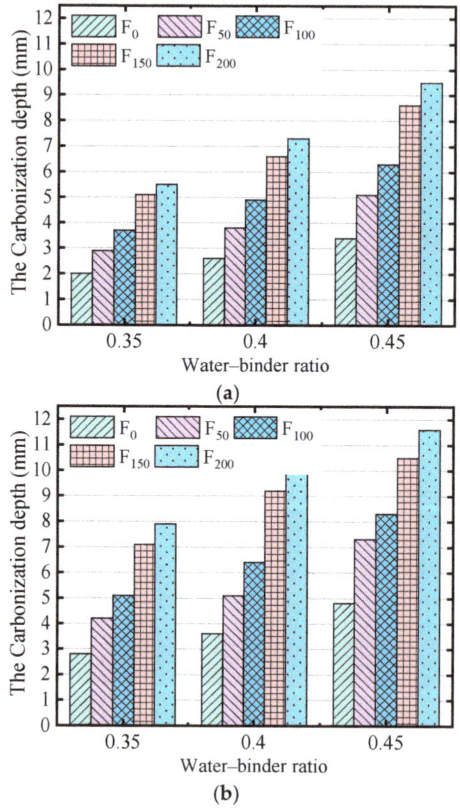

Figure 6. Relationship between water–binder ratio and the depth of carbonation for (**a**) 14 days of carbonation and (**b**) 28 days of carbonation.

Figure 6 shows that, under the action of FTCs, the carbonization depth of the concrete increases with an increase in the water–binder ratio, and the greater the number of FTCs and the higher the water–binder ratio, the faster the growth rate of carbonization depth. The carbonization depths of the concrete with water–binder ratios of 0.35, 0.40, and 0.45 for 28 days were 2.8, 3.6, and 4.1 mm, respectively. After 200 FTCs, the 28-day carbonization depths were 7.8, 10.1, and 11.6 mm, increased by 1.79 times, 1.81 times, and 1.83 times, respectively. The larger the water–binder ratio, the more pores inside the concrete and the faster the carbonization reaction rate. By contrast, in the case of a certain unit of water consumption, when the water–binder ratio is increased, the amount of cementing material is reduced, the fluidity is increased, and more tiny pores and channels are formed after the evaporation of water during the hardening of the concrete. Therefore, increasing the water–binder ratio increases the porosity and pore connectivity of concrete and promotes the diffusion rate of H_2O and CO_2 in concrete, and the carbonization reaction is accelerated.

3.1.4. Effect of Freeze–Thaw Damage on Carbonation Depth of Concrete

In combination with a single carbonation test and carbonation test data collected under freeze–thaw conditions, the influence of freeze–thaw damage on concrete carbonation performance is calculated and analyzed. The influence coefficient of the FTC on the carbonation performance of concrete is defined as λ_F and is expressed as follows [38]:

$$\lambda_F = d_{F+C}/d_C, \tag{2}$$

where d_{F+C} is the carbonization depth of the concrete after the FTC, d_C is the carbonization depth of the concrete under carbonization, and λ_F is the influence coefficient of the FTC on the carbonation depth of the concrete. $\lambda_F > 1$ indicates that the FTC promotes carbonization, while $\lambda_F < 1$ indicates that the FTC inhibits the development of carbonization depth to a certain extent. Figure 7 shows the influence coefficients of the three water–binder ratios on the carbonation performance of concrete under the action of FTCs.

Figure 7 shows that, after carbonization under freeze–thaw conditions, the influence coefficients of carbonization performance are all greater than 1 and increase with an increase in the number of FTCs. This result shows that the FTC has a promoting effect on carbonization, and the higher the number of FTCs, the more evident the promoting effect. With an increase in the number of cycles, the porosity inside the concrete may increase, and the connectivity between the pore size and pores also increases, providing convenient conditions for the diffusion of CO_2 in the concrete and accelerating the occurrence of the carbonization reaction. The research of H. Kuosa [39] and Yang Li [33,40,41] confirmed that the carbonation depth of concrete increased with the increase in FTCs, which was attributed to the increase in surface cracking and internal microcracks of concrete caused by FTCs and the increase in porosity, which accelerated the carbonation reaction.

As shown in Figure 7a, 100 FTCs led to increases of 86%, 85%, and 82% in the carbonization depth after 7, 14, and 28 days of carbonization, respectively, and 200 FTCs led to increases of 179%, 175%, and 172% in the carbonization depth for the same carbonization durations. As shown in Figure 7b, 100 FTCs led to increases of 83%, 79%, and 78% in the carbonization depth after 7, 14, and 28 days, respectively, and 200 FTCs led to increases of 189%, 181%, and 180% in the carbonization depth for the same carbonization durations. As shown in Figure 7c, 100 FTCs led to increases of 88%, 85%, and 73% in the carbonization depth after 7, 14, and 28 days, respectively, while 200 FTCs led to increases of 179%, 178%, and 142% in the carbonization depth for the same carbonization durations. Therefore, under the same number of FTCs, the influence coefficient gradually decreases with an increase in carbonization age, which indicates that the influence of the freeze–thaw cycle on carbonization gradually weakens with an increase in carbonization age. With an increase in carbonization age, the products of the carbonization reaction fill part of the pores in the concrete, resulting in a decrease in porosity, which affects the carbonization reaction rate. Therefore, the carbonization reaction rate decreases in the later stage, and the influence of the FTCs on carbonization decreases.

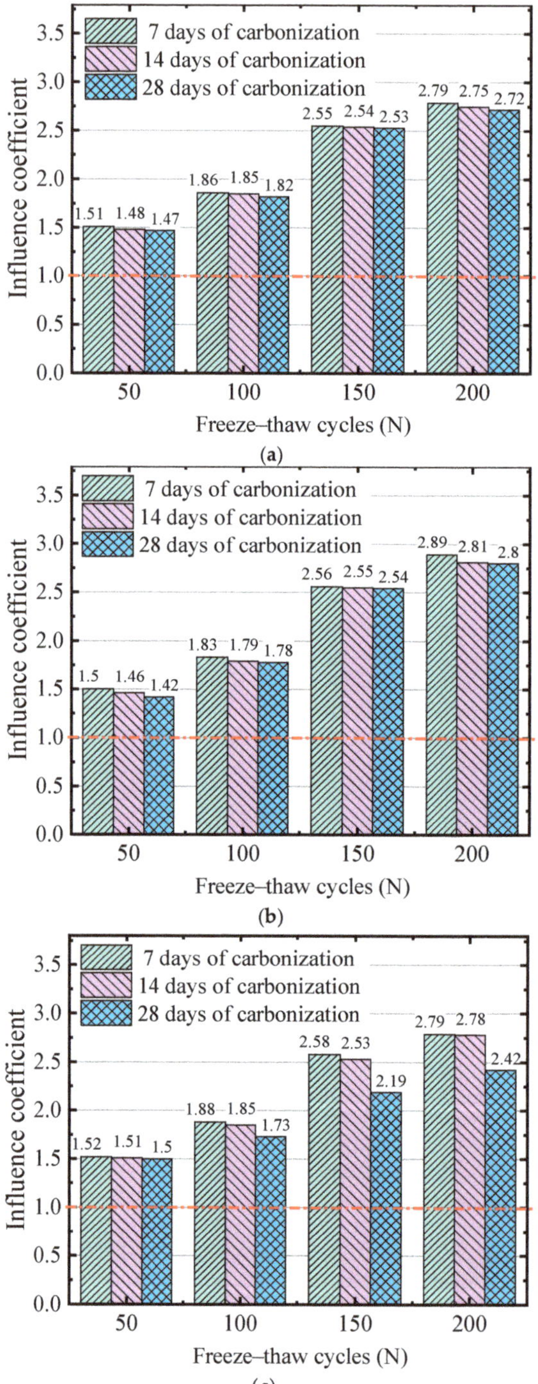

Figure 7. Relationship between the number of FTCs and influence coefficient of carbonization performance for (**a**) 0.35 water–binder ratio, (**b**) 0.40 water–binder ratio, (**c**) 0.45 water–binder ratio.

3.2. Relative Dynamic Modulus of Elasticity

The changes in the RDEM of concrete with different water–cement ratios after carbonation under FTC conditions are shown in Figure 8.

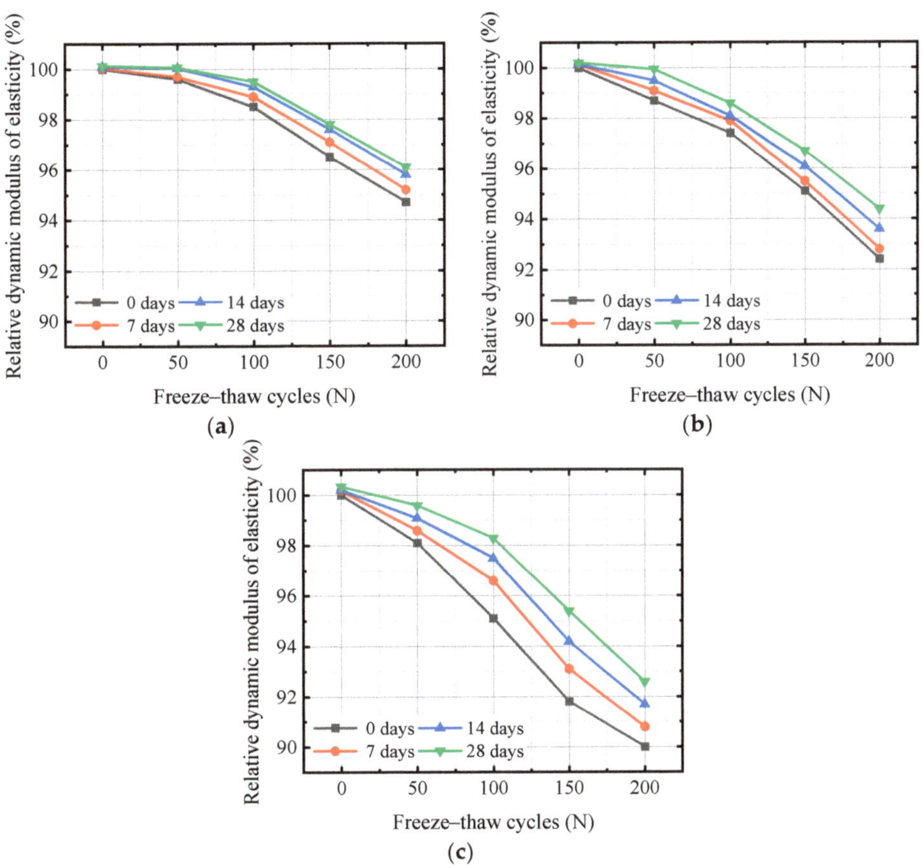

Figure 8. Relationship between the number of FTCs and the RDEM for (**a**) 0.35 water–binder ratio, (**b**) 0.40 water–binder ratio, (**c**) 0.45 water–binder ratio.

Figure 8 shows that the RDEM of concrete after carbonation under freeze–thaw conditions decreases with an increase in the number of FTCs. As the water–cement ratio increases, the RDEM of concrete continuously decreases, and the larger the water–cement ratio, the larger the decrease. For unfrozen specimens, the RDEM after carbonization is greater than 100%. After 200 FTCs, the RDEMs of the concrete specimens with 0.35, 0.40, and 0.45 water–cement ratios without carbonation were 94.7%, 92.4%, and 90%, respectively. After 28 days of carbonation, they were 96.1%, 94.4%, and 92.6%, respectively. The carbonation effect at 28 days resulted in increases of 1.48%, 2.16%, and 2.89% in the RDEM. These results indicate that carbonation improves the RDEM of concrete, restrains the expansion of freeze–thaw degradation characteristics, and alleviates the damage caused by FTCs to concrete. This conclusion is consistent with the research results in the literature.

Under the action of FTCs, the internal structure of concrete gradually transforms from dense to porous, and microcracks gradually expand and extend, leading to an increase in porosity. In a rapid carbonation environment, CO_2 with strong permeability enters the interior of concrete through pores, undergoes a neutralization reaction with the $Ca(OH)_2$ in the pores, and alleviates the damage to the RDEM under FTCs.

3.3. Compressive Strength

The compressive strength of concrete with three water–binder ratios after carbonation under freeze–thaw damage conditions is shown in Figure 9. As shown in the figure, as the carbonization age increases, the compressive strength of concrete increases, and the higher the water–cement ratio, the faster the growth rate of compressive strength after carbonization. For the group of specimens that underwent 0 FTCs, after 28 days of carbonization, the compressive strength of the specimens with 0.35, 0.40, and 0.45 water–binder ratios increased from 56.5, 49.4, and 36.8 MPa to 64.6, 57.1, and 49.5 MPa, respectively, and increased to 114.34%, 115.35%, and 134.51% of their respective compressive strength before carbonization. Thus, the compressive strength of concrete shows a trend of first increasing fast and then increasing slowly with an increase in carbonization age. After 14 days of carbonization, the increase in strength was relatively slow. During the carbonation reaction of concrete, calcium hydroxide reacts with CO_2 to generate $CaCO_3$, which increases the volume of the generated product relative to the reactant and plays a particular filling role in the pore structure, making the pore structure denser and improving the strength of the concrete. The diffusion rate of CO_2 in concrete directly affects the rate of the carbonation reaction.

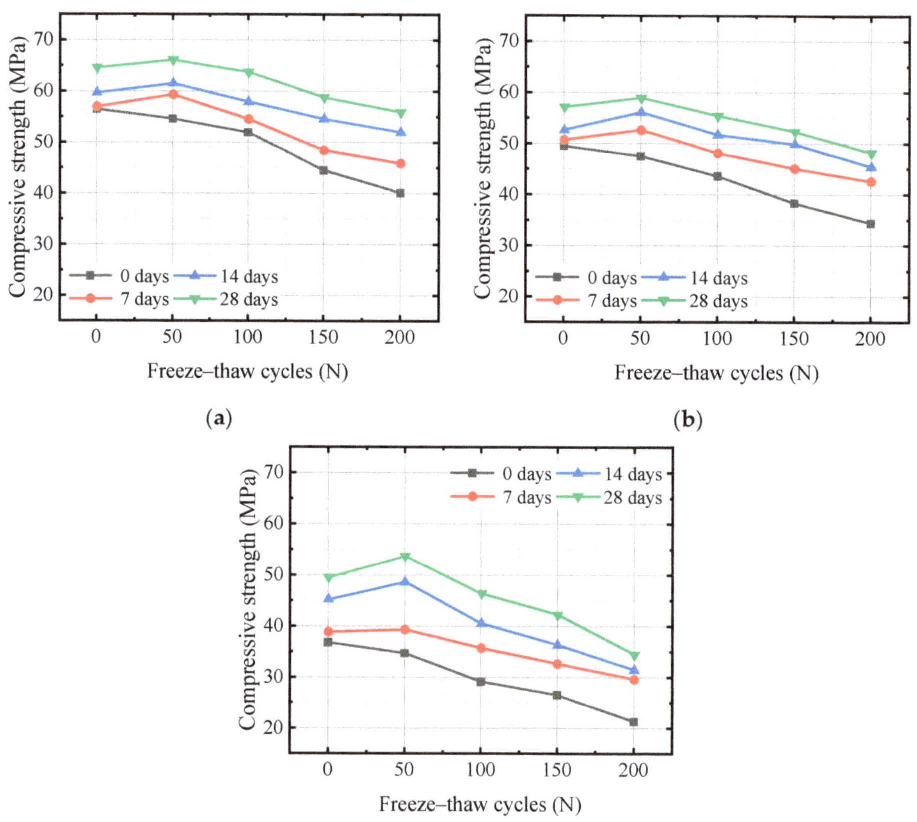

Figure 9. Relationship between carbonization age and compressive strength for (**a**) 0.35 water–binder ratio, (**b**) 0.40 water–binder ratio, (**c**) 0.45 water–binder ratio.

Combining the subfigures of Figure 9 reveals that carbonization can alleviate the damage to concrete strength caused by FTCs. The increase in compressive strength after 7 days of carbonization was greater than the damage to the compressive strength caused by 50 FTCs. After 100 FTCs, the carbonization reaction was insufficient to compensate for the strength loss caused by freeze–thaw damage. For the specimens that underwent 200 FTCs, the 28 days of carbonization resulted in a 39.33% increase in the compressive strength of the 0.35 water–binder ratio concrete compared with the concrete, a 40.12% increase in the compressive strength of the 0.40 water–binder ratio concrete, and a 61.5% increase in the compressive strength of the 0.45 water–binder ratio concrete. Concrete specimens with a large water–binder ratio have a large internal porosity and relatively loose pore structure, resulting in a higher diffusion rate of carbon dioxide, which leads to a faster carbonation reaction rate and a larger content of calcium carbonate generated. The rate of filling dense pores with the generated product is also relatively faster, resulting in an accelerated rate of strength improvement. Therefore, the higher the water–binder ratio of concrete, the faster the strength growth rate after carbonization. Reference [4] also confirmed that the water–binder ratio is the main reason that affects the carbonation performance of concrete, and carbonation improves the compactness and mechanical properties of concrete under FTCs.

Carbonization products fill some pores, which blocks the pores' connectivity. The diffusion of CO_2 is blocked, and the carbonization reaction speed slows down, which leads to slow strength growth after 14 days of carbonization. After 50 FTCs, the damage to the compressive strength was relatively small, and the increase in concrete strength caused by carbonation was greater than the damage caused by FTC. After 100 FTCs, the damage to the concrete caused by FTCs was evident, and the increase in the compressive strength caused by the carbonation reaction was not enough to compensate for the damage to the concrete compressive strength.

3.4. Microstructure Evolutions

3.4.1. Micromorphology

SEM images of concrete samples with different carbonization ages at a magnification of 2000 are shown in Figure 10.

As shown in Figure 10c, the number of pores further decreases, indicating that the concrete becomes denser after carbonation. A comprehensive comparison of the micro changes in the surface of concrete reveals that, with an increase in the number of carbonation days, the pore structure of concrete undergoes considerable changes; the smoothness of the concrete surface is enhanced, the integrity is increased, and the number of small pores is reduced. The reason for this is that, during carbonation, the $Ca(OH)_2$ in the concrete reacts with CO_2 and the carbonization product $CaCO_3$, which fills the interior of the concrete, thereby reducing the porosity of the internal structure of the concrete, increasing the compactness of the concrete, and improving its strength.

SEM analysis was conducted on concrete specimens that underwent 100 FTCs, 200 FTCs, and carbonation for 0 and 28 days. The microstructure of the samples at a magnification of 2000 is shown in Figure 11.

Figure 11a shows that the concrete that underwent 100 FTCs has spherical fly ash particles without complete hydration on its surface before carbonization. Under the freeze–thaw action, cracks appear, exhibiting more pores, and the surface is rugged with many thin-layered and spherical $Ca(OH)_2$ crystals. Figure 11b shows a sample that underwent 100 FTCs and 28 days of carbonization. The surface still shows many cracks and pores, but compared with Figure 11a, the hydration products on the surface are considerably reduced, the width of the cracks is narrowed, and the structure is complete and dense. During the carbonization reaction, CO_2 reacts with $Ca(OH)_2$ to generate $CaCO_3$, which consumes a portion of $Ca(OH)_2$ and flattens the structure. The reaction products fill some pores and cracks, and various hydration products connect to form a whole, making the structure dense, reducing the porosity of the concrete, and improving its strength. Figure 11c shows that, after 200 FTCs, microcracks further increase, the aggregate detaches from the slurry, gaps are visible at the cross-section, and the surface becomes loose and porous. As shown in Figure 11d, carbonization causes the structure to become more complete and denser than that under 200 FTCs with 0 days of carbonization, and the carbonization reactant $CaCO_3$ can be seen attached to the pore wall at the edge of the crack. However, freeze–thaw damage remains relatively evident, and the carbonization products are not enough to fill the cracks caused by FTCs. This result explains the decrease in the RDEM, ultrasonic wave velocity, splitting tensile strength, and compressive strength from a microscopic perspective.

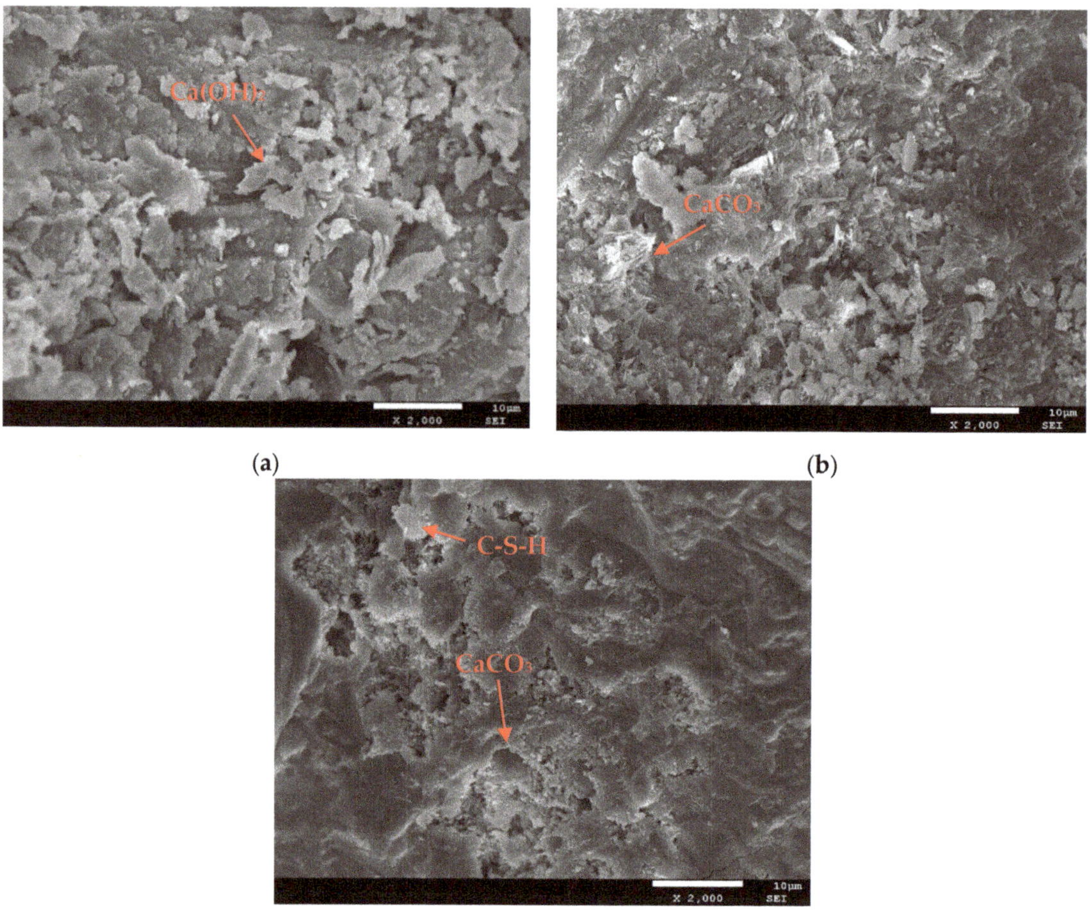

(a) (b) (c)

Figure 10. Electron microscope scanning of concrete after carbonization: (**a**) carbonization for 0 days, (**b**) carbonization for 14 days, (**c**) carbonization for 28 days.

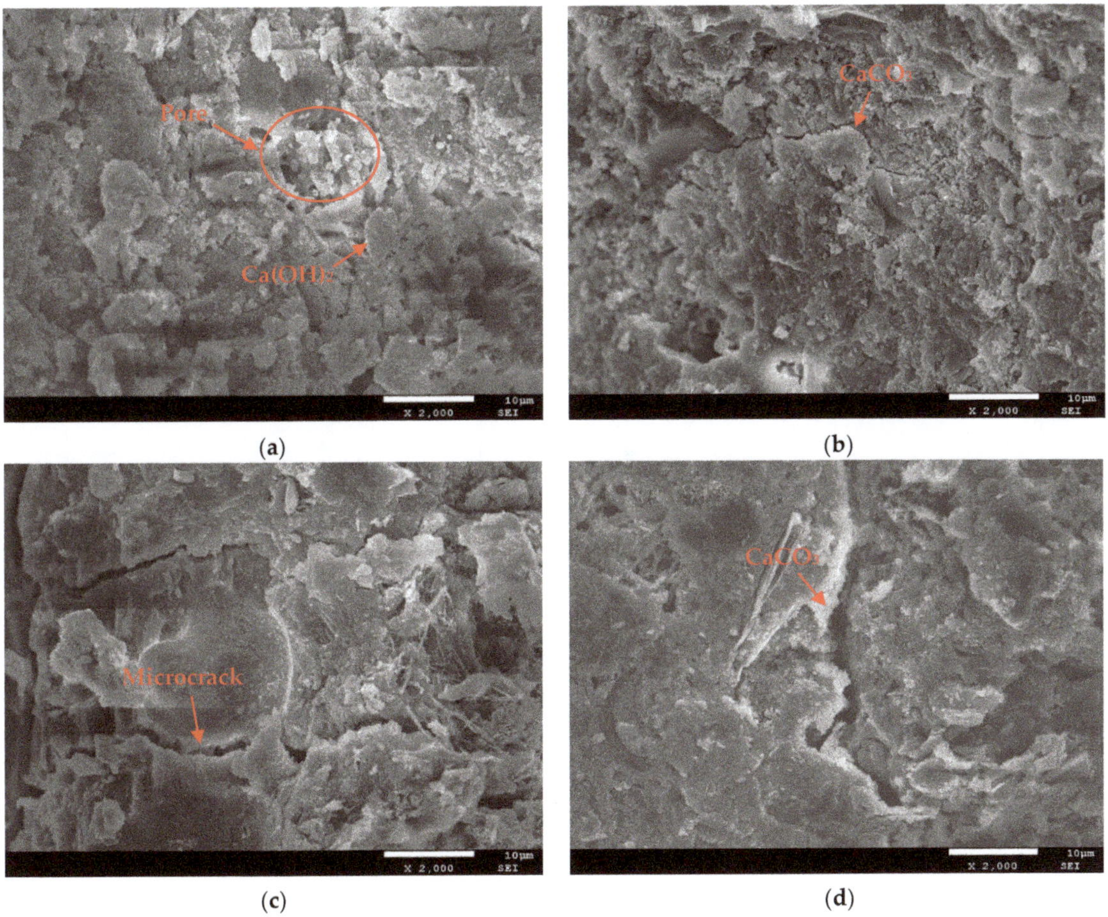

Figure 11. Electron microscope scanning of concrete after carbonization under freeze–thaw conditions: (**a**) carbonization for 0 days under 100 FTCs, (**b**) carbonization for 28 days under 100 FTCs, (**c**) carbonization for 0 days under 200 FTCs, (**d**) carbonization for 28 days under 200 FTCs.

3.4.2. Pore Structure

Porosity

In order to investigate the influence mechanism of FTCs on the carbonation of concrete, three water–binder ratios of concrete were subjected to 0, 100, and 200 FTCs and then placed in a carbonation box to undergo accelerated carbonation for 14 and 28 days. Finally, NMR analysis was performed on the samples to obtain their porosity. The relationship between the carbonation age and porosity of concrete under FTCs is shown in Figure 12.

Figure 12 shows that, under freeze–thaw damage conditions, the porosity of the carbonated concrete increases with an increase in the number of FTCs and the water–cement ratio. As the carbonization age increases, the porosity gradually decreases, and after 14 days of carbonization, the rate of porosity change slows down. Freeze–thaw damage leads to an increase in the porosity of concrete, affecting the diffusion of CO_2 inside the concrete and accelerating the carbonation process. The higher the number of FTCs, the greater the acceleration effect. As the water–binder ratio increases, more small pores and channels are formed by water evaporation after concrete hardening. The internal porosity of the concrete is relatively high, and the structure is relatively loose, promoting the diffusion of water and CO_2 in the concrete, thereby accelerating the rate of carbonation reaction and generating more $CaCO_3$. As the carbonization age increases, CO_2 reacts with $Ca(OH)_2$ to form $CaCO_3$, which has a particular filling effect on the pore structure and reduces the porosity. However, after a carbonization reaction period (14 days), as the internal pores become denser, the generated carbonization product $CaCO_3$ fills some of the pores, playing a blocking role, which hinders the diffusion process of CO_2 and reduces the rate of porosity change.

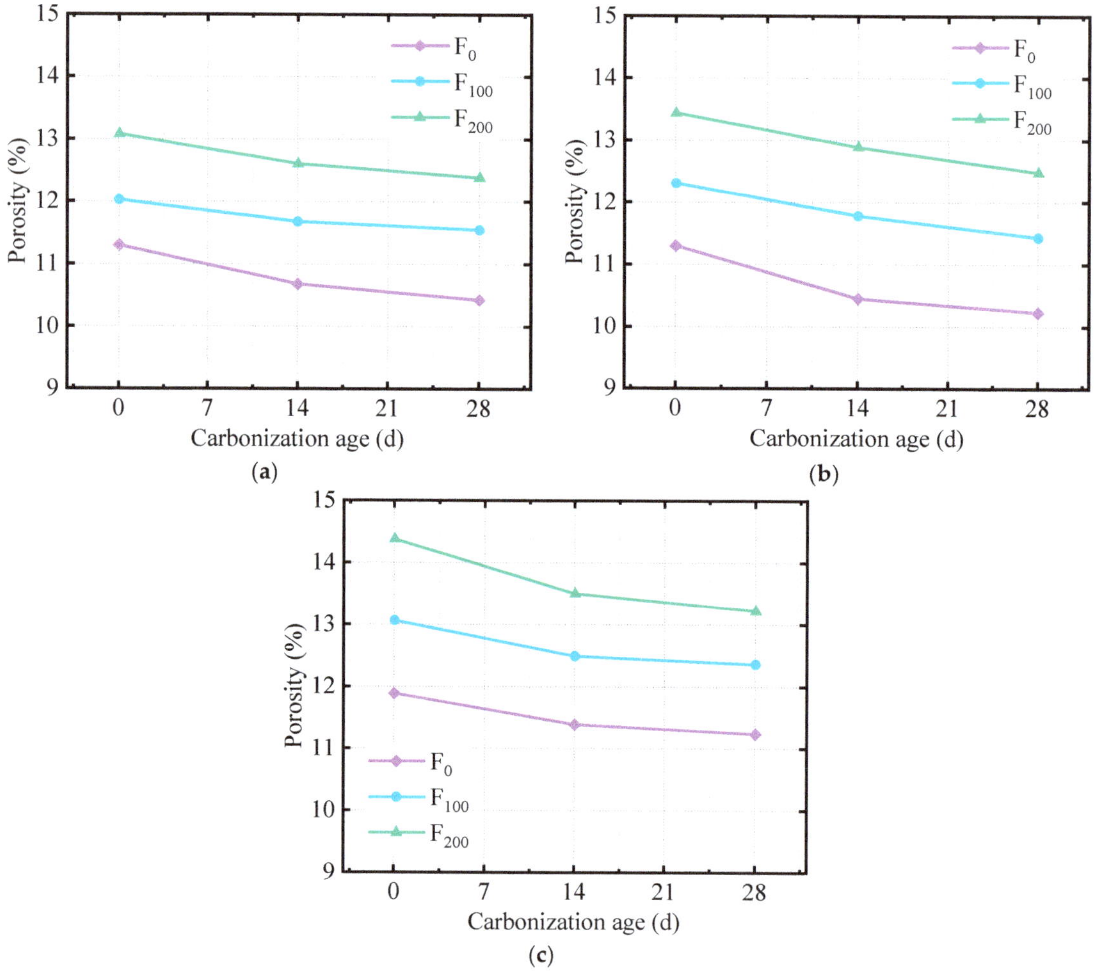

Figure 12. Porosity after carbonation under freeze–thaw conditions for (**a**) 0.35 water–binder ratio, (**b**) 0.40 water–binder ratio, (**c**) 0.45 water–binder ratio.

Pore Size Distribution

The porosity, PSD, and pore connectivity of concrete directly affect the transport characteristics and mechanical properties of water, gas, and ions in concrete. As PSD can better evaluate the comprehensive performance of hydrated concrete, the concrete specimens that underwent 0 and 28 days of carbonization after 0 and 200 FTCs were selected as samples. A magnetic resonance analyzer was used to analyze the influence of carbonization on the PSD of concrete under FTC damage conditions, as shown in Figure 13.

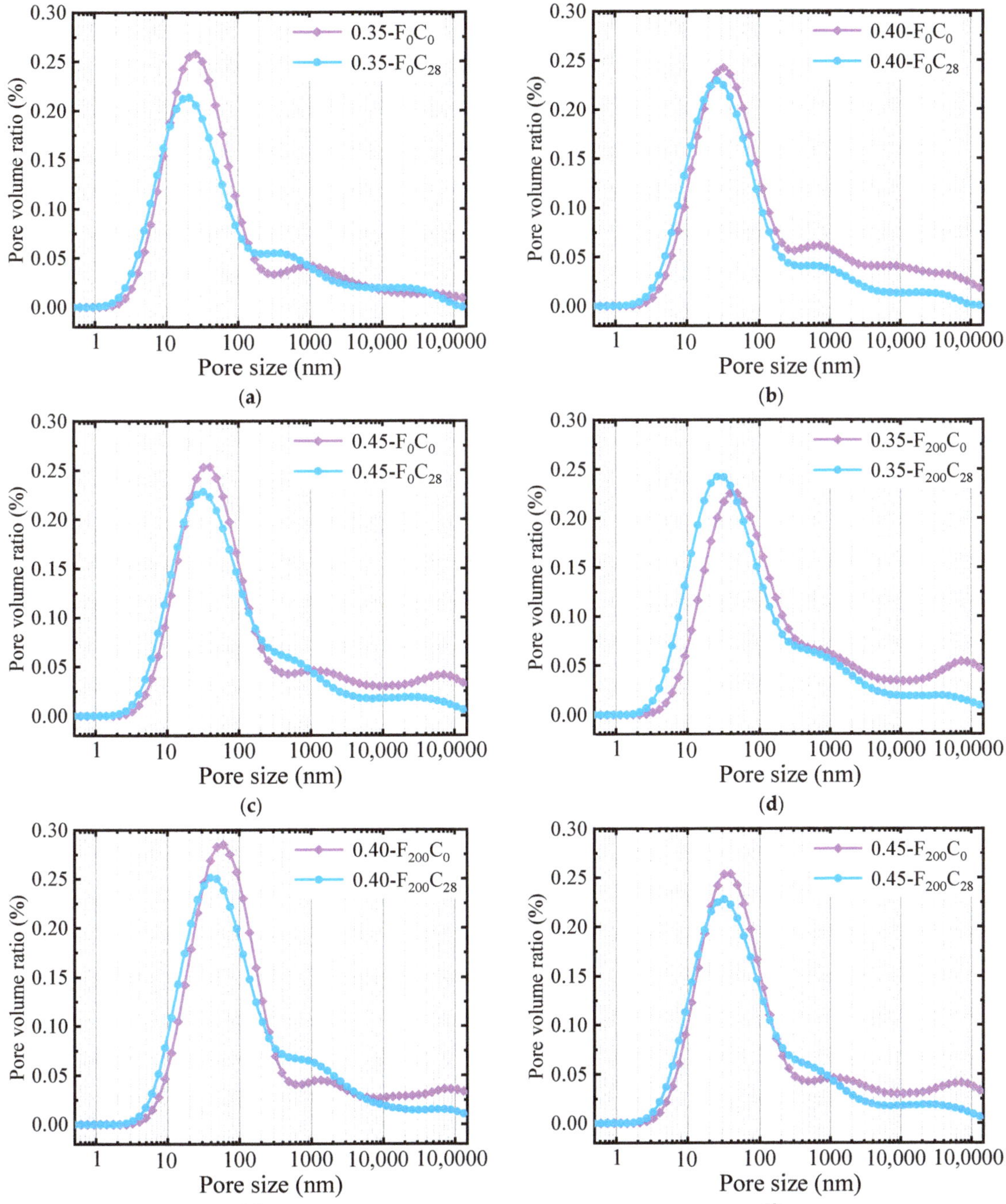

Figure 13. PSD before and after carbonization under FTC conditions for (**a**) 0.35 water–binder ratio, 0 FTCs, (**b**) 0.40 water–binder ratio, 0 FTCs, (**c**) 0.40 water–binder ratio, 0 FTCs, (**d**) 0.35 water–binder ratio, 200 FTCs, (**e**) 0.40 water–binder ratio, 200 FTCs, (**f**) 0.40 water–binder ratio, 200 FTCs.

Figure 13 shows that carbonization changes the PSD of the concrete. Whether it occurs before or after the FTC, the average pore diameter and the most available pore diameter decrease, and the total porosity decreases. Figure 13a–c shows that, for noncarbonized specimens that underwent 0 FTCs, the most probable pore sizes for the concrete with 0.35, 0.40, and 0.45 water–binder ratios were 24.07, 31.77, and 36.50 nm, respectively. After 28 days of carbonization, the most probable pore sizes decreased to 20.45, 25.8, and 29.48 nm, respectively, decreases of 15.04%, 18.79%, and 19.23%. The greater the water–binder ratio, the more the most probable pore size decreased. Figure 13d–f shows that, after 200 FTCs and before carbonization, the maximum pore sizes of the concrete with the three water–binder ratios before carbonization were 41.94, 59.35, and 73.09 nm, respectively, and the maximum pore sizes were reduced to 21.44, 41.94, and 44.96 nm after 28 days of carbonization—decreases of 25.04%, 29.33%, and 38.49%. Carbonization reduces the most probable pore size of concrete before and after the FTC, and the greater the water–binder ratio, the greater the change value. The carbonization reaction produces $CaCO_3$, which adheres to the pore wall, resulting in a decrease in the pore diameter participating in the reaction, changing the pore diameter distribution, and reducing the most probable pore size of the concrete. The larger the water–binder ratio, the larger the initial porosity, and the greater the damage under the FTCs, which speeds up the diffusion rate of CO_2 in concrete and accelerates the carbonization reaction rate. Therefore, the larger the water–binder ratio, the faster the reduction rate of the most probable pore size.

To further explore the influence mechanism of the micropore structure of carbonized concrete on macroproperties under FTC conditions, on the basis of the results of NMR analysis and referencing the literature [42,43] on pore structure classification, the pore structure of carbonized concrete under freeze–thaw conditions was classified, and its PSD was statistically analyzed. Different types of pore distribution are plotted in Figure 14.

Figure 14 shows that, after 28 days of carbonization, the gel pores ($d < 10$ nm), transition pores (10 nm $\leq d <$ 100 nm), and macropores ($d >$ 1000 nm) changed considerably. The content of gel pores and macropores decreased, and the content of transition pores increased. For capillary pores (100 nm $\leq d <$ 1000 nm), a slight fluctuation occurred, but the change was not considerable. The carbonization reaction affects all four kinds of pores, but the evolution law of pores states that the reaction degree is different. CO_2 gradually diffuses from the surface to the interior of the concrete pores under the effect of concentration difference and reacts with $Ca(OH)_2$ in the pores to generate $CaCO_3$, which is attached to the pore wall, resulting in a smaller pore size and lower porosity. The pore diameter of the gel becomes smaller and cannot be detected by the NMR analyzer; thus, the gel pore content decreases. The considerable decrease in macropore content is due to the large surface area of macropores, which generate more $CaCO_3$ in full contact with CO_2, resulting in a considerable reduction in pore size. The increase in transitional pore content is due to the fact that, although carbonization occurs in the range of 10 nm $\leq d <$ 100 nm, the carbonization reaction speed is faster in pore sizes of 100 nm $\leq d <$ 1000 nm and $d >$ 1000 nm, and the pore size decreases more, ultimately leading to an increase in pore size content in the range of 10 nm $\leq d <$ 100 nm.

Figure 14a,c show that, under the action of the FTC, the percentage of transition pores with a pore size of 10 nm $\leq d <$ 100 nm decreases, the percentage of macropores with $d >$ 1000 nm increases, and the amplitude of gel pores with $d <$ 10 nm and capillary pores with a pore size of 100 nm $\leq d <$ 1000 nm is not evident, indicating that the FTC has no considerable effect on gel pores. The FTC causes the change in pore size when $d \leq$ 100 nm, which ultimately increases macropores and decreases transition pores.

Combining the results of the FTC and carbonization testing, including RDEM and compressive strength values, reveals that the reason for the increase in porosity and strength after carbonization is the decrease in macropore content with $d >$ 1000 nm and the increase in transitional pore content with a pore size of 10 nm $\leq d <$ 100 nm caused by carbonization. Controlling the content of large pores with $d >$ 1000 nm and increasing the content of transition pores with a pore size of 10 nm ≤ d < 100 nm can improve the carbonation durability of concrete under FTC conditions.

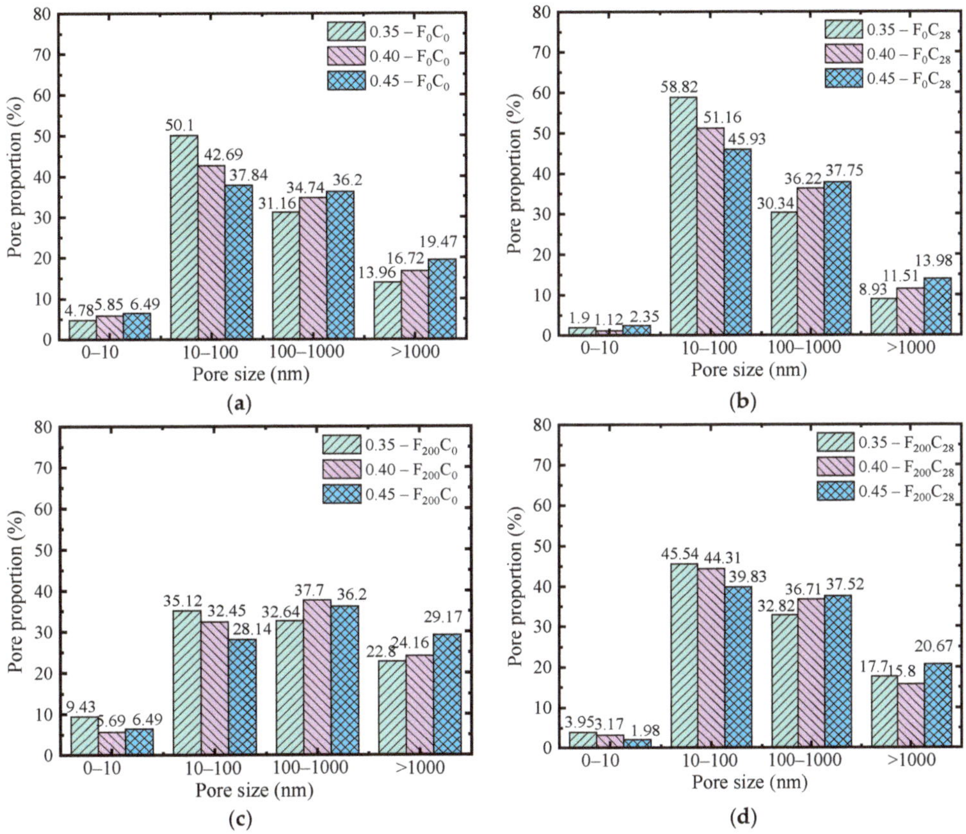

Figure 14. PSD of specimens with different corrosion rates: (**a**) 0 FTCs, carbonization for 0 days, (**b**) 0 FTCs, carbonization for 28 days, (**c**) 200 FTCs, carbonization for 0 days, (**d**) 200 FTCs, carbonization for 28 days.

4. Discussion

On the basis of the gray relational analysis model, the influence of pore structure on the RDEM, carbonation depth, and compressive strength of concrete under the combined action of FTCs and carbonation is discussed. Through fitting analysis between porosity and carbonation depth, the relationship between pore structure and carbonation depth is obtained, and the influence mechanism of the FTC on carbonation depth is discussed from macro to micro perspectives.

4.1. Analysis of Influencing Factors Based on Gray Correlation Model

The gray correlation analysis method measures the correlation program between factors based on their development trend. The degree of correlation directly reflects the degree of influence of various factors in the system on the target value, indicating the degree of correlation between two systems or factors [44,45]. By calculating the correlation between the target value (reference sequence) and the influencing factors (comparison series), the correlation is sorted to identify the main factors that affect the target value. By conducting data calculations on the various factors that must be analyzed, the correlation can be found in a random sequence of factors, and the main influencing factors can be identified. The general calculation steps for factor analysis using the gray correlation degree are as follows:

(1) Establish reference and comparison series.

$X_0(k) = \{X_0(1), X_0(2), X_0(3), \cdots, X_0(n)\}$ is set as the system reference series and $X_i(k) = \{X_i(1), X_i(2), X_i(3), \cdots, X_i(m)\}$ as the comparison sequence of m-many related factors in the model. Six indicators reflecting the durability of concrete against carbonation erosion, i.e., the RDEM, carbonation depth, porosity, compressive strength, tensile compressive strength ratio, and carbonation performance influence coefficient, are used as reference sequences for the model. Three factors that affect carbonization durability, i.e., the water–binder ratio, FTCs, and carbonization age, are set as comparative series.

(2) Initialize each sequence as follows:

$$X_j(k) = \frac{X_j(k)}{\max_k X_j(k)}, \tag{3}$$

where $j = 0, 1, 2, \cdots, m$.

(3) Calculate the correlation coefficient.

The correlation between the comparison sequence and the reference sequence is as follows:

$$\xi_i(k) = \frac{\min_i \min_k |X_0(k) - X_i(k)| + \rho \max_i \max_k |X_0(k) - X_i(k)|}{|X_0(k) - X_i(k)| + \rho \max_i \max_k |X_0(k) - X_i(k)|} \tag{4}$$

In the formula, $\xi_i(k)$ is the correlation coefficient between element k of comparison sequence X_i and element k of reference sequence X_0; ρ is the resolution coefficient, where $\rho \in [0, 1]$, usually taken as 0.5.

(4) Solve the correlation degree.

The correlation degree is calculated according to the following equation:

$$\gamma_i = \frac{1}{n} \sum_{k=1}^{n} \xi_i(k) \tag{5}$$

In the equation, γ_i is the correlation between the comparison sequence X_i and the reference sequence X_0.

(5) Sort association order.

On the basis of the degree of correlation, the order of the impact of each factor on the target value is determined. The magnitude of the correlation degree determines the importance of the influencing factors; the larger the correlation degree value, the closer the development trend of X_i and X_0 is, and the greater the effect of X_i on X_0.

4.1.1. Correlation Analysis of Macro-Influencing Factors

Five durability indexes of carbonation resistance of reactive concrete, such as RDEM, carbonation depth, porosity, compressive strength, and carbonation performance influence coefficient, are taken as the model reference sequence. The water–binder ratio, FTCs, carbonation age, and three other factors that affect carbonation durability are set as a comparison series. According to the calculation method of grey correlation degree, the macro-influencing factors are analyzed by grey correlation degree, and the results are shown in Table 2.

Table 2. Calculation results of grey correlation degree of macro-influencing factors.

Influencing Factor	RDEM	Carbonization Depth	Porosity	Compressive Strength	Influence Coefficient of Carbonization Performance
water–binder ratio	0.879	0.656	0.885	0.783	0.702
FTCs	0.579	0.659	0.576	0.568	0.704
carbonation age	0.547	0.774	0.550	0.610	0.588

As shown in Table 2, the influence of water–binder ratio, freeze–thaw cycles, and carbonation age on the RDEM and porosity of concrete after carbonation under freeze–thaw conditions is as follows: water–binder ratio > FTCs > carbonation age, and the water–binder ratio has the greatest influence. The correlation coefficients of the water–binder ratio to RDEM and porosity are 0.879 and 0.855, which show a strong correlation. The order of influence of different factors on carbonation depth is carbonation age (0.774) > FTCs (0.659) > water–binder ratio (0.656). For the compressive strength, the influence of different factors is as follows: water–binder ratio (0.783) > carbonation age (0.61) > FTCs (0.568). The water–binder ratio and the number of FTCs have the greatest influence on the carbonation performance, followed by carbonation age. Thus, under the combined action of FTC and carbonation, the carbonation age has the greatest influence on carbonation depth, and the water–binder ratio has the greatest influence on other macro properties of concrete, while the carbonation age and FTCs have relatively little influence. Properly increasing the water–binder ratio can improve the carbonation resistance of concrete under freeze–thaw conditions.

4.1.2. Correlation Analysis of Influencing Factors of Pore Structure

According to the calculation method of the gray correlation degree, the RDEM, carbonization depth, and compressive strength are taken as the reference sequence of the model, and the gel hole (d < 10 nm), transition hole (10 nm ≤ d ≤ 100 nm), capillary hole (100 nm ≤ d ≤ 1000 nm), and large hole (d > 1000 nm) are set as the comparison series for gray correlation degree analysis. Table 3 shows the calculation results of the correlation degree coefficients.

Table 3. Calculation results of correlation degree coefficient of pore structure parameters.

Pore Diameter	RDEM	Carbonization Depth	Compressive Strength
Gel pore (d < 10 nm)	0.597	0.539	0.547
Transition pore (10 nm ≤ d ≤ 100 nm)	0.818	0.643	0.907
Capillary pore (100 nm ≤ d ≤ 1000 nm)	0.890	0.653	0.749
Macro pore (d > 1000 nm)	0.714	0.595	0.607

As shown in Table 3, the capillary pores (100 nm ≤ d ≤ 1000 nm) and transition pores (10 nm ≤ d ≤ 100 nm) have a great influence on the carbonization depth under freeze–thaw conditions, and the correlation coefficients are 0.653 and 0.643. The order of influence on the RDEM is capillary hole > transition hole > macropore > gel hole. The capillary hole has the greatest influence, and the correlation coefficient exceeds 0.8, showing a strong correlation. The order of influence on compressive strength is transition pore > capillary pore > macropore > gel pore, and the correlation coefficients are 0.907, 0.749, 0.607, and 0.547, respectively. After carbonization under freeze–thaw conditions, capillary pores (100 nm ≤ d ≤ 1000 nm) have the greatest influence on the RDEM and carbonization depth, followed by transition pores (10 nm ≤ d ≤ 100 nm), large pores (d > 1000 nm), and gel pores (d < 10 nm). Under freeze–thaw conditions, the effect of gel pores on the macroscopic properties of concrete is small. Reducing the content of large pores and increasing the content of transition pores and capillary pores can improve the compressive strength, RDEM, and carbonization properties of concrete.

4.2. Relationship Between Carbonization Depth and Pore Structure

The relationship between the carbonation depth, porosity, and most probable pore size of concrete after carbonation under freeze–thaw conditions is shown in Figure 15. The relationship between the carbonation depth and porosity of the concrete demonstrates a nonlinear positive correlation after carbonation for 14 days and 28 days, with correlation coefficients of 0.978 and 0.959, respectively. The relationship between the carbonation depth and the most probable pore size reveals a linear positive correlation, with correlation coefficients of 0.988 and 0.984, respectively, showing excellent correlations.

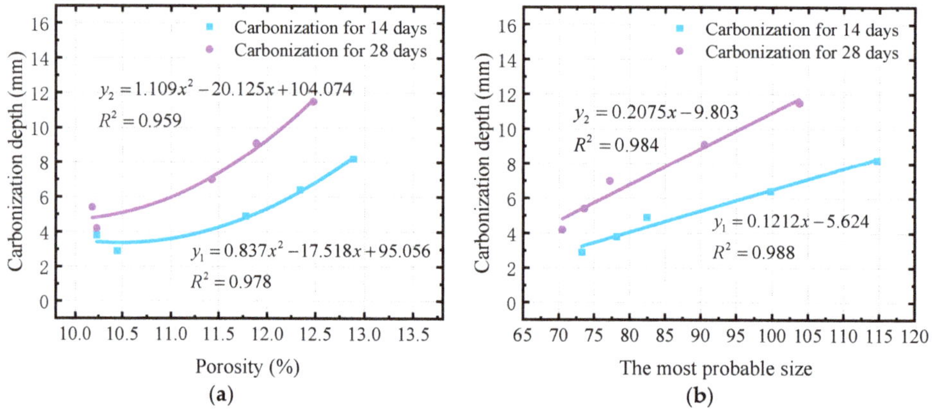

Figure 15. Relationship between carbonation depth and (**a**) porosity, (**b**) the most probable size.

4.3. Effect Mechanism of Freeze–Thaw Cycle on Carbonation

The carbonation mechanism of concrete under freeze–thaw damage conditions is basically the same as that in general atmospheric environments. At the early stage of the FTC, microcracks occur in volume-expanded concrete due to freezing. With an increase in the number of FTCs, the internal microcracks gradually develop, the mortar layer on the surface of the concrete spalls, and the structure becomes loose and porous. The porosity inside the concrete also changes greatly, and the number and pore size of the pores increase, which is conducive to the diffusion of CO_2 and speeds up the carbonization reaction. As the carbonization reaction continues, CO_2 and $Ca(OH)_2$ inside the concrete react chemically, resulting in $CaCO_3$ attached to the inner pore wall of the concrete, which fills the pores inside the concrete to a certain extent, resulting in a decrease in the porosity inside the concrete, an increase in the compactness of the concrete, and an increase in strength. With the later process of carbonization, the pore structure of concrete is blocked by the generated $CaCO_3$, resulting in an increase in the internal compactness of the concrete and a decrease in pore volume, which is the reason for the decrease in porosity. At the same time, this process prevents external CO_2 from entering the concrete pores and undergoing the carbonizing reaction. Thus, after 14 days of carbonization, the strength increase is relatively slow, which is the reason for the low carbonization reaction rate in the later stage.

The FTC accelerates the carbonation reaction and reduces the durability of concrete, which is also confirmed by the research of TIAN Li [46] and NIU Ditao [38]. When freeze–thaw and carbonation phenomena are observed in the actual service environment, predicting service life and design durability based only on the carbonation law determined by a single carbonation factor in the laboratory is unreasonable. The combined effects of the FTC, carbonation, and other factors must be considered.

5. Conclusions

In this study, freeze–thaw cycle and carbonation experiments were conducted, and the relationship between carbonation depth and the microstructure of concrete mixed with fly ash under FTCs was analyzed. The results of this study are summarized in the concluding remarks below.

(1) The freeze–thaw cycle, as the power source of concrete damage, increases the porosity of concrete, which creates favorable conditions for the carbonation reaction and thus accelerates the process. Carbonization improves the compressive strength of concrete under freeze–thaw damage. The compressive strength of concrete specimens with water–binder ratios of 0.35, 0.40, and 0.45 increased by 39.3%, 40.2%, and 61.4%, respectively, after 200 FTCs.
(2) The carbonation depth of concrete increases with the number of FTCs and the water–binder ratio. Carbonization improves the RDEM of concrete, inhibits the expansion of freeze–thaw deterioration characteristics, and alleviates the damage to the concrete caused by the FTC. Carbonization changes the microstructure of concrete, reducing the porosity and the most probable pore size of concrete under freeze–thaw damage conditions.
(3) The content of macropores with d > 1000 nm decreases, and the content of transition pores with a pore size of 10 nm ≤ d < 100 nm increases after carbonation under freeze–thaw damage conditions, which is the direct reason for the decrease in porosity and the improvement in strength. Therefore, controlling the content of macropores with d > 1000 nm and increasing the content of transition pores with a pore size of 10 nm ≤ d < 100 nm can improve the carbonation durability of fly ash concrete under freeze–thaw conditions.
(4) Gray relational analysis showed that the carbonation age has the greatest influence on carbonation depth, and the water–binder ratio has the greatest influence on other macroproperties of concrete. Thus, properly reducing the water–binder ratio can improve the carbonation resistance of concrete under freeze–thaw conditions. The capillary hole (100 nm ≤ d ≤ 1000 nm) has the greatest influence on the RDEM and carbonation depth, followed by the transition hole (10 nm ≤ d ≤ 100 nm), and then the macropore (d > 1000 nm), and the gel hole (d < 10 nm) has the least influence.
(5) The relationship between the carbonation depth and pore structure of concrete shows a nonlinear positive correlation between carbonation depth and porosity under freeze–thaw conditions and a linear positive correlation with the most probable pore size.

Author Contributions: Conceptualization, G.T. and B.T.; formal analysis, Z.Q. and J.C; investigation, J.C. and B.C; methodology, B.C.; project administration, S.Z.; resources, B.T.; writing—review and editing, S.Z. All authors have read and agreed to the published version of the manuscript.

Funding: This work was financially supported by the National Natural Science Foundation of China (nos. 51879145 and 52109158) and the Seventh Geological Brigade of Hubei Geological Bureau Science and Technology project (DQKJ2023-10).

Institutional Review Board Statement: Not applicable.

Informed Consent Statement: Not applicable.

Data Availability Statement: The original contributions presented in this study are included in the article. Further inquiries can be directed to the corresponding author.

Conflicts of Interest: The authors declare no conflicts of interest.

References

1. Quan, H.Z. Research on durability of high volume fly ash concrete. *Appl. Mech. Mater.* **2012**, *174–177*, 380–383. [CrossRef]
2. Wang, R.; Zhang, Q.; Li, Y. Deterioration of concrete under the coupling effects of freeze–thaw cycles and other actions: A review. *Constr. Build. Mater.* **2022**, *319*, 126045. [CrossRef]
3. Li, J.; Yang, H.; Wu, H. Evaluation of Concrete Carbonation Based on a Fiber Bragg Grating Sensor. *Micromachines* **2024**, *15*, 29. [CrossRef] [PubMed]
4. Yang, Q.; Liu, M.; Li, J.; Chen, X. Study on the degradation models based on the experiments considering the coupling effect of freeze-thaw and carbonation. *Structures* **2024**, *64*, 106659. [CrossRef]

5. Zhang, D.; Yang, Q.; Mao, M.; Li, J. Carbonation performance of concrete with fly ash as fine aggregate after stress damage and high temperature exposure. *Constr. Build. Mater.* **2020**, *242*, 118125. [CrossRef]
6. Dhanesh, S.; Kumar, K.S.; Maruthur, P.; Rejumon, R.; Usmansha, G.S. Experimental investigation of strength of Aramid kelvar and chopped carbon reinforced concrete beam. *Mater. Today: Proc.* **2021**, *45*, 1269–1273. [CrossRef]
7. Wang, X.; Yang, Q.; Peng, X.; Qin, F. A Review of Concrete Carbonation Depth Evaluation Models. *Coatings* **2024**, *14*, 386. [CrossRef]
8. Sun, X.; Cui, Y.; Chen, J.; Yi, S.; Li, X.; Chen, L. MFRWA: A Multi-Frequency Rayleigh Wave Approximation Method for Concrete Carbonation Depth Evaluation. *Buildings* **2024**, *14*, 1821. [CrossRef]
9. Moghaddas, S.A.; Nekoei, M.; Golafshani, E.M.; Nehdi, M.; Arashpour, M. Modeling carbonation depth of recycled aggregate concrete using novel automatic regression technique. *J. Clean. Prod.* **2022**, *371*, 133522. [CrossRef]
10. Qu, F.; Xia, W.; Sun, C.; Hou, H.; Huang, B.; Wang, G.; Hu, S. Modeling carbonation depth of recycled aggregate concrete containing chlorinated salts. *Constr. Build. Mater.* **2024**, *430*, 136478. [CrossRef]
11. Xiao, Q.H.; Li, Q.; Guan, X.; Zou, Y.X. Prediction model for carbonation depth of concrete subjected to freezing-thawing cycles. *IOP Conf. Ser. Mater. Sci. Eng.* **2018**, *322*, 022048. [CrossRef]
12. You, Q.; Ren, G.; Fraedrich, K.; Kang, S.; Ren, Y.; Wang, P. Winter temperature extremes in China and their possible causes. *Int. J. Clim.* **2013**, *33*, 1444–1455. [CrossRef]
13. Yang, Y.; Chang, W. Analysis of Spatial and Temporal Distribution and Changes in Extreme Climate Events in Northwest China from 1960 to 2021: A Case Study of Xinjiang. *Sustainability* **2024**, *16*, 4960. [CrossRef]
14. Liu, F.; You, Z.; Yang, X.; Wang, H. Macro-micro degradation process of fly ash concrete under alternation of freeze-thaw cycles subjected to sulfate and carbonation. *Constr. Build. Mater.* **2018**, *181*, 369–380. [CrossRef]
15. Zhao, R.; Yuan, Y.; Cheng, Z.; Wen, T.; Li, J.; Li, F.; Ma, Z.J. Freeze-thaw resistance of Class F fly ash-based geopolymer concrete. *Constr. Build. Mater.* **2019**, *222*, 474–483. [CrossRef]
16. Ma, Z.; Zhu, F.; Ba, G. Effects of freeze-thaw damage on the bond behavior of concrete and enhancing measures. *Constr. Build. Mater.* **2019**, *196*, 375–385. [CrossRef]
17. Zhang, P.; Cong, Y.; Vogel, M.; Liu, Z.; Müller, H.S.; Zhu, Y.; Zhao, T. Steel reinforcement corrosion in concrete under combined actions: The role of freeze-thaw cycles, chloride ingress, and surface impregnation. *Constr. Build. Mater.* **2017**, *148*, 113–121. [CrossRef]
18. Zhang, S.; Chen, B.; Tian, B.; Lu, X.; Xiong, B. Effect of Fly Ash Content on the Microstructure and Strength of Concrete under Freeze–Thaw Condition. *Buildings* **2022**, *12*, 2113. [CrossRef]
19. Zhang, S.; Tian, B.; Chen, B.; Lu, X.; Xiong, B.; Shuang, N. The Influence of Freeze–Thaw Cycles and Corrosion on Reinforced Concrete and the Relationship between the Evolutions of the Microstructure and Mechanical Properties. *Materials* **2022**, *15*, 6215. [CrossRef]
20. Papadakis, V.G. Effect of supplementary cementing materials on concrete resistance against carbonation and chloride ingress. *Cem. Concr. Res.* **2000**, *30*, 291–299. [CrossRef]
21. Heidari, M.; Salaudeen, S.; Arku, P.; Acharya, B.; Tasnim, S.; Dutta, A. Development of a mathematical model for hydrothermal carbonization of biomass: Comparison of experimental measurements with model predictions. *Energy* **2021**, *214*, 119020. [CrossRef]
22. Van den Heede, P.; De Belie, N. Durability based life cycle assessment of concrete with supplementary cementitious materials exposed to carbonation. *Int. Conf. Sustain. Struct. Concr. Proc.* **2015**, I-13–I-24.
23. Heede, P.V.D.; De Belie, N. A service life based global warming potential for high-volume fly ash concrete exposed to carbonation. *Constr. Build. Mater.* **2014**, *55*, 183–193. [CrossRef]
24. Li, S.; Chen, Y.; Tang, C.; Wang, J.; Liu, R.; Wang, H. Experimental and Theoretical Study on Carbonization Coefficient Model of NS/SAP Concrete. *Buildings* **2022**, *12*, 2227. [CrossRef]
25. Kruse, A.; Grandl, R. Hydrothermal: 3. Kinetic model, Chemie-Ingenieur-Technik. *Chem. Ing. Tech.* **2015**, *87*, 449–456. [CrossRef]
26. Tan, Y.; Ma, C.; Zhao, B.; Xiong, W.; Chen, X.; Yu, J. Study and Microanalysis on the Effect of the Addition of Polypropylene Fibres on the Bending Strength and Carbonization Resistance of Manufactured Sand Concrete. *Polymers* **2023**, *15*, 2139. [CrossRef] [PubMed]
27. Ou, G.; Xin, Z.; Chengkai, Z. Prediction models of the carbonization depth of recycled concrete. *Zhongguo Kuangye Daxue Xuebao/J. China Univ. Min. Technol.* **2015**, *44*, 54–58.
28. Zhang, J.-S.; Cheng, M.; Zhu, J.-H. Carbonation Depth Model and Prediction of Hybrid Fiber Fly Ash Concrete. *Adv. Civ. Eng.* **2020**, *2020*, 1–11. [CrossRef]
29. Chen, Z.; Hu, Y.C.; Zhao, Y.F.; Yu, B. Multi—Factor Computation Model of Concrete Carbonation Depth Based on Material Parameters in Standard Carbonization Environment. *Bull. Chin. Ceram. Soc.* **2019**, *38*.
30. Li, G.F.; Shen, X.D. A Study of the deterioration law and mechanism of aeolian-sand powder concrete in the coupling environments of freeze-thaw and carbonization. *J. Ceram. Soc. Jpn.* **2019**, *127*, 551–563. [CrossRef]
31. Yang, L.; Zhu, Z.; Sun, H.; Huo, W.; Zhang, J.; Wan, Y.; Zhang, C. Durability of waste concrete powder-based geopolymer reclaimed concrete under carbonization and freeze–thaw cycles. *Constr. Build. Mater.* **2023**, *403*, 133155. [CrossRef]
32. Niu, J.G.; Yan, L.; Zhai, H.T. Study on the influence of freeze-thaw on the carbonation property of fly ash concrete. *Appl. Mech. Mater.* **2013**, *357–360*, 939–943. [CrossRef]

33. Li, Y.; Wang, R.; Zhao, Y. Effect of coupled deterioration by freeze-thaw cycle and carbonation on concrete produced with coarse recycled concrete aggregates. *J. Ceram. Soc. Jpn.* **2017**, *125*, 36–45. [CrossRef]
34. Zhang, R.; Du, L.; Li, Z.; Lin, P.; Cao, F. Design Method For Cemented Sand and Gravel Mix Proportion Using Jaw-Crushed Material. *J. Eng. Sci. Technol. Rev.* **2023**, *16*, 92–99. [CrossRef]
35. *GB/T 50082-2009*; Standard for Test Methods of Long-Term Performance and Durability of Ordinary Concrete. Ministry of Construction of the People's Republic of China: Beijing, China, 2009.
36. Zhang, D.; Wang, Y.; Ma, M.; Guo, X.; Zhao, S.; Zhang, S.; Yang, Q. Effect of Equal Volume Replacement of Fine Aggregate with Fly Ash on Carbonation Resistance of Concrete. *Materials* **2022**, *15*, 1550. [CrossRef] [PubMed]
37. Li, B.; Tian, Y.; Zhang, G.; Liu, Y.; Feng, H.; Jin, N.; Jin, X.; Wu, H.; Shao, Y.; Yan, D.; et al. Author Correction: Comparison of detection methods for carbonation depth of concrete. *Sci. Rep.* **2024**, *14*, 5442. [CrossRef] [PubMed]
38. Niu, D.; Xiao, Q.; Zhu, W. Concrete damage and neutralization under coupling effect of carbonation and freeze-thaw cycles. *J. Wuhan Univ. Technol. Mater. Sci. Ed.* **2012**, *27*, 353–357. [CrossRef]
39. Kuosa, H.; Ferreira, R.; Holt, E.; Leivo, M.; Vesikari, E. Effect of coupled deterioration by freeze–thaw, carbonation and chlorides on concrete service life. *Cem. Concr. Compos.* **2014**, *47*, 32–40. [CrossRef]
40. Li, Y.; Wang, R.; Li, S.; Zhao, Y.; Qin, Y. Resistance of recycled aggregate concrete containing low- and high-volume fly ash against the combined action of freeze–thaw cycles and sulfate attack. *Constr. Build. Mater.* **2018**, *166*, 23–34. [CrossRef]
41. Rao, M.J.; Dong, Y.; Yang, H.Q.; Li, M.X.; Yu, Z. Influence of carbonation and freeze-thaw on macro-properties of concrete. *J. Phys. Conf. Ser.* **2016**, *755*, 011001. [CrossRef]
42. Zhang, K.; Zhou, J.; Yin, Z. Experimental study on mechanical properties and pore structure deterioration of concrete under freeze–thaw cycles. *Materials* **2021**, *14*, 6568. [CrossRef] [PubMed]
43. Chen, B.; Guan, B.; Lu, X.; Tian, B.; Li, Y. Thermal conductivity evolution of early-age concrete under variable curing temperature: Effect mechanism and prediction model. *Constr. Build. Mater.* **2022**, *319*, 126078. [CrossRef]
44. Zhou, H.; Mi, T.; Zhao, C.; Liang, Z.; Fang, T.; Wang, F.; Zhou, Y. Identification Model of Fault-Influencing Factors for Dam Concrete Production System Based on Grey Correlation Analysis. *Appl. Sci.* **2024**, *14*, 4745. [CrossRef]
45. Zhou, M.; Dong, W. Grey relativity correlations between the pore structures and compressive strength of aeolian sand concrete undergoing carbonation and freeze-thaw cycles. *J. Build. Eng.* **2023**, *77*, 107515. [CrossRef]
46. Tian, L.; Chen, J.; Zhao, T. Durability of lining concrete of subsea tunnel under combined action of freeze-thaw cycle and carbonation. *J. Wuhan Univ. Technol. Mater. Sci. Ed.* **2012**, *27*, 779–782. [CrossRef]

materials

MDPI

Article

Integrated Use of Furnace Bottom Ash as Fine Aggregate and Cement Replacement for Sustainable Mortar Production

Waiching Tang [1,*], Ali M. Onaizi [1], Sagheer A. Onaizi [2,3], Umer Sajjad [1] and Yanju Liu [4]

1 School of Architecture and Built Environment, The University of Newcastle, University Drive, Callaghan, NSW 2308, Australia; ali.onaizi@uon.edu.au (A.M.O.); umer.sajjad@newcastle.edu.au (U.S.)
2 Department of Chemical Engineering, King Fahd University of Petroleum and Minerals, Dhahran 31261, Saudi Arabia; onaizi@kfupm.edu.sa
3 Interdisciplinary Research Center for Hydrogen Technologies and Carbon Management, King Fahd University of Petroleum and Minerals, Dhahran 31216, Saudi Arabia
4 Global Centre for Environmental Remediation, The University of Newcastle, University Drive, Callaghan, NSW 2308, Australia; yanju.liu@newcastle.edu.au
* Correspondence: patrick.tang@newcastle.edu.au; Tel.: +61-(2)49217246

Abstract: Recycling fly ash (FA) and furnace bottom ash (FBA) help with reducing greenhouse gas emissions, conserving natural resources, and minimizing waste accumulation. However, research on recycling FBA is progressing more slowly compared to FA. This research aims to investigate the combined use of FBA as a replacement for both fine aggregate and cement and its influence on the performance of mortar. The findings indicated that incorporating 25% FBA as a fine aggregate replacement and 10% or 20% ground FBA (GFBA) as a cement replacement significantly enhanced compressive strength after 28 and 56 days. Flexural strength was comparable to control mortar at 28 days and superior at 56 days. However, increasing the FBA content beyond 25% as a fine aggregate replacement reduced workability and increased porosity, which negatively affected mechanical performance and water absorption. Microstructural analyses revealed denser and more compact structures in the mortar with combined FBA replacement for both fine aggregate and cement, specifically 25% as a fine aggregate replacement and 10% and 20% as cement replacements. Optimal performance was noted in mixtures with Ca/Si and Ca/Al ratios within the ranges of 1.8–1.5 and 0.24–0.19, respectively. Trace element leaching analysis has not shown significant differences between GFBA, FA, and OPC. Regarding environmental impact assessment, using FBA as a fine aggregate replacement did not show a significant reduction in CO_2 emissions, but replacing cement with FBA reduced emissions remarkably. Generally, using FBA as a replacement for both fine aggregate and cement in mortar enhances compressive and flexural strengths at optimal levels, promotes sustainability by reducing landfill waste and CO_2 emissions, and supports cleaner production practices despite some workability challenges.

Keywords: furnace bottom ash; cement replacement; fine aggregate replacement; mechanical properties; environmental impact

1. Introduction

Due to the economic advantages of coal, many countries still rely on the use of coal as the main source of electrical power generation [1,2]. The global proven coal reserves exceeded 1074 billion tonnes in 2020, enough to last more than 130 years at present global production rates [3–5]. According to the Statistical Review of World Energy by BP 2022 [6], global coal production increased from 6087.1 Mt in 2005 to 8185.6 Mt in 2012, with an annual growth rate of around 262 Mt%. The peak production was 8256.1 Mt in 2013, but this then declined to 7476.0 Mt in 2016 due to climate change awareness. Production rose again from 2017, reaching 8172.6 Mt in 2021, driven by economic growth in developing countries like China, India, and ASEAN. Australia ranks third globally, after the US and Russia, with

150.227 billion tonnes, about 14% of global reserves [5]. Coal demand is expected to rise this decade, increasing Coal Combustion Residues (CCPs) [5,7]. In the case of Australia, since May 2022, the Federal Environment Minister has approved four new coal mines, and there are 25 additional proposals for new or expanded coal mines currently waiting for federal government approval [8]. That means that the demand on coal is expected to continue rising at least during this decade [9]. Every year, Australian coal-fired power stations produce 12 million tonnes of ash from burning coal [10]. Coal ash accounts for 18% of Australia's entire solid waste generation, which represents 500 kg per capita annually [10]. According to the World of Coal Ash (WOCA), Asian nations generate over 66% of global coal waste, followed by Europe and the Americas [11,12]. China produces around 395 Mt of coal waste, the US 118 Mt, and India 105 Mt [12]. China and the US recycle about 80% of coal fly ash, while India recycles 78% [13]. Most recycled quantities are used in the building and construction sectors, though specific data on recycled FBA is lacking [12].

Traditional disposal methods, involving open land areas or water bodies, exacerbate environmental pollution through air, water, and soil contamination. Therefore, devising effective solutions to effectively manage these vast amounts of coal ash is essential for sustainable development in Australia. Similar to fly ash (FA), recycling furnace bottom ash (FBA) meets sustainability goals by decreasing greenhouse gas (GHG) emissions, conserving natural resources, and minimizing waste generation. However, it is noteworthy that research progress on FBA is lagging behind that of FA. Zhou et al. [14] identify three main reasons why researchers tend to prioritize FA over FBA: (1) FBA production is significantly lower than FA, comprising only about 20–30% of the total coal ash generated; (2) FBA exhibits less leaching of toxic elements compared to FA; and (3) the heterogeneous nature of FBA complicates its recycling. Nonetheless, appropriate treatments such as sieving and grinding can mitigate many of the quality issues currently hindering FBA recycling.

In the case of using FBA as a sand replacement, most previous studies have noted a decrease in the workability of cementitious composites, and the reduction in workability increases with higher levels of replacement [15–19]. When FBA is utilized as a cement substitute, previous studies have shown contradictory results. Some research [20,21] reported an increase in workability, while others [22–25] observed a decrease. It is crucial to highlight that the impact of FBA on workability is influenced by several factors, including the physical and morphological properties of FBA, the mix design, and the replacement ratios [7]. Additionally, the irregular shape of FBA increases friction, and its porosity leads to higher water absorption, further reducing the mixture's workability [15,17,26]. Also, the workability of concrete and mortar containing FBA as a cement substitute is affected by multiple factors, such as the grinding method, fineness and gradation, type and source of the coal ash, and replacement ratios [7,20,21,25,27,28].

Similarly, previous studies have reported inconsistent findings regarding the compressive strength of mortar and concrete incorporating FBA as a fine aggregate replacement, especially in the early stages. For example, Hasim et al. [29] and Rafieizonooz et al. [30] found that concrete mixes with FBA ranging from 50 to 100% as fine aggregate replacements showed similar compressive strength to that of the control specimen after 28 curing days. In contrast, other studies [17,18,31] reported a reduction in compressive strength for all replacement levels (from 20 to 100%) and at all curing days. Several factors significantly influence strength development, including the replacement level, the type of sand being replaced, and the treatment applied to the FBA before use. Soman et al. [32], Nanda and Rou [19], and Rafieizonooz et al. [18] observed similar trends in flexural and splitting tensile strengths. They reported a slight initial reduction in both flexural and splitting tensile strengths when FBA was used as a fine aggregate replacement, but these strengths improved with extended curing time.

When using FBA as cement replacement, several previous studies [20,21,23,24,27,33–39] have presented inconsistent findings regarding the compressive strength of cementitious composites containing 5–40% FBA as a cement replacement at early stages. However, most of those studies demonstrated acceptable outcomes in later stages. The compressive strength fluctuation generally ranged from a reduction of 30% to an increase of up to 10% at 28 days compared with the control samples. It is worth mentioning that the majority of studies revealed that the compressive strength performance at replacement levels of 10 and 15% was mostly similar to that of the control samples at 28 days, whereas a reduction was evident at higher replacement levels (e.g., at 30%). A similar trend to that of compressive strength was observed for flexural and splitting tensile strengths [24,25,34]. Despite the above-mentioned studies, the utilization of FBA for fine aggregate and cement replacement is still controversial, requiring more in-depth and well-designed studies to resolve the contradictory observations. While previous studies have explored the use of FBA as either cementitious materials or fine aggregate separately, this study focuses on its dual use as both a fine aggregate and cement replacement—a relatively underexplored area. This dual replacement approach aims to enhance the sustainability of mortar. Additionally, the results of the available studies are still contradictory, necessitating further research and investigation. This research aims to determine the optimal replacement ratios and their effects on compressive and flexural strength, workability, porosity, and environmental impact, reinforcing insights into the feasibility and benefits of FBA recycling in cementitious composite production. The significance of this study aligns with several Sustainable Development Goals (SDGs), including SDGs 12 and 13 for responsible consumption and production and SDG 9 for sustainable industrialization and infrastructure innovation. By incorporating FBA in cleaner production applications, particularly in the construction materials industry, we address the challenges of handling and disposing of FBA. This promotes recycling practices, reduces health and social risks from landfill contamination, and contributes to a healthier environment while significantly reducing carbon emissions in the construction sector.

2. Materials and Methods

2.1. Materials Preparation

Traditional ordinary Portland cement (OPC) was obtained from Boral company, Sydney, Australia, conforming to AS 3972 [40] for Type GP and FBA from Vale Point Power Station in NSW were used as the raw materials. To prepare FBA for use as a cement substitute, the FBA was oven-dried for 24 h at 105 °C and sieved using a 2.36 mm sieve. Then, ball milling was conducted using an Across International PQ series planetary ball mill, with the ball milling mode set to unidirectional. A constant speed of 450 rpm was maintained for a grinding period of 30 min. The chemical compositions of OPC and ground FBA (GFBA) were then determined by XRF, as shown in Table 1. OPC primarily consists of CaO (70.91%), with $Al2O_3$, SiO_2, and Fe_2O_3 percentages of 4.64%, 15.30%, and 3.4%, respectively. In contrast, GFBA is mainly composed of SiO_2, Al_2O_3, and Fe_2O_3, which collectively represent around 84.9% of its total mass. The findings from the XRF analysis were reinforced by the EDS analysis shown in Figure 1, which indicates SiO_2, Fe_2O_3, and Al_2O_3 as the dominant oxides in GFBA. It is evident that GFBA complies with the ASTM C618 [41] requirements for Type F pozzolanic materials, especially fly ash (FA), as well as the AS/NZS 3582.1: 2016 [42] standards. The mean sum of its pozzolanic content (SiO_2 + Al_2O_3 + Fe_2O_3) exceeds 70%, achieving 86.7% of the total weight ratio in its chemical composition. Additionally, its calcium content is 2.40%, which falls within the ASTM C618 and AS/NZS 3582.1: 2016 requirements for Type F. Specifically, ASTM C618 sets a maximum of 5% calcium content, while AS/NZS 3582.1: 2016 stipulates it should not exceed 10% (25% for New Zealand specifications) for the three grades (I, II, and Special grade). Figure 2 shows the XRD patterns of FBA, with a semi-quantitative XRD analysis conducted using the EMPYREAN diffractometer system (Cu Kα 1.54 Å) and PANalytical Data Collector software v.4.7. The analysis covered a 2θ range from 10.02° to 79.95°, with a step size of 0.0390° and scan step time of 195.585 s. HighScore Plus software v5.2, along with the ICDD database, was used for peak fitting and matching/searching, enabling semi-quantification of the sample's mineralogical composition. The XRD analysis showed that GFBA primarily comprises silicon oxide (60%), mullite (27%), Feldspar (9%), and others constituents such as Hematite-Proto and Iron Cobalt.

Table 1. Chemical compositions (wt%) of the utilized materials.

	CaO	Al_2O_3	SiO_2	Fe_2O_3	SO_3	Cl	K_2O	Na_2O	TiO_2	MnO	LOI
OPC	69.77	4.31	17.04	3.73	3.63	0.16	0.83	-	0.24	0.17	2.23
GFBA	2.40	15.10	62.37	9.30	0.17	0.17	1.45	0.74	0.58	0.25	5.61

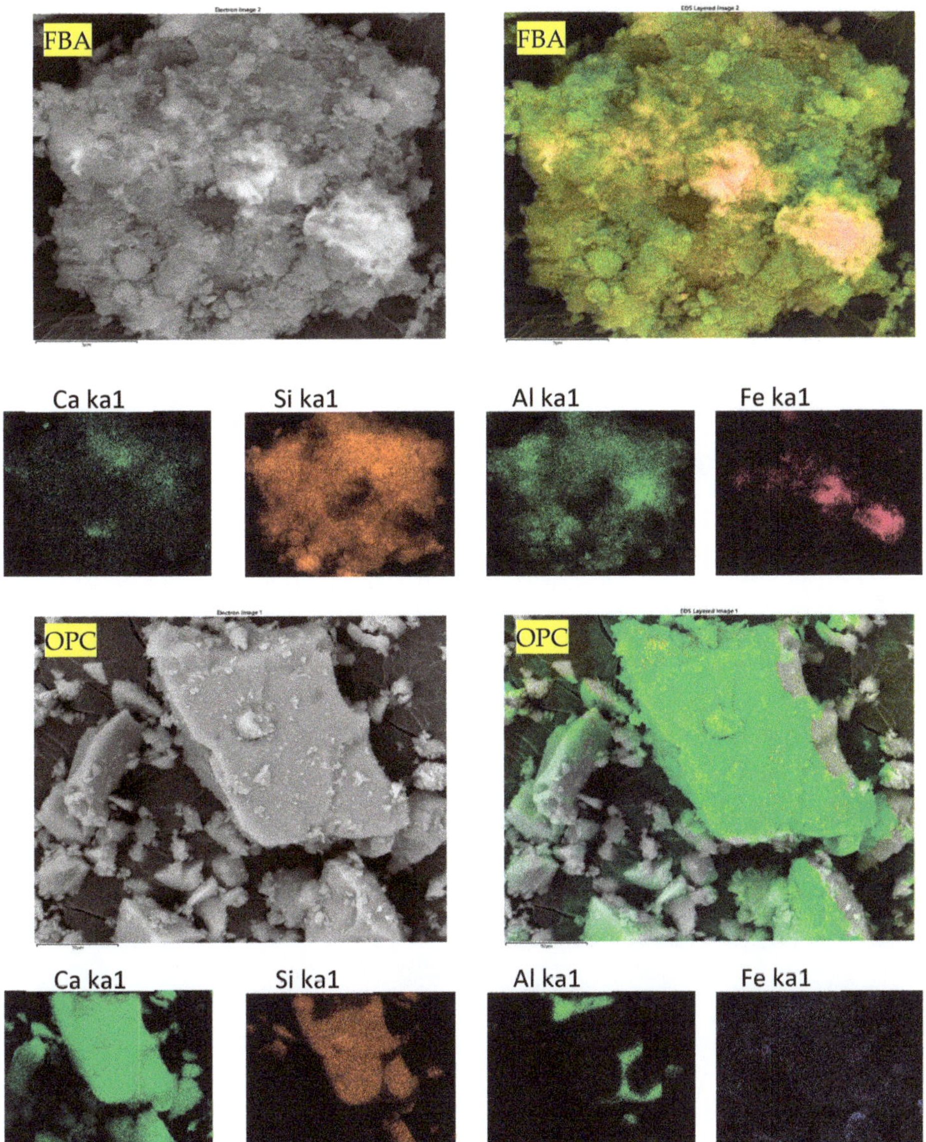

Figure 1. SEM and EDS images of GFBA and OPC.

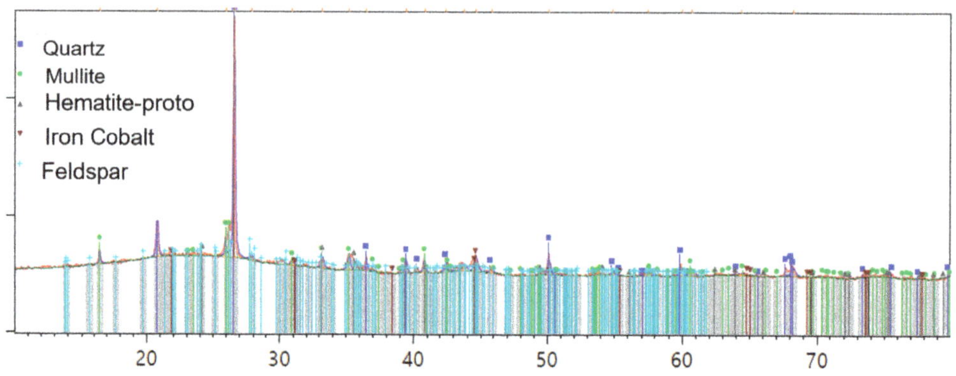

Figure 2. XRD patterns of ground FBA.

Figure 3 illustrates the particle size distribution, while Table 2 presents the physical properties of OPC and GFBA. The properties of particle size distribution and specific surface areas were determined using MasterSizer2000 (Malvern Panalytical Ltd., Malvern, Worcestershire, UK) with water as the dispersant medium, as both materials dissolve easily in water. Additionally, a helium gas pycnometer was used to determine specific gravity. Three samples of each material were measured to determine specific gravity, particle size distribution, and specific surface area, and the averages were taken. The densities of OPC and GFBA were 3.12 and 2.45 g/cm^3, respectively, while the specific surface areas were 8650 and 9857 cm^2/g.

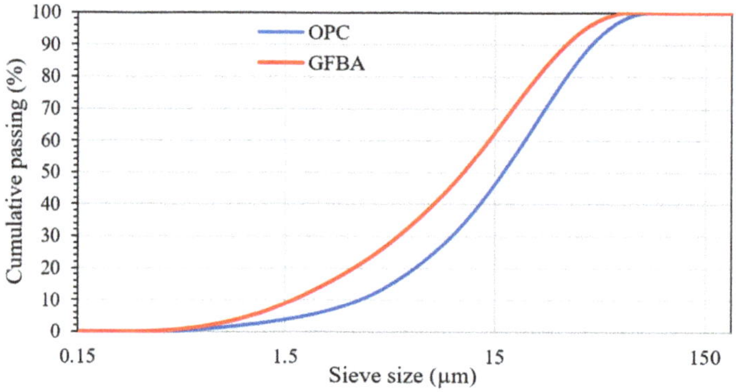

Figure 3. Particle size distribution of binder components.

Table 2. Physical properties of OPC and GFBA.

Result Transform Type	GFBA	Ordinary Portland Cement
Specific gravity g/cm^3	2.45	3.12
Volume weighted mean (μm)	6.17	20.37
Fineness % (passing on 45 μm)	100	82.12
Uniformity	1.28	0.75
Specific surface area (cm^2/g)	9857	8650
Surface weighted mean	4.68	6.94
d (0.1)	1.74	3.37
d (0.5)	13.76	16.48
d (0.9)	58.92	43.44

The fine aggregate consisted of a combination of river sand and fine crushed stones in a 1:1 ratio, sourced locally and sieved through a 4.75 mm sieve. For FBA utilized as a replacement for fine aggregate, it was oven-dried for 24 h at 105 °C and sieved through a 4.75 mm sieve. The particle size distribution of FBA and the fine aggregate, compared to the upper and lower limits of AS 2758.1 [43], is presented in Figure 4. The physical properties of FBA as a fine aggregate alternative and the fine aggregate are shown in Table 3. The fineness moduli of the combinations of fine aggregates and FBA were 4.7 and 4.4, respectively. The specific gravity and water absorption of FBA and the fine aggregate were also determined in accordance with AS 1141.5 [44], while the densities and moisture content were measured according to AS 1141.4 [45] and AS 3583.2 [46], respectively. The water absorption of the combinations of fine aggregates and FBA were 1.2% and 11.30, respectively, while bulk densities were 1650 kg/m^3 and 950 kg/m^3, respectively.

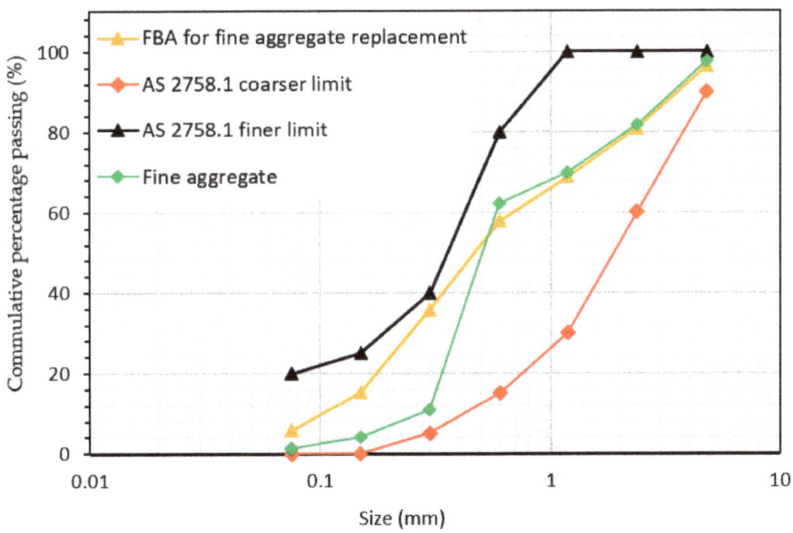

Figure 4. Particle size distribution of FBA and fine aggregates.

Table 3. Physical properties of FBA and fine aggregates.

Property	Unit	Fine Aggregate	FBA	Method
Water absorption	%	1.20	11.30	AS 1141.5 [44]
Clay-fine silt content	%	2.1	3.7	AS 1141.33 [47]
Specific gravity	g/cm^3	2.50	1.45	AS 1141.5 [44]
Compacted bulk density	g/cm^3	1.65	0.95	AS 3583.2 [46]
Uncompacted bulk density	g/cm^3	1.41	0.89	AS 3583.2 [46]
Fineness modulus	/	4.7	4.4	/

2.2. *Mix Proportions and Testing Setup*

Table 4 displays the mix proportions for the proposed mortars. Firstly, four proportions of fine aggregate replacement by FBA were considered: 25%, 50%, 75%, and 100% by volume. Three essential properties of the mortar were utilized to determine the optimal percentage, including flowability, compressive strength, and water absorption. Assessing these properties was crucial for selecting the optimal percentage of FBA as a fine aggregate alternative in the mixture. These parameters were chosen because compressive strength is the primary indicator of mechanical performance, water absorption affects durability, and flowability influences rheological properties. This stage is crucial for achieving the desired balance between workability, structural integrity, and durability of the mortar, ensuring that the FBA-based mortar meets the required performance standards for construction applications. After determining the optimal level of FBA as a fine aggregate replacement, the chosen proportion was used to proceed with mortar groups using FBA as cement replacement. Three different mix formulations were developed to investigate the mortar performance with optimal FBA as fine aggregate (i.e., 25%), and 10%, 20% and 30% GFBA as cement replacements. The water-to-binder (w/b) ratio was fixed at 0.6 for all mixtures. The mortar preparation and sampling were conducted following AS/NZS 2350.12 [48]/AS 2701 [49]. 2015 specifications. Figure 5a,b shows actual photos of prepared samples. Mortar flowability (Figure 5c) was determined through the flow table test according to AS 2701 [49], while compressive strength (Figure 5d) and flexural strength (Figure 5e) were determined according to AS/NZS 2350.11 [50] and AS 2350.12 [48].

Table 4. Mortar mix proportions (kg/1 m^3).

Mixture	Sand	FBA as Sand	GFBA as Cement	Cement	w/c
CR0	1537.5	0	0	467.5	0.6
FBA25	1153.2	234.4	0	467.5	0.6
FBA50	1041.7	604.2	0	467.5	0.6
FBA75	520.8	906.3	0	467.5	0.6
FBA100	0.0	1208.3	0	467.5	0.6
F10	1153.2	234.4	46.75	420.75	0.6
F20	1153.2	234.4	93.50	374.00	0.6
F30	1153.2	234.4	140.24	327.25	0.6

Note: FBA refers to the sieved FBA to be use for fine aggregate replacement. GFBA refers to the ground FBA to be used as cement replacement.

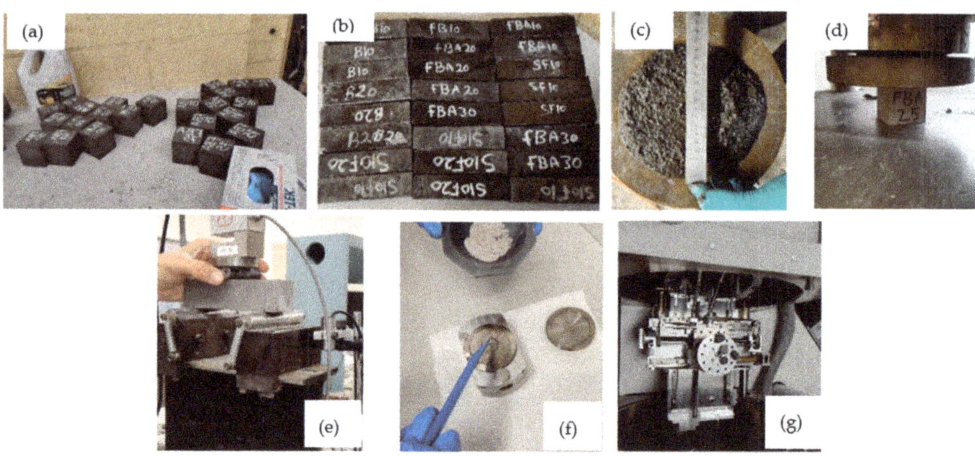

Figure 5. Actual photos for prepared samples and testing setup, where (**a**): mortar cubs, (**b**): mortar prisms, (**c**): flow table test, (**d**): compressive strength test setup, (**e**): flexural strength test setup, (**f**) preparing powder for XRD, and (**g**) SEM setup.

For the XRD analysis, small portions of the interior specimen tested for compressive strength were taken and ground until they passed through a 125 nm sieve, then filled into the testing disk as shown in Figure 5f. For SEM/EDS mapping, interior pieces from the tested specimen were taken and mounted on carbon tape as shown in Figure 5g. The samples were analyzed using an environmental SEM (ESEM) setup to accommodate the high-pressure chamber conditions without coating. To determine the water absorption, the mortar specimens were removed from the water tank at 28 days and kept in the lab atmosphere until the saturated-surface-dry (SSD) condition was achieved. Then, the samples were weighted, and the mass was recorded as m1. Then the mortar samples were dried in the oven at 105 °C for 24 h. After that, the specimens were cooled, the mass was recorded as m^2, and the water absorption was calculated as:

$$WA(\%) = \frac{m_1 - m_2}{m_2} \times 100 \quad (1)$$

2.3. Environmental Impact Assessment

2.3.1. Analysing Emissions

The embodied carbon dioxide (on a mass basis, i.e., CO_2-e kg/m^3 of mortar) for each virtual mix was calculated according to the contribution from each of its constituents, using the values given in Table 5, and described by Purnell and Black [51] and Al-Naqkeeb [52], according to the following equation:

$$CO_2\text{-e kg/m}^3 = \sum (m_i \times f_i) \quad (2)$$

where m_i is the mortar material i in kg/m^3 and f_i is the global warming parameter equivalent to material i production. The f_i values were obtained from Daracon group database information on materials embodied emissions sourced from Infrastructure Sustainability Materials, as shown in Table 5. The value of fi for GFBA was modified based on the treatment conditions applied to the GFBA.

Table 5. Global warming parameter values.

Material	Portland Cement	GFBA for Cement Replacement	Fine Aggregate	FBA as Fine Aggregate Rep.
kg CO_2-e/m^3	0.982	0.121	0.00789	0.00758

2.3.2. Leachability: Inductively Coupled Plasma Test

The analysis was performed in duplicate, and a standard solution calibration curve was established with a slope of 0.99 and an R-square value of 0.9999 according to AS 4439.3 [53]. The continuing calibration verification (CCV) was checked, ensuring recovery ranged from 93.64% to 100%. For quality assurance and quality control (QA/QC) purposes, a standard reference soil was analysed. Fe, Al, K, Ca, Mg and Na were analysed using Inductively Coupled Plasma Optical Emission Spectroscopy (ICP-OES), while the others were analysed using Inductively Coupled Plasma Mass Spectrometry (ICP-MS). The Limits of Reporting (LOR) for ICP-MS and ICP-OES were 0.1 ppb and 0.1 ppm, respectively.

3. Results and Discussion

3.1. Mortar Flowability

Figure 6 illustrates the results of a flow table test to evaluate the influence of FBA as a fine aggregate and partial cement replacement on the flowability of mortar. The results indicate that the mix with no replacement (CR0) exhibited the highest flowability; when FBA was included as a fine aggregate replacement, the mortar flowability decreased. The reduction in mortar flowability increased as the replacement level increased, with reduction percentages of 1.6%, 4.9%, 7.0% and 10.8% corresponding to the inclusion of 25%, 50%, 75%, and 100% of FBA as a fine aggregate replacement, respectively. These reductions are attributable to the high-water absorption of FBA compared to natural fine aggregate, as presented in Section 2.1. The irregular and rough shape of the FBA particles may have had an effect on increasing intermolecular friction and reducing the kinetic energy of the mortar, thus reducing its flowability [54,55]. The results are consistent with observations reported elsewhere [15,17], where the workability reduction tended to increase as the level of replacement increased. In the case of the inclusion of GFBA as a partial cement replacement, the F10 mix showed a 4.9% decrease in flow value compared to CR0, identical to the FBA25 mix. The F20 and F30 mixes experienced a 3.8% and 5.4% reduction compared to CR0 and 2.2% and 3.8% compared to the FBA25 mix, respectively. The decrease in flow diameter values compared to the control samples could have attributed to the rough and irregular surface of GFBA (Figure 1) compared to cement particles. The rough and irregular shape of the GFBA particles may have had an effect of increasing intermolecular friction, unlike the semi-smooth shape of cement particles, thus reducing the mortar flowability. Furthermore, the high fineness of GFBA particles as shown in Table 2 means they to tend to fill the inner area of the mixture particles, thus reducing the area available for particles dynamics; therefore, an excessive amount of water is needed to lubricate the mixture to give it enough flowability [56]. These findings are consistent with several studies that have reported decreased flow values in mixes containing GFBA, such as the studies by Abdulmatin et al. [23], Abbas [24], and Aydin [25]. These researchers found that the reduction in mortar flowability gradually increased with increased GFBA content.

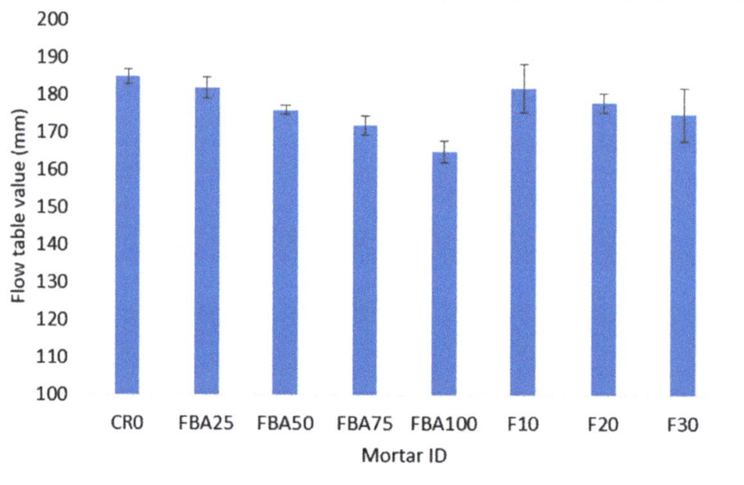

Figure 6. Flow diameter of designed mortars.

3.2. Mortar Densities

Figure 7 presents the densities of the prepared mortar samples measured at 7, 28, and 56 days, with varying percentages of FBA as fine aggregate and cement replacements. From Figure 7, the density decline is obvious in the case of using FBA as a fine aggregate. For the FBA0 mix, which had no FBA replacement, the densities were 2.34, 2.34, and 2.32 kg/m^3 at 7, 28, and 56 days, respectively. The FBA25 mix has densities of 2.20, 2.19, and 2.17 kg/m^3 at the respective time points, showing a similar gradual decrease. This trend was consistent across the mixes with higher FBA contents; for instance, the FBA100 mix, in which the fine aggregate was fully replaced by FBA, recorded the lowest densities across all the time points—1.81, 1.78, and 1.80 kg/m^3. The reduction in the mortar density with the increase in FBA content might be due to the lower density of FBA compared to the fine aggregate it replaces. In the case of the inclusion of GFBA as partial cement replacement, there was no regular trend in densities. For example, F10 showed a 0.6% higher density compared to FBA25 at 28 days. This could be attributed to the filling effect, leading to a more compact matrix. The reduction extent of densities was lower than that of using FBA as a fine aggregate. However, the F20 mortar showed similar densities to those of FBA25, while F30 showed lower densities by around 1%. Compared to previous studies, Mangi et al. [57] and Oruji et al. [27] found that mortar densities decreased with increasing replacement levels of cement by GFBA. They attributed this to the lower specific gravity of GFBA compared to Portland cement. Similarly, in the case of fine aggregate replacement, several studies [58–60] indicated a decrease in mortar density when FBA was used as a sand substitute. This is due to the lower specific gravity of FBA, which ranges from

1.3 to 2.2 g/cm^3, as summarized by Onaizi et al. [7], compared to approximately 2.5 to 2.65 for conventional aggregates. These findings are consistent with the results of this study, which observed a clear reduction in mortar densities with increased fine aggregate replacement due to the lower density of FBA (1.45 g/cm^3) compared to 2.5 g/cm^3 for the natural fine aggregate.

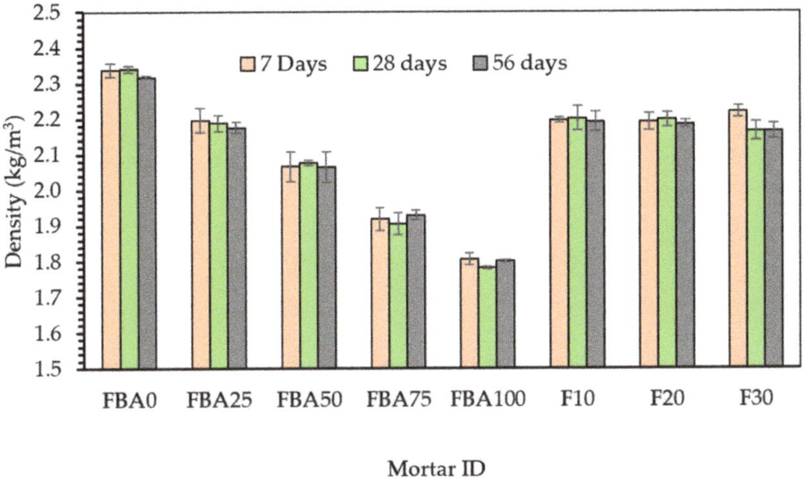

Figure 7. Densities of the designed mortars.

3.3. Compressive Strength

Figure 8 depicts the compressive strength of the mortar samples with various replacement levels of FBA over time, specifically at 7, 28, and 56 days. The control mix (CR0), without any FBA replacement exhibited an initial compressive strength of 25.4 MPa at 7 days, which substantially increased to 34.9 MPa at 28 days and reached 38.9 MPa at 56 days. In contrast, the FBA25 mix, with a 25% replacement of fine aggregate by FBA, started with a higher initial strength of 27.2 MPa at 7 days, surpassed CR0 at 28 days by 7.2%, and achieved the highest compressive strength among all mixes at 56 days with 42.3 MPa. For the FBA50, FBA75, and FBA100 mixes, there was a slight decrease in strength at all time points compared to CR0, and the strength reduction increased with increase in fine aggregate replacement level. However, FBA50 achieved higher strength by 5.9% compared to the control sample without FBA. This could be attributed to the partial pozzolanic reaction between the fine particles of FBA and $Ca(OH)_2$, resulting in the enhancement of the early strength [61]. This potentially occurs with all mixes containing fine FBA as sand; however, at high replacement levels, the negative effect of FBA porosity and weak FBA particles likely outweighed that improvement. When 10% and 20% GFBA were used as cement replacements with 25% of FBA as a fine aggregate replacement, there was more enhancement of compressive strength. The compressive strength of both mixes, F10 and F20, surpassed those of CR0 and FBA25 at 28 and 56 days. This indicates the effectiveness of co-incorporation of GFBA as partial replacements for both cement and fine aggregate. This improvement might be correlated to improvement of microstructures, which provides a more compacted and stronger bonding possibility between the binder and the aggregate. Meanwhile, the finer particles and those used as a substitute for cement reacted with the $Ca(OH)_2$ produced from the initial cement hydration, providing additional hydrates [62]. The relevant studies [63,64] indicate that the improvement in the strength of cementitious mixtures is enhanced by the growth of AFt, which significantly improves the microstructures of the matrices. This was specifically observed in this study for the F10 and F20 samples (as discussed in Section 3.6). These results align with previous observations reported in the literature. For instance, a study [65] indicated that compressive strength associated with a 15% replacement level of FBA surpassed that of control specimens by approximately 17% at 28 days. Similarly, another study [66] reported that 10% GFBA increased compressive strength by around 10% at 28 days. Moreover, another study [24] noted comparable compressive strength to control specimens at 28 days for specimens made with 10% and 20% FBA. These findings suggest that finely ground coal bottom ash can improve the compressive strength of mortars by up to 20% by mass substitution compared to a control mortar without GFBA. However, these findings contradict some earlier studies [20,23,67] that reported a reduction in mortar strength regardless of the replacement ratios. It was suggested that the lower early-stage strength gain was due to the lower activity of the GFBA, which retards the cement hydration process [7,59]. However, it is important to note that these previous studies focused on the use of GFBA as a cement substitute, and there are currently no results available on the combined replacement of fine aggregate and cement.

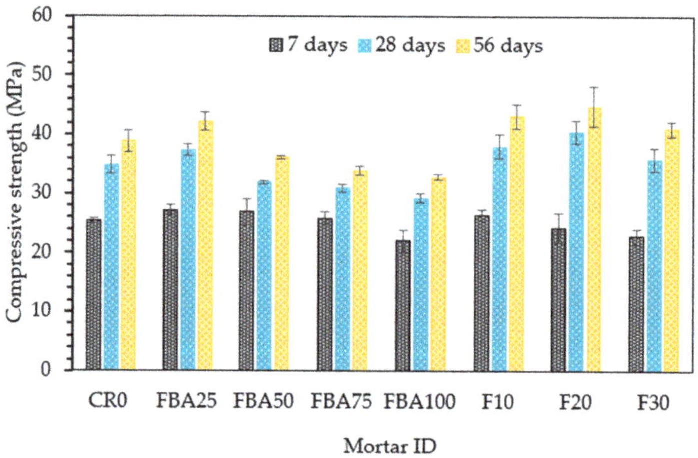

Figure 8. Compressive strength performance of FBA-based mortars.

3.4. Flexural Strength

Figure 9 illustrates the flexural strength of the designed FBA-based mortar samples at 28 and 56 days. All samples containing FBA (25% of FBA as fine aggregate and 20% and 30% as cement replacements) demonstrated lower strength compared to CR0 at 28 days. For example, FBA25 demonstrated a 13.3% reduction in strength compared to CR0 at 28 days, while it showed similar flexural strength to that of CR0 at 56 days. The FBA25 sample exhibited a substantial improvement of 20.51%, with flexural strength increasing from 3.9 MPa to 4.7 MPa over the same period. At 28 days, F10 closely matched the control sample CR0, indicating comparable early strength, while F20 and F30 showed slightly lower flexural strength. After 56 days, both F10 and F20 showed significant strength increases, with F10 (5.2 MPa) being stronger than both CR0 (8.3%) and FBA25 (10.6%), proving that co-incorporating GFBA as a partial replacement for fine aggregate and cement was more beneficial. These findings are consistent with those reported in other studies [68–70], which demonstrated that the optimal GFBA replacements are lower than 25% and the enhancement in flexural strength usually appears after 56 days. This finding is relatively compliant with what was recorded in previous studies [24,25,34] that recorded a comparable flexural strength for control specimens at low replacement levels, with a reduction in flexural strength starting to appear at increased GFBA replacement amounts above 20%. For example, Abbas et al. [24] found that specimens with 10% FBA had comparable flexural strength to control samples, while higher FBA content resulted in reduced strength. Similarly, Kurama and Kaya [34] reported a 10.8% increase in flexural strength at 56 days with 10% FBA as a cement replacement.

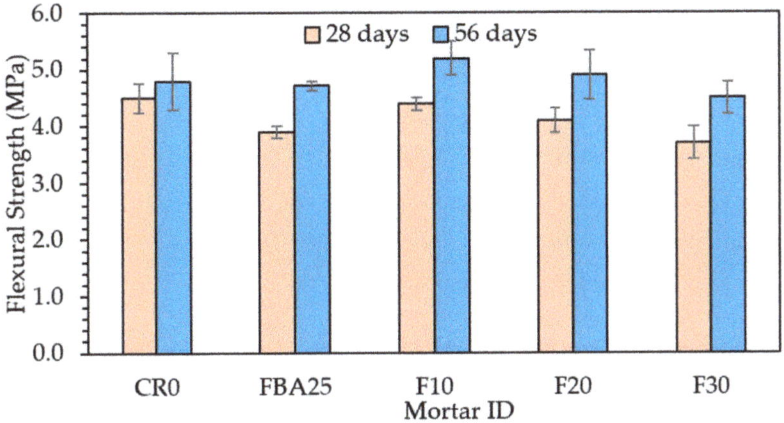

Figure 9. Flexural strength of the designed FBA-based mortars.

3.5. Water Absorption

Figure 10 shows absorbed water percentages for mortar samples with FBA as partial alternatives of fine aggregate and cement at 28 days. The water absorption rate of CR0 was 4.61%. When FBA is used as fine aggregate replacement, the percentages of absorbed water increased, and the increment was boosted by increasing the replacement level. Starting with FBA25, the increment was 8.2% compared to that of CR0. This trend reversed slightly with FBA75, where water absorption decreased from 5.98 for FBA50 to 5.54% for FBA75. The higher water absorption rate was observed for the mortar sample with 100% replacement of the fine aggregate with FBA; for this sample, the water absorption was 6.58%, which was higher than that of CR0 by 42.7%. The increase in water absorption with increasing FBA content could be attributed to the intrinsic properties of FBA, particularly its porosity, which might contribute to a greater overall porosity of mortar and thus higher water absorption. However, with the co-incorporation of FBA, 25% as fine aggregate replacement and 10%, 20%, and 30% GFBA as partial cement replacements, the water absorption rates were slightly lower than the FBA25 sample. This improvement in water absorption resistance can be attributed to both the pozzolanic reaction and the filling action by the fine particles. It can be claimed that the high fineness boosted the pozzolanic reactivity of GFBA particles due to the availability of higher surface area per volume, which resulted in the C-S-H gel filling the pores, hence, reducing water absorption [39]. This finding contradicts that reported by Bheel et al. [65], who reported that the water absorption regularly decreased with an increase in the FBA content for all mixture groups. The water absorption decreased from 3.8% (for the control sample) to 1.98%, 1.75%, 1.6%, and 1.52% when cement was replaced with 10%, 20%, 30%, and 40% FBA, respectively. However, it is worth noting that the majority of studies [71,72] conducted on the use of FBA as an alternative to fine aggregate recorded an increase in water absorption due to the high porosity of FBA particles compared to natural aggregate.

The relationship between the compressive strength and water absorption of the designed mortars at 28 days is illustrated in Figure 11. The water absorption was found to be inversely proportional to compressive strength, whereby the water absorption dropped from 6.6 to 4.5% and the strength increased from 29.2 to 40.5 MPa. The exponential regression method was applied to correlate the experimental data with the R^2 value of 0.8002, indicating a significantly good confidence in the relationships.

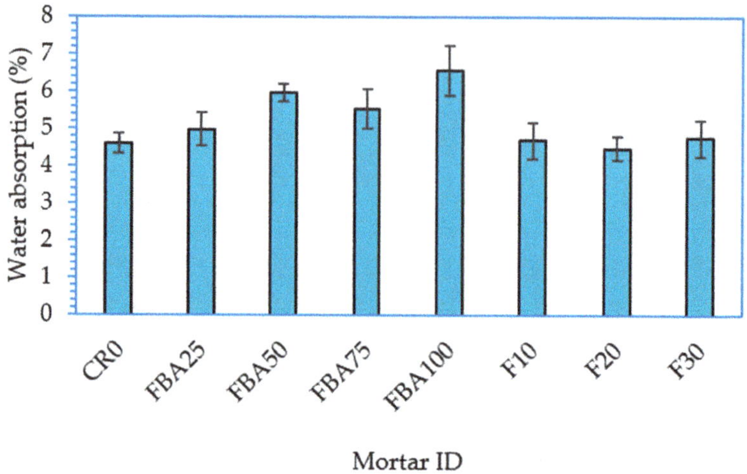

Figure 10. Water absorption of the designed FBA-based mortars at 28 days.

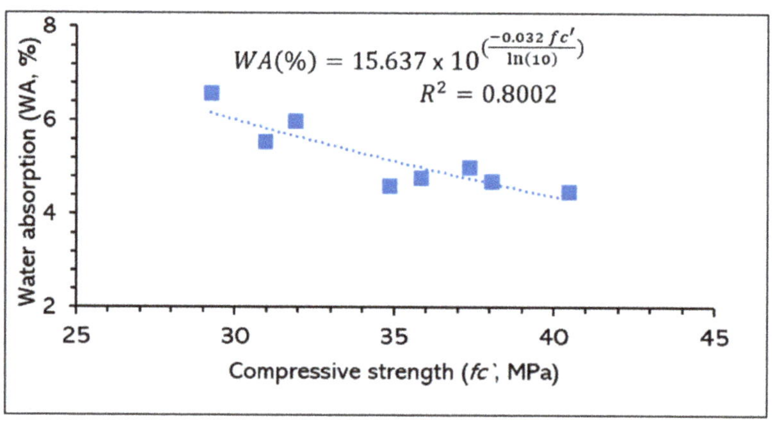

Figure 11. Relationship between compressive strength and water absorption of designed mortars.

3.6. Microstructure Analysis and Discussion

Figure 12 shows the SEM images of the CR0, FBA25, F10, F20, F30, and FBA100 mortars at 28 days. FBA25 shows noticeable pores in some regions within the matrix, potentially due to the presence of the more porous FBA particles. Overall, FBA25 has a more uniform microstructure than CR0. The number of these pores increased with increasing FBA content as a replacement for fine aggregate, as illustrated in Figure 12f for the FBA100 sample. When 25% FBA was used as fine aggregate replacement and 10%, 20%, and 30% ground FBA were used as partial cement alternatives, the microstructure became even more compact and uniform. This could be attributed to the finer particles of GFBA filling the pores and cracks in the mortar matrices, positively impacting the mechanical performance of F10, F20, and F30 samples.

The ability of FBA to absorb water may also play a role in reducing the available water-to-cement ratio (w/c) in the mixtures, enhancing compactness of the mortar matrixes and interfacial transition zones. Wang et al. [73] indicated that in cementitious composites, an excess amount of water beyond what is necessary for cement hydration leads to increased voids in the hardened pastes. This may explain the homogeneity of the microstructures in F10 and F20 compared to CR0. However, with the increased content of FBA as a replacement for fine aggregate, as in the FBA100 sample, or as a substitute for cement, as in the F30 sample, the negative effects of high porosity and a weaker pozzolanic reaction compared to cement gradually negated the positive effects of filling and controlling the water-to-cement ratio available for hydration and early pozzolanic reaction.

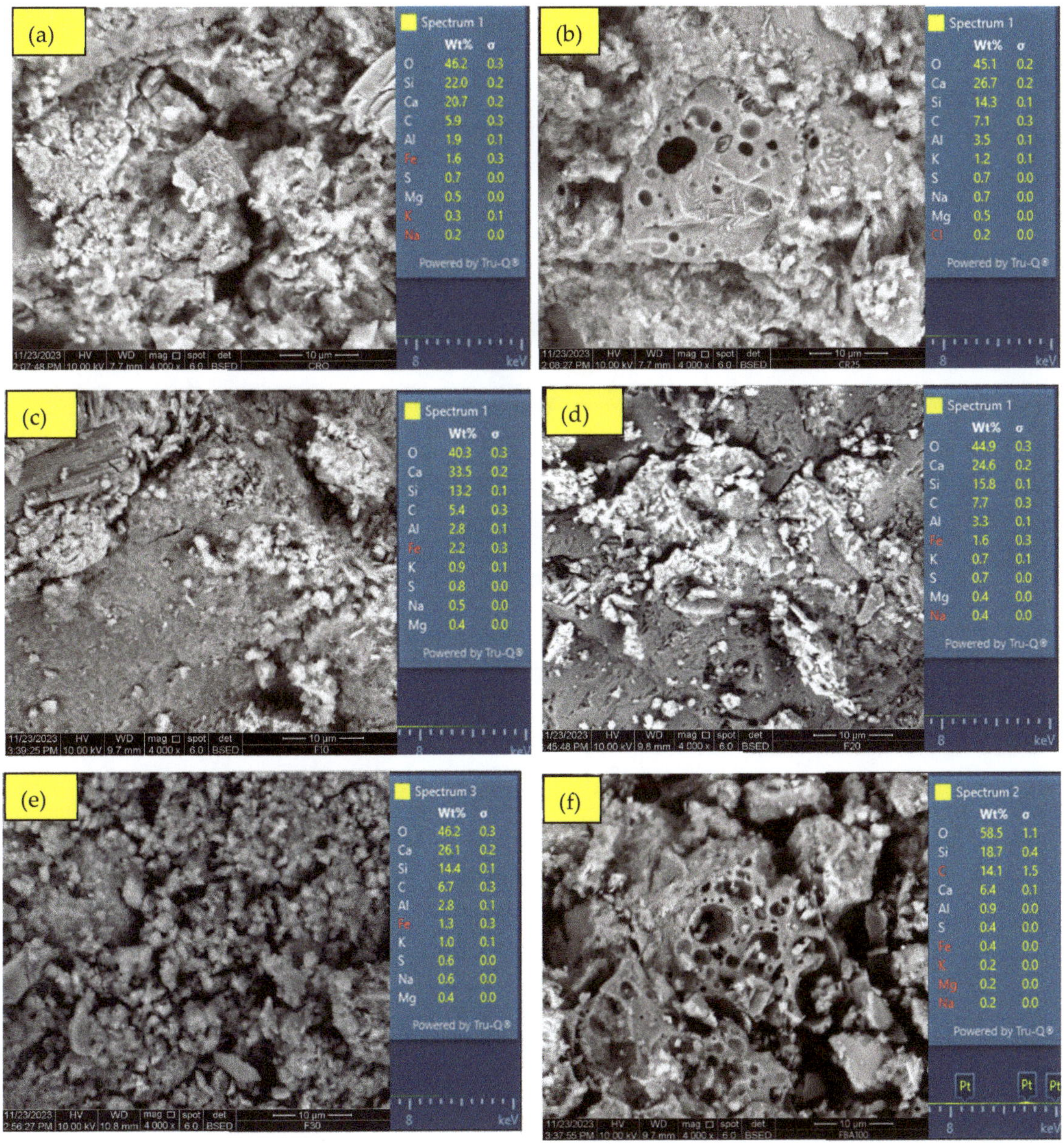

Figure 12. SEM imaging and sectional EDS element mapping of mortar, where (**a**) CR0; (**b**) FBA25; (**c**) F10; (**d**) F20; (**e**) F30; and (**f**) FBA100.

The average elemental mapping from scanned 500-micron areas is also shown in Figure 12. The EDS mapping demonstrated that Ca decreased with increasing level of FBA either as a fine aggregate or cement alternative. In contrast, the concentrations of Si and Al increase with the increased inclusion of FBA. Figure 13 shows the ratios of Ca/Si, Ca/Al/Si/Al, and modifiers/formers network elements. The highest Ca/Si ratio was recorded for control mortar (CR0), where it was 2.54, which then dropped to 1.87 when 25% of FBA was added as a fine aggregate alternative. The drop in Ca/Si continued with the increasing inclusion of FBA, where it reached 0.94 for F30 and 0.34 for FBA100. When correlating the Ca/Si ratio to strength performance, it can be concluded that the ideal range for the Ca/Si ratio to obtain optimum strength is between 2 and 1, as shown in Table 6.

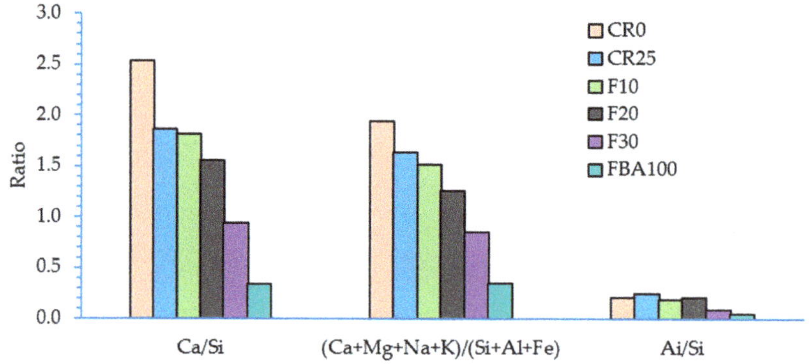

Figure 13. Ca/Si, Ai/Si, and (Ca + Mg + Na + K)/(Si + Al + Fe) ratios in the designed mortars.

Table 6. Summary of Ca/Si, Ca/Al, Ai/Si, and (Ca + Mg + Na + K)/(Si + Al + Fe) and the corresponding compressive strength.

Mortar	Compressive Strength (MPa)	Ca/Si	Ca/Al	Ai/Si	(Ca + Mg + Na + K)/(Si + Al + Fe)
CR0	34.85	2.54	11.96	0.21	1.94
FBA25	37.37	1.87	7.63	0.24	1.63
FBA100	29.25	1.81	9.32	0.19	1.52
F10	38.07	1.56	7.45	0.21	1.26
F20	40.47	0.94	10.89	0.09	0.85
F30	35.86	0.34	7.11	0.05	0.35

Hydration of pure Portland cement results in a C–S–H phase with a Ca/Si ratio of approximately 1.7 [74], which decreases with increasing replacement by SCMs. As SiO_2 replacement increases, the Ca/Si ratio of C–S–H decreases significantly, ranging from 0.67 to 2.0 [75,76]. Kunther et al. [77] demonstrated that compressive strength increases with decreasing Ca/Si ratio across all samples and ages. Low Ca/Si C–S–H pastes (Ca/Si = 0.83 and 1.00) showed significant strength increases over time, whereas high Ca/Si C–S–H pastes showed only small increases. After 3 months, the compressive strength of the Ca/Si = 0.83 binder was more than double that of the highest Ca/Si ratio binder. However, the range of Ca/Si ratio in this study was between 0.83 and 1.5. Two hypotheses explain this correlation [77]:

1. Low Ca/Si C–S–H hydrates have denser microstructures, leading to lower porosities and higher compressive strengths, as lower Ca/Si ratios result in higher surface areas.
2. The density and molar volume of the C–S–H phase are correlated with the Ca/Si ratio; lower Ca/Si ratios result in lower molar volumes and higher densities, contributing to higher strengths.

Despite the consensus among researchers regarding the impact of both Ca/Si and Al/Si ratios on the mechanical properties of cementitious composites, the nature of this relationship remains debated. Some studies [78,79] suggest that the values of the modulus of elasticity and creep increase with the rise in Ca/Si ratio, which might explain the higher flexural strength of CR0 compared to other samples. Conversely, other studies indicate optimal Ca/Si ratios, deviations from which negatively affect mechanical properties. For instance, Wang et al. [80] found that the typical Ca/Si ratio, when incorporating fly ash, is around 1.4, while Garcia's results [81] suggest a value closer to 1.8. Wang [80] also noted that the optimal Al/Si ratio is approximately 0.25, which is close to the values observed for the mortar samples with optimal performance obtained in this study: 0.24, 0.19, and 0.21 for FBA25, F10, and F20 samples, respectively. This ratio decreased to 0.09 and 0.05 for the F30 and FBA100 samples, respectively, and this decrease falls within the range defined by Wang et al. [80] as 0.5 to 0.05. Similarly, the Ca/Si ratio also declined to 0.94 and 0.34 for the F30 and FBA100 samples, respectively.

It is also important not to ignore the impact of the ratios of network modifiers to network formers due to their role in modifying pore alkalinity and participating in various complex hydrate chains. In this study, a consistent decrease in the ratio of network modifiers to network formers was observed with the increased incorporation of GFBA. This decrease might have resulted from the reduction of calcium and the increase in amorphous silica in the mixtures containing GFBA. Similar to the cases of Ca/Si and Al/Si ratios, this decrease can be beneficial to mechanical performance up to a certain point, as it involves the consumption of $Ca(OH)_2$ in forming hydrate chains. However, a sharp decrease might lead to lower pore solution alkalinity, which hinders the dissolution of pozzolan particles. Previous studies indicate that increased alkali content accelerates cement hydration, enhancing strength development [82–84]. This acceleration occurs because alkali cations in the fresh cement mixture's liquid phase accelerate C3A hydration by releasing Ca^{2+} cations [84–86]. It is also known that increasing glass network modifiers boosts the alkaline medium, enhancing dissolution of glass phase activation [87,88]. However, some studies have shown that higher alkali content can reduce concrete compressive strength [82,83,86]. This reduction is attributed to a porous microstructure and the lower strength of alkali-containing C–S–H gel in the hardened mixture [84,89,90]. This highlights the need for further research to optimize the balance between glass network modifiers and formers and the conditions that control them.

Figure 14 shows the types of crystalline phases formed in the formulated mortars. As observed, six crystalline phases were detected: quartz, portlandite, calcite, calcium silicate hydrate, AFt, and monosulfate (AFm). The figure indicates an increase in the intensity of monosulfate peaks with the increase in the level of cement replacement by GFBA (Supplementary material), especially at a 30% replacement level. This increase was due to the abundance of alumina dissolved from the GFBA, which can react with ettringite in the presence of gypsum to form monosulfate. A slight decrease in AFm peaks was also observed in the outperforming samples such as F10 and FBA25, which may be attributed to its conversion to AFt. This may be due to the chemical instability of AFm, as it may react with calcium carbonate in the presence of appropriate humidity to turn into AFt [91]. Calcite was also observed, with its peaks primarily distributed between 50° and 27°. Its presence can be attributed to the relative instability of $Ca(OH)_2$, which undergoes accelerated carbonation under atmospheric conditions. The quartz peaks were nearly identical in the mortar samples CR0, FBA25, F10, F20, and F30, but their intensity was more pronounced in the mortar containing a 100% replacement level of fine aggregate with FBA.

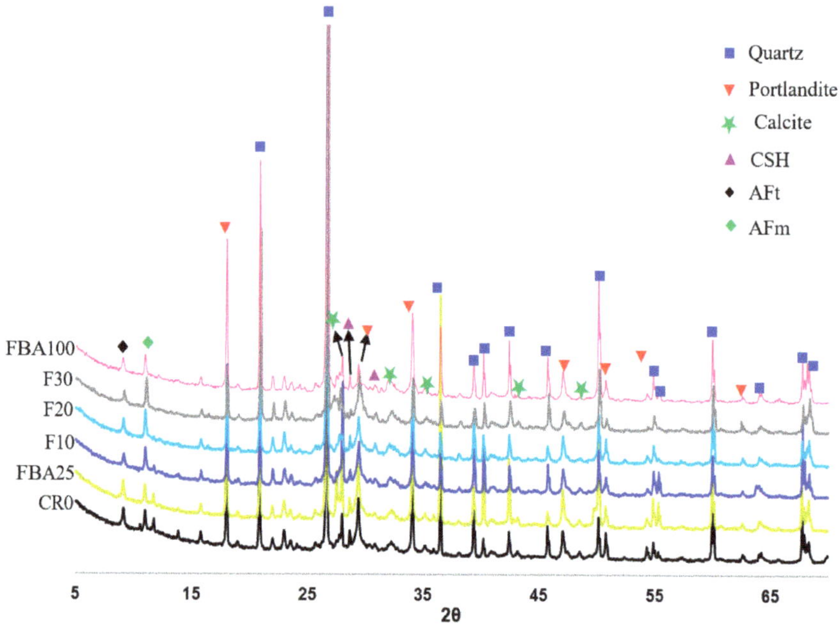

Figure 14. XRD patterns of the designed mortars.

3.7. Environmental Impact Assessment

3.7.1. Leachability

Table 7 provides a comparative analysis of the leachability of trace elements from GFBA, OPC, and FA. The results indicate that nickel was detected in FBA but was absent in both FA and OPC, with its concentration exceeding the upper limits specified by regulations, including those of the EPA NSW, as depicted in Figure 15. Additionally, GFBA exhibited measurable concentrations of Mn, Zn, Cr, Fe, Co, Cu, Mo, Ag, Sn, and Ba, with some elements having lower concentrations than those found in FA or OPC, or both, as shown in Table 7. When comparing concentrations with the regulatory upper limits for trace elements shown in Figure 15, it becomes evident that GFBA surpasses the allowable limits for Ba, Se, Cr, and Ni. This issue is not confined to GFBA alone; certain trace elements in FA and OPC also exceeded the permissible limits established by environmental regulations. For instance, OPC exhibited concentrations of Si, Ba, Cr, and V that were higher than the allowed limits. Further research is required to understand the potential impact of these trace elements on environmental health.

However, previous studies have shown that although FBA and FA exhibit higher concentrations of some trace elements compared to relevant standards, including them in concrete bodies can immobilize their leachability. For example, Sutcu et al. [92] and Eliche-Quesada et al. [93] investigated the leaching concentrations from FBA-based fired bricks and discovered that the brick body effectively immobilized the heavy metals. Similarly, Rafieizonooz et al. [30] noted that the leaching of heavy metals from concrete containing up to 100% FBA as a sand replacement was lower than the original FBA concentration and within the US EPA's SW-846 andard limits [94]. This indicates that the concrete body immobilized the heavy metals sufficiently, rendering them negligible. From the above, it can be concluded that the inclusion of FBA in concrete composites, similar to FA, significantly reduced the concentration of leached heavy metals compared to using FBA or FA alone.

Table 7. Concentrations (in mg/kg) of detected elements in FA, OPC, and GFBA.

Element	FA	GFBA	OPC
Be	0.59	0.35	0.43
V	11.9	6.2	22.4
Cr	NA	610	11.8
Mn	56	251	216
Fe	6273	44152	32292
Co	0.48	6.21	1
Ni	NA	750.86	NA
Cu	1.47	38.5	5.52
Zn	NA	NA	1.88
As	1.88	0.39	0.93
Se	0.91	0.46	0.34
Sr	43.2	34.1	68.6
Mo	2.36	23	1.08
Ag	0.05	0.24	0.2
Cd 111	0.06	0.04	0.48
Cd 114	0.1	0.09	0.46
Sn	2.89	4.71	3.21
Sb	0.81	0.2	0.58
Ba	1158	1599	205
Pb	0.6	0.3	0
Al	5428	9519	14029
K	2239	2630	3237
Ca	17962	16357	15144
Mg	978	2388	3870
Na	594	2085	3594

Note: NA indicates that the element was not detected.

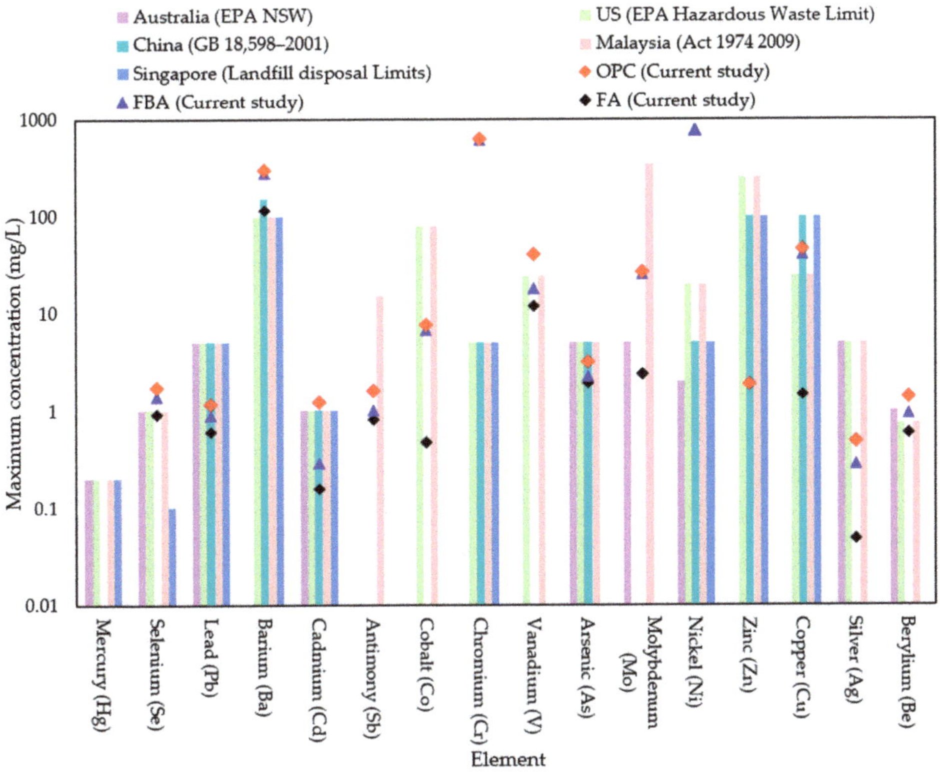

Figure 15. Trace element concentrations from FBA leachate and the maximum limits allowed in the relevant standards/regulations to classify solid waste as non-hazardous waste. The maximum allowable limits for Australia [95], the US [94] China [96], Malaysia [97], and Singapore [98,99] are shown in the figure.

3.7.2. Equivalent Emissions

Figure 16 illustrates the equivalent carbon emissions for each mortar mix. It was observed that replacing fine aggregate with FBA did not result in a significant change in the amount of embodied carbon for the FBA25, FBA50, FBA75, and FBA100 mixes compared to the control sample CR0. This is attributed to global warming parameter values of natural fine aggregate (0.00789) and FBA aggregate (0.00758). The reduction rates were 0.26%, 0.53%, 0.80%, and 1.1% when 25%, 50%, 75%, and 100% of the fine aggregate was replaced with FBA, respectively. When the replacement ratio of fine aggregate with FBA was fixed at 25% and cement was replaced with 10%, 20%, and 30% GFBA, the equivalent carbon dioxide decreased from 469.87 kg/m^3 to 428.36, 386.85, and 345.33 kg/m^3, respectively. This corresponds to reductions of approximately 9.1%, 17.9%, and 26.7%, respectively, compared to CR0. This is consistent with the conclusions of previous studies, which indicated that CO_2 emissions decreased linearly with increasing FBA content as a cement replacement [100,101].

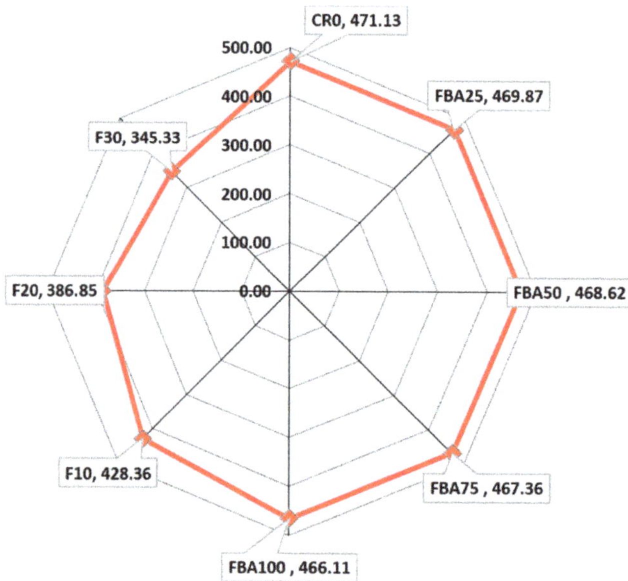

Figure 16. Embodied emissions (in kg/m^3) from each mix.

4. Conclusions

In this study, the impact of co-incorporating FBA as a replacement for fine aggregate and cement on mortar performance was examined, focusing on flowability, compressive strength, flexural strength, water absorption, and microstructural characteristics. The combined incorporation of 25% FBA as a fine aggregate replacement and 10% or 20% GFBA as a cement replacement significantly improved the compressive strength of the mortar after 28 and 56 days of curing, surpassing the control samples and sample with 25% of FBA as a fine aggregate replacement, though no significant variation was observed at 7 days. When the replacement level of fine aggregate exceeded 50%, notable decreases in compressive strength were observed.

The flexural strength of mixtures with 25% FBA as a fine aggregate replacement alone and 25% FBA combined with 30% GFBA as cement replacement was similar to conventional mortar at 56 days, with a slight decrease at 28 days. However, samples with 25% FBA as fine aggregate and 10% and 20% GFBA as cement replacement showed flexural strength similar to traditional mortar after 28 days and improved strength after 56 days. Water absorption rates increased significantly with the inclusion of FBA as a fine aggregate replacement, while there was no noticeable difference in water absorption with 10%, 20%, or 30% GFBA as a cement replacement. There was a notable trend of increased compressive strength with decreasing water absorption rates.

Microstructural analyses revealed denser and more compact structures in mortars with combined FBA replacement, particularly those with 25% fine aggregate and 10% and 20% GFBA as cement replacements, positively impacting their strength performance and water absorption characteristics. EDS analysis showed consistent decreases in Ca/Si and Ca/Al with increasing GFBA content. XRD analysis identified six crystalline phases in the mortars: quartz, portlandite, calcite, calcium silicate hydrate, AFt, and AFm.

In terms of environmental impact, replacing fine aggregate with FBA did not significantly change the amount of embodied carbon. However, fixing the fine aggregate replacement at 25% and replacing cement with 10%, 20%, and 30% GFBA resulted in reductions of approximately 9.1%, 17.9%, and 26.7%, respectively. GFBA exhibited lower concentrations of Mn, Zn, Cr, Fe, Co, Cu, Mo, Ag, Sn, and Ba compared to FA or OPC but exceeded the permissible limits for Ba, Se, Cr, and Ni according to regulatory standards such as the EPA NSW and US EPA for Hazardous Waste. However, based on relevant investigations conducted on concrete and bricks based on FBA, the leaching rates were lower and within permissible ranges. Further research is needed in this aspect.

Supplementary Materials: The following supporting information can be downloaded at: https://www.mdpi.com/article/10.3390/ma17153834/s1.

Author Contributions: Conceptualization, W.T. and A.M.O.; methodology, W.T., A.M.O., U.S. and Y.L.; software, A.M.O.; validation, W.T., A.M.O. and S.A.O.; formal analysis, W.T., A.M.O., U.S. and Y.L.; investigation, W.T., A.M.O., U.S. and Y.L.; resources, W.T.; data curation, W.T., A.M.O. and S.A.O.; writing—original draft preparation, A.M.O.; writing—review and editing, W.T., A.M.O., S.A.O., U.S. and Y.L; visualization, W.T. and A.M.O.; supervision, W.T. and Y.L.; project administration, W.T.; funding acquisition, W.T. All authors have read and agreed to the published version of the manuscript.

Funding: This research was funded by the University of Newcastle and SmartCrete CRC, grant number (21.R1.0042).

Institutional Review Board Statement: Not applicable.

Informed Consent Statement: Not applicable.

Data Availability Statement: The raw data supporting the conclusions of this article will be made available by the authors on request.

Conflicts of Interest: The authors declare that they have no known competing financial interests or personal relationships that could have appeared to influence the work reported in this paper.

References

1. Zhao, Y.; Duan, Y.; Liu, Q.; Cui, Y.; Mohamed, U.; Zhang, Y.; Ren, Z.; Shao, Y.; Yi, Q.; Shi, L. Life cycle energy-economy-environmental evaluation of coal-based CLC power plant vs. IGCC, USC and oxy-combustion power plants with/without CO_2 capture. *J. Environ. Chem. Eng.* **2021**, *9*, 106121. [CrossRef]
2. Shahsavari, A.; Akbari, M. Potential of solar energy in developing countries for reducing energy-related emissions. *Renew. Sustain. Energy Rev.* **2018**, *90*, 275–291. [CrossRef]
3. Heidrich, C.; Feuerborn, H.-J.; Weir, A. Coal combustion products: A global perspective. *VGB PowerTech* **2013**, *22*, 46–52.
4. World Coal Association. Coal Statistics. 2014. Available online: https://www.bp.com/content/dam/bp/business-sites/en/global/corporate/pdfs/energy-economics/statistical-review/bp-stats-review-2022-full-report.pdf (accessed on 27 June 2024).
5. Bp Statistical Review of World Energy. 2022. Available online: https://www.bp.com/content/dam/bp/business-sites/en/global/corporate/pdfs/energy-economics/statistical-review/bp-stats-review-2022-full-report.pdf (accessed on 27 June 2024).
6. Looney, B. Statistical Review of World Energy 2020, 69th ed. 2020. Available online: https://www.bp.com/content/dam/bp/business-sites/en/global/corporate/pdfs/energy-economics/statistical-review/bp-stats-review-2020-full-report.pdf (accessed on 27 June 2024).
7. Onaizi, A.M.; Tang, W.; Amran, M.; Liu, Y.; Sajjad, U.; Alhassan, M. Towards increased adoption of furnace bottom ash as sustainable building materials: Characterization, standardization, and applications. *J. Build. Eng.* **2024**, *82*, 108274. [CrossRef]
8. Research TAI. Coal Mine Tracker. 2023. Available online: https://australiainstitute.org.au/initiative/coal-mine-tracker/ (accessed on 23 June 2024).

9. Staltari, D. The Ash Development Association of Australia. 2020; pp. 41–47. Available online: https://www.accc.gov.au/system/files/public-registers/documents/Submission%20by%20Ash%20Development%20Association%20of%20Australia%20-%2018.06.20%20-%20PR%20-%20AA1000517%20Boral.pdf (accessed on 27 June 2024).
10. Millington, B. Coal ash has become one of Australia's biggest waste problems—And a solution is being ignored. *ABC Newcastle*, 10 March 2019.
11. Mohammed, S.A.; Koting, S.; Katman, H.Y.B.; Babalghaith, A.M.; Abdul Patah, M.F.; Ibrahim, M.R.; Karim, M.R. A review of the utilization of coal bottom ash (Cba) in the construction industry. *Sustainability* **2021**, *13*, 8031. [CrossRef]
12. Singh, N.; Bhardwaj, A. Reviewing the role of coal bottom ash as an alternative of cement. *Constr. Build. Mater.* **2020**, *233*, 117276. [CrossRef]
13. Marinina, O.; Nevskaya, M.; Jonek-Kowalska, I.; Wolniak, R.; Marinin, M. Recycling of coal fly ash as an example of an efficient circular economy: A stakeholder approach. *Energies* **2021**, *14*, 3597. [CrossRef]
14. Zhou, H.; Bhattarai, R.; Li, Y.; Si, B.; Dong, X.; Wang, T.; Yao, Z. Towards sustainable coal industry: Turning coal bottom ash into wealth. *Sci. Total Environ.* **2022**, *804*, 149985. [CrossRef] [PubMed]
15. Singh, M.; Siddique, R. Effect of coal bottom ash as partial replacement of sand on workability and strength properties of concrete. *J. Clean. Prod.* **2016**, *112*, 620–630. [CrossRef]
16. Jamaluddin, N.; Hamzah, A.F.; Ibrahim, M.H.W.; Jaya, R.P.; Arshad, M.F.; Abidin, N.E.Z.; Dahalan, N.H. Fresh properties and flexural strength of self-compacting concrete integrating coal bottom ash. *MATEC Web Conf.* **2016**, *47*, 1010. [CrossRef]
17. Kim, H.K.; Lee, H.K. Use of power plant bottom ash as fine and coarse aggregates in high-strength concrete. *Constr. Build. Mater.* **2011**, *25*, 1115–1122. [CrossRef]
18. Rafieizonooz, M.; Mirza, J.; Salim, M.R.; Hussin, M.W.; Khankhaje, E. Investigation of coal bottom ash and fly ash in concrete as replacement for sand and cement. *Constr. Build. Mater.* **2016**, *116*, 15–24. [CrossRef]
19. Nanda, B.; Rout, S. Properties of concrete containing fly ash and bottom ash mixture as fine aggregate. *Int. J. Sustain. Eng.* **2021**, *14*, 809–819. [CrossRef]
20. Kim, H.-K. Utilization of sieved and ground coal bottom ash powders as a coarse binder in high-strength mortar to improve workability. *Constr. Build. Mater.* **2015**, *91*, 57–64. [CrossRef]
21. Mangi, S.A.; Wan Ibrahim, M.H.; Jamaluddin, N.; Arshad, M.F.; Mudjanarko, S.W. Recycling of coal ash in concrete as a partial cementitious resource. *Resources* **2019**, *8*, 99. [CrossRef]
22. Onaizi, A.M.; Tang, W.; Liu, Y. Co-grinding treatment for developing integrated-properties SCMs from basic oxygen furnace slag and furnace bottom ash: A step toward synthesis advanced SCMs (Under review). *Case Stud. Constr. Mater.* **2024**, *20*, e03163. [CrossRef]
23. Abdulmatin, A.; Tangchirapat, W.; Jaturapitakkul, C. An investigation of bottom ash as a pozzolanic material. *Constr. Build. Mater.* **2018**, *186*, 155–162. [CrossRef]
24. Abbas, S.; Arshad, U.; Abbass, W.; Nehdi, M.L.; Ahmed, A. Recycling untreated coal bottom ash with added value for mitigating alkali–silica reaction in concrete: A sustainable approach. *Sustainability* **2020**, *12*, 10631. [CrossRef]
25. Aydin, E. Novel coal bottom ash waste composites for sustainable construction. *Constr. Build. Mater.* **2016**, *124*, 582–588. [CrossRef]
26. Argiz, C.; Moragues, A.; Menéndez, E. Use of ground coal bottom ash as cement constituent in concretes exposed to chloride environments. *J. Clean. Prod.* **2018**, *170*, 25–33. [CrossRef]
27. Oruji, S.; Brake, N.A.; Nalluri, L.; Guduru, R.K. Strength activity and microstructure of blended ultra-fine coal bottom ash-cement mortar. *Constr. Build. Mater.* **2017**, *153*, 317–326. [CrossRef]
28. Al Biajawi, M.I.; Embong, R.; Muthusamy, K.; Ismail, N.; Obianyo, I.I. Recycled coal bottom ash as sustainable materials for cement replacement in cementitious Composites: A review. *Constr. Build. Mater.* **2022**, *338*, 127624. [CrossRef]
29. Hasim, A.M.; Shahid, K.A.; Ariffin, N.F.; Nasrudin, N.N.; Zaimi, M.N.S. Properties of high volume coal bottom ash in concrete production. *Mater. Today Proc.* **2022**, *48*, 1861–1867. [CrossRef]
30. Rafieizonooz, M.; Salim, M.R.; Mirza, J.; Hussin, M.W.; Khan, R.; Khankhaje, E. Toxicity characteristics and durability of concrete containing coal ash as substitute for cement and river sand. *Constr. Build. Mater.* **2017**, *143*, 234–246. [CrossRef]
31. Singh, M.; Siddique, R. Compressive strength, drying shrinkage and chemical resistance of concrete incorporating coal bottom ash as partial or total replacement of sand. *Constr. Build. Mater.* **2014**, *68*, 39–48. [CrossRef]
32. Soman, K.; Sasi, D.; Abubaker, K.A. Strength properties of concrete with partial replacement of sand by bottom ash. *Int. J. Innov. Res. Adv. Eng.* **2014**, *1*, 2163–2349.
33. Mangi, S.A.; Ibrahim, M.H.W.; Jamaluddin, N.; Arshad, M.F.; Jaya, R.P. Short-term effects of sulphate and chloride on the concrete containing coal bottom ash as supplementary cementitious material. *Eng. Sci. Technol. Int. J.* **2019**, *22*, 515–522. [CrossRef]
34. Kurama, H.; Kaya, M. Usage of coal combustion bottom ash in concrete mixture. *Constr. Build. Mater.* **2008**, *22*, 1922–1928. [CrossRef]
35. Arun, N.R.; Singh, P.; Gupta, S. Utilisation of ground bottom ash in concrete. *Mater. Today Proc.* **2020**, *32*, 663–669. [CrossRef]
36. Cheah, C.B.; Liew, J.J.; Khaw Le Ping, K.; Siddique, R.; Tangchirapat, W. Properties of ternary blended cement containing ground granulated blast furnace slag and ground coal bottom ash. *Constr. Build. Mater.* **2022**, *315*, 125249. [CrossRef]
37. Khan, R.A.; Ganesh, A. The effect of coal bottom ash (CBA) on mechanical and durability characteristics of concrete. *J. Build. Mater. Struct.* **2016**, *3*, 31–42. [CrossRef]

38. Lo, F.-C.; Lee, M.-G.; Lo, S.-L. Effect of coal ash and rice husk ash partial replacement in ordinary Portland cement on pervious concrete. *Constr. Build. Mater.* **2021**, *286*, 122947. [CrossRef]
39. Mangi, S.A.; Wan Ibrahim, M.H.; Jamaluddin, N.; Arshad, M.F.; Ramadhansyah, P.J. Effects of ground coal bottom ash on the properties of concrete. *J. Eng. Sci. Technol.* **2019**, *14*, 338–350.
40. *AS 3972*; General Purpose and Blended Cements. Standards Australia: Sydney, Australia, 2010.
41. *ASTM C. C618-15*; Standard Specification for Coal Fly Ash and Raw or Calcined Natural Pozzolan for Use in Concrete. ASTM International: West Conshohocken, PA, USA, 2015.
42. *AS/NZS 3582.1:2016*; Supplementary Cementitious Materials, Part 1: Fly Ash. Standards Australia: Sydney, Australia, 2016.
43. *AS 2758.1:2014*; Aggregates and Rock for Engineering Purposes, Part 1: Concrete Aggregates. 2014.
44. *AS 1141.5-2000*; Method for Sampling and Testing Aggregates, Method 5: Particle Density and Water Absorption of Fine Aggregate. Standards Australia: Sydney, Australia, 2000.
45. *AS 1141.4-2000*; Methods for Sampling and Testing Aggregates, Method 4: Bulk Density of Aggregate. Standards Australia: Sydney, Australia, 2000.
46. *AS 3583.2-1991*; Methods of Test for Supplementary Cementitious Materials for Use with Portland Cement, Method 2: Determination of Moisture Content. Standards Australia: Sydney, Australia, 1991.
47. *AS 1141.33:2015*; Methods for Sampling and Testing Aggregates, Method 33: Clay and Fine Silt (Settling Method). Standards Australia: Sydney, Australia, 2015.
48. *AS 2350.12-2006*; Methods of Testing Portland, Blended and Masonry Cements, Method 12: Preparation of a Standard Mortar and Moulding of Specimens. Standards Australia: Sydney, Australia, 2006.
49. *AS 2701-2001*; Methods of Sampling and Testing Mortar for Masonry Construction. Standards Australia: Sydney, Australia, 2001.
50. *AS/NZS 2350.11:2006*; Methods of Testing Portland, Blended and Masonry Cements, Method 11: Compressive Strength. Standards Australia: Sydney, Australia, 2006.
51. Purnell, P.; Black, L. Embodied carbon dioxide in concrete: Variation with common mix design parameters. *Cem. Concr. Res.* **2012**, *42*, 874–877. [CrossRef]
52. Al-Naqkeeb, A.S. Relationship between Mix Design, Concrete Performance and Carbon Footprint. 2020. Available online: https://www.google.com/url?sa=t&source=web&rct=j&opi=89978449&url=https://etheses.whiterose.ac.uk/27244/1/A.%2520Al-Naqkeeb.pdf&ved=2ahUKEwjQzuXm9byHAxXnhlYBHYDxMV0QFnoECBsQAQ&usg=AOvVaw3yWBeReldPdCm1lKOnEoTK (accessed on 26 June 2024).
53. *AS 4439.3:2019*; Wastes, Sediments and Contaminated Soils. Standards Australia: Sydney, Australia, 2019.
54. Hisseine, O.A.; Tagnit-Hamou, A. Development of ecological strain-hardening cementitious composites incorporating high-volume ground-glass pozzolans. *Constr. Build. Mater.* **2020**, *238*, 117740. [CrossRef]
55. Arowojolu, O.; Fina, J.; Pruneda, A.; Ibrahim, A.; Mahmoud, E. Feasibility study on concrete performance made by partial replacement of cement with nanoglass powder and fly ash. *Int. J. Civ. Eng.* **2019**, *17*, 1007–1014. [CrossRef]
56. Onaizi, A.M.; Lim, N.H.A.S.; Huseien, G.F.; Amran, M.; Ma, C.K. Effect of the addition of nano glass powder on the compressive strength of high volume fly ash modified concrete. *Mater. Today Proc.* **2022**, *48*, 1789–1795. [CrossRef]
57. Mangi, S.A.; Ibrahim, M.H.W.; Jamaluddin, N.; Shahidan, S.; Arshad, M.F.; Memon, S.A.; Jaya, R.P.; Mudjanarko, S.W.; Setiawan, M.I. Influence of ground coal bottom ash on the properties of concrete. *Int. J. Sustain. Constr. Eng. Technol.* **2018**, *9*, 26–34. [CrossRef]
58. Baite, E.; Messan, A.; Hannawi, K.; Tsobnang, F.; Prince, W. Physical and transfer properties of mortar containing coal bottom ash aggregates from Tefereyre (Niger). *Constr. Build. Mater.* **2016**, *125*, 919–926. [CrossRef]
59. Ankur, N.; Singh, N. Performance of cement mortars and concretes containing coal bottom ash: A comprehensive review. *Renew. Sustain. Energy Rev.* **2021**, *149*, 111361. [CrossRef]
60. Hashemi, S.S.G.; Mahmud, H.B.; Djobo, J.N.Y.; Tan, C.G.; Ang, B.C.; Ranjbar, N. Microstructural characterization and mechanical properties of bottom ash mortar. *J. Clean. Prod.* **2018**, *170*, 797–804. [CrossRef]
61. Hamada, H.; Alattar, A.; Tayeh, B.; Yahaya, F.; Adesina, A. Sustainable application of coal bottom ash as fine aggregates in concrete: A comprehensive review. *Case Stud. Constr. Mater.* **2022**, *16*, e01109. [CrossRef]
62. Ali, T.; Buller, A.S.; Abro, F.U.R.; Ahmed, Z.; Shabbir, S.; Lashari, A.R.; Hussain, G. Investigation on Mechanical and Durability Properties of Concrete Mixed with Silica Fume as Cementitious Material and Coal Bottom Ash as Fine Aggregate Replacement Material. *Buildings* **2022**, *12*, 44. [CrossRef]
63. Wang, J.; Wang, Y.; Yu, J.; Xu, L.; Li, M.; Cheng, J.; Li, Z. Effects of sodium sulfate and potassium sulfate on the properties of calcium sulfoaluminate (CSA) cement based grouting materials. *Constr. Build. Mater.* **2022**, *353*, 129045. [CrossRef]
64. Wang, Y.; He, F.; Wang, J.; Wang, C.; Xiong, Z. Effects of calcium bicarbonate on the properties of ordinary Portland cement paste. *Constr. Build. Mater.* **2019**, *225*, 591–600. [CrossRef]
65. Bheel, N.; Khoso, S.; Baloch, M.H.; Benjeddou, O.; Alwetaishi, M. Use of waste recycling coal bottom ash and sugarcane bagasse ash as cement and sand replacement material to produce sustainable concrete. *Environ. Sci. Pollut. Res.* **2022**, *29*, 52399–52411. [CrossRef] [PubMed]
66. Oruji, S.; Brake, N.A.; Guduru, R.K.; Nalluri, L.; Günaydın-Şen, Ö.; Kharel, K.; Rabbanifar, S.; Hosseini, S.; Ingram, E. Mitigation of ASR expansion in concrete using ultra-fine coal bottom ash. *Constr. Build. Mater.* **2019**, *202*, 814–824. [CrossRef]

67. Mousa, A. Utilization of coal bottom ash from thermal power plants as a cement replacement for building: A promising sustainable practice. *J. Build. Eng.* **2023**, *74*, 106885. [CrossRef]
68. Günay, G.; Cihan, M.T.; Güneş, E. Evaluation of mechanical properties and leaching tests results of mortars containing waste bottom ash as replacement of cement. *J. Mater. Cycles Waste Manag.* **2024**, *26*, 2378–2392. [CrossRef]
69. Guan, X.; Wang, L.; Mo, L. Effects of ground coal bottom ash on the properties of cement-based materials under various curing temperatures. *J. Build. Eng.* **2023**, *69*, 106196. [CrossRef]
70. Cheah, C.B.; Liew, J.J.; Kevin, K.L.P.; Siddique, R.; Tangchirapat, W. Influence of milling parameters on the properties of ground coal bottom ash and its blended cement. *Constr. Build. Mater.* **2023**, *363*, 129745. [CrossRef]
71. Rodríguez-Álvaro, R.; González-Fonteboa, B.; Seara-Paz, S.; Rey-Bouzon, E.J. Masonry mortars, precast concrete and masonry units using coal bottom ash as a partial replacement for conventional aggregates. *Constr. Build. Mater.* **2021**, *283*, 122737. [CrossRef]
72. Lakhiar, M.T.; Bai, Y.; Wong, L.S.; Paul, S.C.; Anggraini, V.; Kong, S.Y. Mechanical and durability properties of epoxy mortar incorporating coal bottom ash as filler. *Constr. Build. Mater.* **2022**, *315*, 125677. [CrossRef]
73. Wang, J.; Xie, J.; Wang, Y.; Liu, Y.; Ding, Y. Rheological properties, compressive strength, hydration products and microstructure of seawater-mixed cement pastes. *Cem. Concr. Compos.* **2020**, *114*, 103770. [CrossRef]
74. Richardson, I.G. The nature of C-S-H in hardened cements. *Cem. Concr. Res.* **1999**, *29*, 1131–1147. [CrossRef]
75. Richardson, I.G. Model structures for C-(A)-S-H(I). *Acta Crystallogr. Sect. B Struct. Sci. Cryst. Eng. Mater.* **2014**, *70*, 903–923. [CrossRef]
76. Kumar, A.; Walder, B.J.; Kunhi Mohamed, A.; Hofstetter, A.; Srinivasan, B.; Rossini, A.J.; Scrivener, K.; Emsley, L.; Bowen, P. The Atomic-Level Structure of Cementitious Calcium Silicate Hydrate. *J. Phys. Chem. C* **2017**, *121*, 17188–17196. [CrossRef]
77. Kunther, W.; Ferreiro, S.; Skibsted, J. Influence of the Ca/Si ratio on the compressive strength of cementitious calcium-silicate-hydrate binders. *J. Mater. Chem. A* **2017**, *5*, 17401–17412. [CrossRef]
78. Hu, Z.; Wyrzykowski, M.; Griffa, M.; Scrivener, K.; Lura, P. Young's modulus and creep of calcium-silicate-hydrate compacts measured by microindentation. *Cem. Concr. Res.* **2020**, *134*, 106104. [CrossRef]
79. Furlan, A.P.; Razakamanantsoa, A.; Ranaivomanana, H.; Levacher, D.; Katsumi, T. Shear strength performance of marine sediments stabilized using cement, lime and fly ash. *Constr. Build. Mater.* **2018**, *184*, 454–463. [CrossRef]
80. Wang, T.; Ishida, T.; Gu, R.; Luan, Y. Experimental investigation of pozzolanic reaction and curing temperature-dependence of low-calcium fly ash in cement system and Ca-Si-Al element distribution of fly ash-blended cement paste. *Constr. Build. Mater.* **2021**, *267*, 121012. [CrossRef]
81. García-Giménez, R.; Frias, M.; Vigil de la Villa, R.; Martínez-Ramírez, S. Ca/Si and Si/Al Ratios of Metakaolinite-Based Wastes: Their Influence on Mineralogy and Mechanical Strengths. *Appl. Sci.* **2018**, *8*, 480. [CrossRef]
82. Jawed, I.; Skalny, J. Alkalies in cement: A review: II. Effects of alkalies on hydration and performance of Portland cement. *Cem. Concr. Res.* **1978**, *8*, 37–51. [CrossRef]
83. Odler, I.; Wonnemann, R. Effect of alkalies on portland cement hydration II. Alkalies present in form of sulphates. *Cem. Concr. Res.* **1983**, *13*, 771–777. [CrossRef]
84. Li, Z.; Afshinnia, K.; Rangaraju, P.R. Effect of alkali content of cement on properties of high performance cementitious mortar. *Constr. Build. Mater.* **2016**, *102*, 631–639. [CrossRef]
85. Sant, G.; Kumar, A.; Patapy, C.; Le Saout, G.; Scrivener, K. The influence of sodium and potassium hydroxide on volume changes in cementitious materials. *Cem. Concr. Res.* **2012**, *42*, 1447–1455. [CrossRef]
86. Smaoui, N.; Bérubé, M.A.; Fournier, B.; Bissonnette, B.; Durand, B. Effects of alkali addition on the mechanical properties and durability of concrete. *Cem. Concr. Res.* **2005**, *35*, 203–212. [CrossRef]
87. Hamdan, A.; Hajimohammadi, A.; Rawal, A.; Kim, T. The intrinsic role of network modifiers (Ca versus Mg) in the reaction kinetics and microstructure of sodium silicate-activated CaO-MgO-Al_2O_3-SiO_2 glasses. *Cem. Concr. Res.* **2023**, *164*, 107058. [CrossRef]
88. Kucharczyk, S.; Zajac, M.; Stabler, C.; Thomsen, R.M.; Ben Haha, M.; Skibsted, J.; Deja, J. Structure and reactivity of synthetic CaO-Al_2O_3-SiO_2 glasses. *Cem. Concr. Res.* **2019**, *120*, 77–91. [CrossRef]
89. Way, S.J.; Shayan, A. Early hydration of a Portland cement in water and sodium hydroxide solutions: Composition of solutions and nature of solid phases. *Cem. Concr. Res.* **1989**, *19*, 759–769. [CrossRef]
90. Halaweh, M.A. *Effect of Alkalis and Sulfate on Portland Cement Systems*; University of South Florida: Tampa, FL, USA, 2007.
91. Shen, Y.; Liu, S.; Wang, Y.; Shen, P.; Xuan, D.; Guan, X.; Shi, C. Hydration-hardening properties of low-clinker composite cement incorporating carbonated waste sintering red mud and metakaolin. *Constr. Build. Mater.* **2022**, *354*, 129171. [CrossRef]
92. Sutcu, M.; Erdogmus, E.; Gencel, O.; Gholampour, A.; Atan, E.; Ozbakkaloglu, T. Recycling of bottom ash and fly ash wastes in eco-friendly clay brick production. *J. Clean. Prod.* **2019**, *233*, 753–764. [CrossRef]
93. Eliche-Quesada, D.; Leite-Costa, J. Use of bottom ash from olive pomace combustion in the production of eco-friendly fired clay bricks. *Waste Manag.* **2016**, *48*, 323–333. [CrossRef] [PubMed]
94. *SW-846*; Test Methods for Evaluating Solid Waste, Physical/Chemical Methods. United States Environmental Protection Agency: Washington, DC, USA, 2019.
95. New South Wales Environmental Protection Agency. *Waste Classification Guidelines Part 1: Classifying Waste*; New South Wales Environmental Protection Agency: Parramatta, Australia, 2009.

96. Li, J.; Zhuang, X.; Querol, X.; Font, O.; Moreno, N. A review on the applications of coal combustion products in China. *Int. Geol. Rev.* **2018**, *60*, 671–716. [CrossRef]
97. Berhad, P.A.A. Report on literature study and laboratory testing on characteristics of water treatment plants residuals for DOE consideration to declassify residuals as non-scheduled waste under Environment Quality Act 1974 2009. Available online: https://www.proquest.com/docview/2564444499?pq-origsite=gscholar&fromopenview=true&sourcetype=Dissertations%20&%20Theses (accessed on 26 June 2024).
98. Abdullah, M.H.; Rashid, A.S.A.; Anuar, U.H.M.; Marto, A.; Abuelgasim, R. Bottom ash utilization: A review on engineering applications and environmental aspects. *IOP Conf. Ser. Mater. Sci. Eng.* **2019**, *527*, 12006. [CrossRef]
99. Patra, S.; Whaung, S.T.; Kwan, W.L. Analysis of heavy metals in Incineration Bottom Ash in Singapore and potential impact of pre-sorting on ash quality. *Energy Procedia* **2017**, *143*, 454–459. [CrossRef]
100. Wiranata, D.Y.; Yang, S.-H.; Akgul, C.M.; Hsien, H.Y.; Nugraha, M.Z.P. Use of coal ash cement stabilized material as pavement base material: Laboratory characterization and field evaluation. *Constr. Build. Mater.* **2022**, *344*, 128055. [CrossRef]
101. Ibeto, C.N.; Obiefuna, C.J.; Ugwu, K.E. Environmental effects of concretes produced from partial replacement of cement and sand with coal ash. *Int. J. Environ. Sci. Technol.* **2020**, *17*, 2967–2976. [CrossRef]

materials

MDPI

Article

Engineering Properties and Microstructure of Soils Stabilized by Red-Mud-Based Cementitious Material

Wentao Li [1,2], Ke Huang [1,2], Feng Chen [2], Lihua Li [1,2,*], Yang Cheng [1,2] and Kang Yang [1,2]

1 Key Laboratory of Health Intelligent Perception and Ecological Restoration of River and Lake, Ministry of Education, Hubei University of Technology, Wuhan 430068, China; wli20201027@hbut.edu.cn (W.L.); 102110914@hbut.edu.cn (K.H.); 102210900@hbut.edu.cn (Y.C.); 102110913@hbut.edu.cn (K.Y.)

2 School of Civil Engineering, Architecture and Environment, Hubei University of Technology, Wuhan 430068, China; 102010781@hbut.edu.cn

* Correspondence: researchmailbox@163.com

Abstract: Red mud (RM) is an industrial waste generated in the process of aluminum refinement. The recycling and reusing of RM have become urgent problems to be solved. To explore the feasibility of using RM in geotechnical engineering, this study combined magnesium oxide (MgO) (or calcium oxide (CaO)) with RM as an RM-based binder, which was then used to stabilize the soil. The physical, mechanical, and micro-structural properties of the stabilized soil were investigated. As the content of MgO or CaO in the mixture increased, the unconfined compressive strength (UCS) of the RM-based cementitious materials first increased and then decreased. For the soils stabilized with RM–MgO or RM–CaO, the UCS increased and then decreased, reaching a maximum at RM:MgO = 5:5 or RM:CaO = 8:2. The addition of sodium hydroxide (NaOH) promoted the hydration reaction. The UCS enhancement ranged from 8.09% to 66.67% for the RM–MgO stabilized soils and 204.6% to 346.6% for the RM–CaO stabilized soils. The optimum ratio of the RM–MgO stabilized soil (with NaOH) was 2:8, while that of the RM–CaO stabilized soil (with NaOH) was 4:6. Freeze–thaw cycles reduced the UCS of the stabilized soil, but the resistance of the stabilized soil to freeze–thaw erosion was significantly improved by the addition of RM–MgO or RM–CaO, and the soil stabilized with RM–MgO had better freeze–thaw resistance than that with RM–CaO. The hydrated magnesium silicate generated by the RM–MgO stabilized soil and the hydrated calcium silicate generated by the RM–CaO stabilized soil helped to improve the UCS of the stabilized soil. The freeze–thaw cycles did not weaken the formation of hydration products in the stabilized soil but could result in physical damage to the stabilized soils. The decrease in the UCS of the stabilized soil was mainly due to physical damage.

Keywords: red mud; magnesium oxide; calcium oxide; stabilized soil; freeze–thaw

1. Introduction

Red mud (RM) is an industrial waste produced in the process of aluminum refinement. When 1 ton of alumina is produced, 1–1.5 tons of RM could be generated [1,2], and the annual production of RM worldwide could reach 120 million tons [3]. However, it is difficult for RM to be recycled and reused due to its high alkalinity [4,5], resulting in the consumption of land space, so it has to be disposed. Currently, the disposal storage of RM in China had reached more than one billion tons [6], which causes environmental pollution, groundwater pollution, and land salinization [7]. The recycling and reusing of RM have become urgent problems to be solved and have attracted lots of attention.

RM can be recycled and reused by recovering valuable metals, producing building materials [8], etc. Chavan et al. [9] used RM to partially replace cement to prepare concrete. Their results showed that when RM was used to replace 12% of the cement, the mechanical properties of the concrete were significantly improved. However, as the content of RM increased, the fluidity of the concrete decreased continuously. Some scholars [10] tried to use lime to stabilize kaolin, and after ten years, its UCS value was 21 times that of untreated kaolin. Magnesium oxide and sodium bicarbonate were used to stabilize kaolin and resulted in better strength measurements [11]. Recently, the application of RM as a road base material has emerged as a new way to recycle and reuse RM. Alam et al. [12] attempted to stabilize RM using Na_2SiO_3 and granulated ground blast furnace slag, and Sahile [13] added cement and fly ash. Their results all showed that the unconfined compressive strength (UCS) of stabilized RM was significantly improved and that it had potential as a road base material. Mukiza et al. [14] tried to apply RM in road engineering and found that RM alone had a low UCS. It was found that an amended RM performed better than natural soil as a road base material. These studies all regarded modified RM as a soil material and used it extensively in road engineering.

However, using a large content of RM in road engineering would have a negative impact on the surrounding environment [15–17]. Vigneshwaran et al. [18] found that increasing the content of RM had a negative effect, and RM had a better binding ability with sisal fiber when the content was 10–20%. RM contains active substances, such as SiO_2 and Al_2O_3, that are good raw materials for synthesizing cementitious materials [19]. Zhang et al. [20] mixed RM with slag to make cementitious materials, which produced better mechanical properties. Sreelekshmi et al. [21] used RM, fly ash, and slag as raw materials to prepare cementitious materials. Their research results showed that the RM–fly ash–slag cementitious materials had good processability and strength characteristics. Therefore, using RM as a cementitious material to stabilize poor soil is another way to recycle and reuse RM. A two-year study found that the addition of 5% by weight of RM did not negatively affect plants and soil microorganisms, which indicated that RM could be used as a soil binder to stabilize soils [22]. To explore whether RM could replace more soil, it was found that RM mixed with 2% lime showed a greater strength in 20% of the soil than in 15% of the soil, which ensured that more soil could be replaced by RM after being treated with lime [23]. Some scholars have mixed RM with lime and then used it to stabilize expansive soils. It was found to improve the geotechnical properties of the soil [24]. Ma et al. [25] reported that the addition of an appropriate content of RM could effectively improve the UCS of stabilized soil. Suo et al. [26] investigated the mechanical properties of heavy-metal-contaminated soil stabilized by RM and desulfurized gypsum and found that after adding a stabilizer, the UCS of the contaminated soil increased by about 130–260%. In addition, in practice, temperature (i.e., freeze–thaw) is one of the main factors causing a deterioration in the performance of road base material [27–30]. However, few studies have investigated the capability of RM-based cementitious material (or binder) stabilized soil to resist freeze–thaw cycles.

Therefore, this study used magnesium oxide (MgO) (or calcium oxide (CaO)) to activate RM and then applied RM–MgO (or RM–CaO) to treat soil. The mechanical properties of RM-based cementitious materials were investigated, and the influence of freeze–thaw cycles on the physical and mechanical properties as well as the microstructure of alkali-activated RM-based cementitious material stabilized soil was explored. Nuclear magnetic resonance (NMR) tests, X-ray diffraction (XRD) tests, and scanning electron microscopy (SEM) were used to reveal the stabilization mechanism of the soil treated with the RM-based materials under freeze–thaw cycles.

2. Experimental Materials and Methods

2.1. Experimental Materials

The plastic limit and liquid limit of the kaolin used in this study were 34.6% and 53.6%, respectively. It was provided by Guangzhou Chuangke New Material Technology Co., Ltd. (Guangzhou, China). RM was provided by Henan Porun New Material Co., Ltd. (Gongyi, China). Before testing, RM was dried in a 105 °C oven and passed through a 2 mm sieve. MgO was provided by Hebei Zhiyan Alloy Material Co., Ltd. (Xingtai, China), with an effective MgO content of >75%. CaO was provided by Gongyi Hengnuo

Filter Material Co., Ltd. (Gongyi, China), with an effective CaO content of >90%. Sodium hydroxide (NaOH) was provided by Hebei Zhiyan Alloy Material Co., Ltd. (Xingtai, China), with a purity of >99%. Table 1 provides the chemical composition of the test soil and the chemical binders used in this study. Photographs of the experimental materials are shown in Figure 1.

Table 1. Chemical composition of soil and binders (% by weight).

Composition	CaO	SiO_2	Fe_2O_3	Al_2O_3	MgO	Na_2O	Others	Loss on Ignition
Kaolin	0.06	53.59	0.89	43.24	0.05	0.07	2.10	0
RM	8.22	27.33	26.08	20.95	1.23	9.61	6.58	0
MgO	5.62	10.23	0.03	6.34	76.72	0.27	0.79	0
CaO	92.58	1.87	0.36	1.30	3.10	0.15	0.62	0.02

(**a**) Kaolin (**b**) Red mud (**c**) MgO (**d**) CaO (**e**) NaOH

Figure 1. Experimental materials.

2.2. Experimental Design

A flowchart (Figure 2) has been inserted in this study to better show our methodology. Red mud is first combined with magnesium oxide (or calcium oxide) to form a cementitious material, which is then used to stabilize kaolin. Specimens of both the cementitious material and the stabilized soil were used for UCS testing. Finally, with the addition of NaOH solution to the stabilizer, the stabilized soil (with NaOH) was used for freeze–thaw testing, UCS testing, and microstructure or SEM testing.

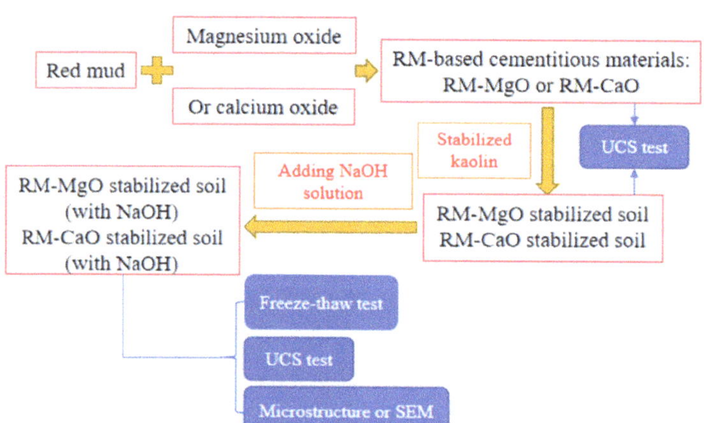

Figure 2. Flowchart.

2.2.1. Mix Design for RM-Based Cementitious Material

RM and MgO (or CaO) were blended with water. After blending them evenly, the mixture was placed into a cubic mold with a length × width × height of 50 mm × 50 mm × 50 mm and stood for 6 h before demolding. After demolding, the specimens were sealed and placed in a standard curing box (with a temperature of 20 °C ± 2 °C and relative humidity of >95%) for curing. Table 2 shows the testing scheme for the experiment:

Table 2. Testing program for RM-based cementitious material.

Test Group No.	Water–Binder Ratio	RM (%)	MgO (%)	CaO (%)	Days of Curing
M-1	0.5	75	25	0	7, 14, 28
M-2		60	40	0	
M-3		43	57	0	
M-4		33	67	0	
L-1		75	0	25	
L-2		60	0	40	
L-3		50	0	50	
L-4		43	0	57	

2.2.2. Soil Stabilization with RM-Based Cementitious Material

To improve the UCS of the specimens, NaOH was added to RM–MgO and RM–CaO stabilized soil [31,32]. As comparison, RM–MgO and RM–CaO without NaOH were used to stabilize the soil. Cylindrical specimens with a size of Φ50 mm × 100 mm were prepared by compaction method. Firstly, RM, MgO (or CaO), kaolin, and water were mixed according to the design mix ratio. After stirring them evenly, compaction was carried out by using a 0.51 kg hammer, which fell freely along a steel bar that was 270 mm long. The mixture was divided into three layers and placed into the mold. Each layer was compacted 30 times with a unit compaction energy of 606.72 kJ × m^{-3}, which is slightly higher than the 605.6 kJ × m^{-3} required by the ASTM standard Proctor test [33]. After compaction was completed, the mold was removed using a jack, and the specimens were sealed with plastic wrap and placed in a standard curing box for curing until the design time. The UCS test was carried out using a universal testing machine with a range of 20 kN. The loading rate was controlled at 1 mm × min^{-1} [34]. The test was stopped immediately after the UCS reached its peak value. The test was performed twice. It was important to note that ASTM D2166M-16 [35] specifies that the specimen size for the UCS test should have a minimum diameter of 30 mm and maximum particle size less than one-tenth of the diameter of the specimen. The height-to-diameter ratio of the specimen should be between 2 and 2.5. In this study, the specimen size met these requirements. The compaction method used in this study was adopted for two reasons. First, it ensured that the size of the compacted specimen was consistent with the specimen size of the UCS test, which improved the accuracy of the test results. Second, it was necessary due to the limited test materials and test conditions in the laboratory.

Figure 3 provides the compaction curve of RM-based cementitious material stabilized soil. As the MgO (or CaO) content increased, that is, as the content of RM in the stabilizer decreased, the optimum moisture content (OMC) of RM–MgO stabilized soil (with or without NaOH) increased. For example, without NaOH solution, as the OMC for the soil stabilized at ratios of RM:MgO = 8:2, 7:3, 6:4, 5:5, and 4:6 was expected to be between the two OMCs at ratios of RM:MgO = 9:1 and 3:7. Compaction tests for the ratios of 8:2, 7:3, 6:4, 5:5, and 4:6 were not conducted. The average OMC values that were obtained from soils at ratios of RM:MgO = 9:1 and 3:7 were used to prepare UCS specimens stabilized at ratios of 8:2, 7:3, 6:4, 5:5, and 4:6. Therefore, the OMC values of RM–MgO (with or without NaOH) stabilized soil were 37% and 38%, respectively. The OMC values of RM–CaO (with or without NaOH) stabilized soil were 39% and 37%, respectively.

Similar to previous research [36], the content of stabilizer was 15% (by the weight of dry soil). The mass ratios of RM, MgO, and CaO in the stabilizer are shown in Table 3.

Table 3. Design of mixing proportions.

Test Group No.	RM (%)	MgO (%)	CaO (%)	NaOH (mol/L)	Days of Curing
RM-1	90	10	0	0	7, 14, 28
RM-2	80	20	0	0	
RM-3	70	30	0	0	
RM-4	60	40	0	0	
RM-5	50	50	0	0	

Table 3. *Cont.*

Test Group No.	RM (%)	MgO (%)	CaO (%)	NaOH (mol/L)	Days of Curing
RM-6	40	60	0	0	
RM-7	30	70	0	0	
NRM-1	90	10	0	1.25	
NRM-2	80	20	0	1.25	
NRM-3	70	30	0	1.25	
NRM-4	60	40	0	1.25	
NRM-5	50	50	0	1.25	
NRM-6	40	60	0	1.25	
NRM-7	30	70	0	1.25	
NRM-8	20	80	0	1.25	
NRM-9	10	90	0	1.25	
NRM-0	0	100	0	1.25	
RL-1	90	0	10	0	7, 14, 28
RL-2	80	0	20	0	
RL-3	70	0	30	0	
RL-4	60	0	40	0	
RL-5	50	0	50	0	
NRL-1	90	0	10	1.25	
NRL-2	80	0	20	1.25	
NRL-3	70	0	30	1.25	
NRL-4	60	0	40	1.25	
NRL-5	50	0	50	1.25	
NRL-6	40	0	60	1.25	
NRL-7	30	0	70	1.25	
NRL-0	0	0	100	1.25	

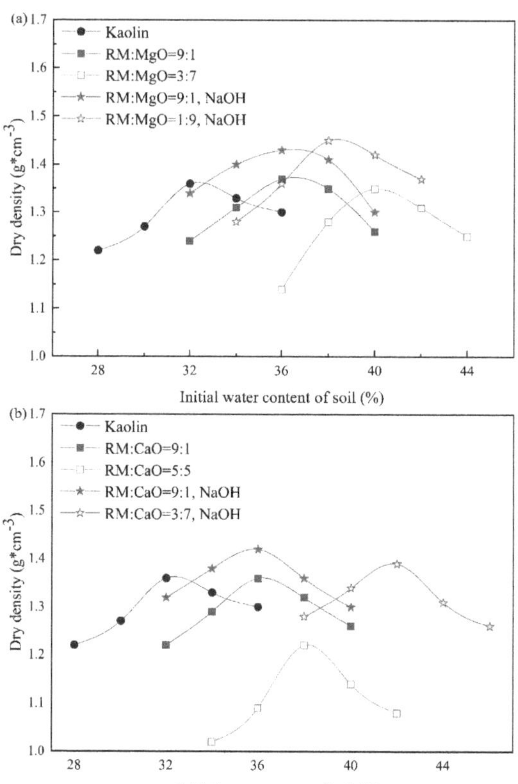

Figure 3. Compaction curves for stabilized soil: (**a**) RM–MgO (or NaOH) stabilized soil; (**b**) RM–CaO (or NaOH) stabilized soil.

2.2.3. Freeze–Thaw Test

According to the reference [37], the freezing temperature was set to −20 °C and the thawing temperature to 20 °C based on the temperature range in China. A complete freeze–thaw cycle consisted of 12 h of freezing and 12 h of thawing. After the specimen was maintained to the specified curing time, it would be vacuumed and saturated in the vacuum saturator. To prevent evaporation of water from the specimens during the freeze–thaw process, they were wrapped in plastic wrap and placed in a test chamber. When the specimens reached the required cycles (0, 3, 6, and 10 times), their mass and volume were measured, which were then used for calculating the mass loss rate and volume change.

(1) Equation (1) was used to calculate the mass loss rate of RM–MgO (or RM–CaO) stabilized soil.

$$\Delta M = (M_0 - M_i)/M_0 \times 100\% \quad (1)$$

where ΔM—the mass loss rate of stabilized soil after i freeze–thaw cycles (%), i—the cycle number, M_0—the initial mass of stabilized soil, and M_i—the mass of stabilized soil after i freeze–thaw cycles.

(2) By measuring the height and diameter of each specimen, the volume of the specimen before and after freeze–thawing was obtained. The method used to calculate the volume change rate of RM–MgO (or RM–CaO) stabilized soil is shown in Equation (2):

$$\Delta V = (V_0 - V_i)/V_0 \times 100\% \quad (2)$$

where ΔV—the volume change rate of stabilized soil after i freeze–thaw cycles (%), i—the cycle number for which a positive value indicates volume shrinkage and a negative value indicates volume expansion, V_0—the initial volume of stabilized soil, and V_i—the volume of stabilized soil after i freeze–thaw cycles.

2.2.4. Microstructure or SEM

After the UCS test was completed, block soil samples were collected and dried in a freeze dryer. The dried samples were ground and passed through a 0.075 mm sieve. The sieved samples were used for XRD tests using a Panalytical Empyrean instrument (Panaco, Almelo, The Netherlands) with a scanning rate of 5°/min. Some unsieved samples were subjected to SEM tests using a Hitachi SU8010 instrument (Hitachi High-Tech, Tokyo, Japan) with resolution of 1 nm/15 kv and 1.3 nm/1 kv. In addition, crushed soil samples with a diameter of 10–15 mm and a height of 20–30 mm were saturated using a vacuum pump. These saturated samples were then used in NMR tests. A PQ-001-Mini nuclear magnetic resonance instrument (Shanghai Newmax Electronic Technology Co., Shanghai, China) with a resonance frequency of 12 MHz, a magnetic induction intensity of 0.52 T, and an effective test range of 60 mm × Φ60 mm was used.

3. Results and Discussion

3.1. UCS of RM-Based Cementitious Material

Figure 4a shows the UCS of the MgO-activated RM cementitious material. As the content of MgO in the mixture increased, the UCS first increased and then decreased. The optimal ratio was RM:MgO = 3:4, with a 7-day UCS of 4.60 MPa. Figure 4b shows the UCS of the CaO-activated RM cementitious material. As the content of CaO increased, the UCS also first increased and then decreased, with the optimal ratio of RM:CaO = 3:2 and a 7-day UCS of 4.22 MPa. This indicated that both MgO and CaO had a significant activation effect on the RM, and under their respective optimal ratios, the UCS of the RM–MgO cementitious material was 5–29% higher than that of the RM–CaO cementitious material. According to Nie et al. [38], it was found that the 7-day UCS of the RM–fly ash cementitious material reached about 2.5 MPa, and the 28-day UCS was only about 3 MPa, both of which were lower than the UCS when the optimal ratio was used in this experiment. This illustrated that MgO and CaO had better effect on the activation of the RM.

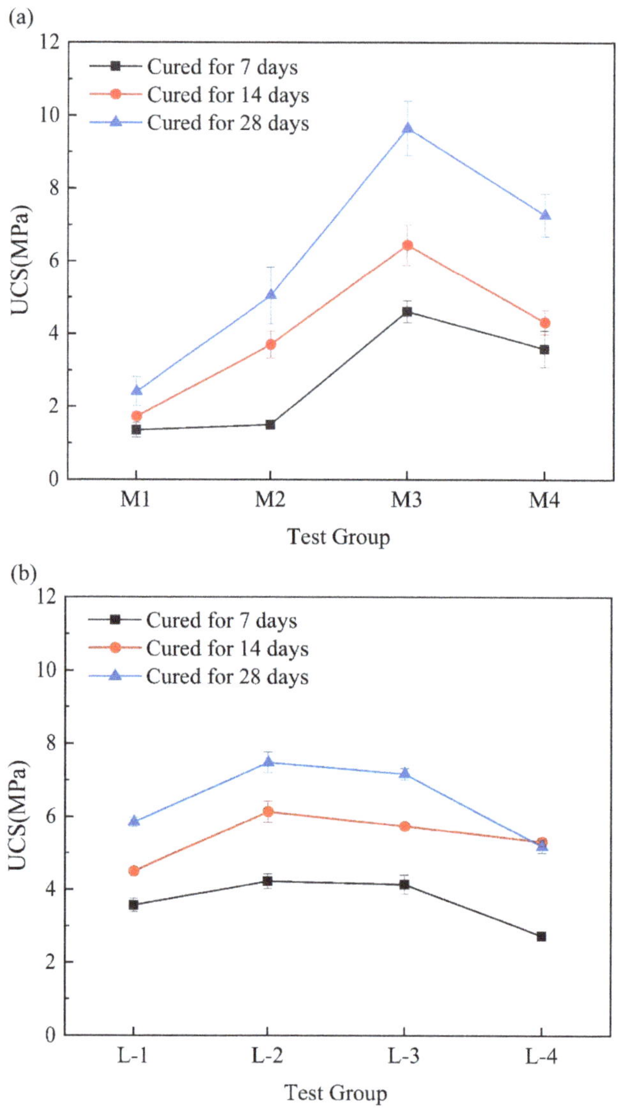

Figure 4. (**a**) UCS of RM–MgO cementitious material; (**b**) UCS of RM–CaO cementitious material.

3.2. UCS of Stabilized Soil

The above results indicated that the RM-based cementitious materials had the potential to stabilize the soil. Therefore, further tests were carried out on the physical and mechanical properties of the RM-based cementitious material stabilized soil.

3.2.1. UCS of RM–MgO (or RM–CaO) Stabilized Soil without NaOH

Figure 5a shows the change in the UCS of the RM–MgO stabilized soil. The UCS of the stabilized soil first increased and then decreased as the content of RM decreased. The 7-day and 14-day UCS values of the stabilized soils were most significantly enhanced when the RM content was 50–60%. At 28 days, the RM-5 (RM:MgO = 5:5) reached a maximum UCS of 1.73 MPa. Figure 5b shows the relationship between the UCS and the content of RM for the RM–CaO stabilized soils. After 7 days of curing, as the content of the RM in the mixture decreased, the UCS gradually increased, and when the mass ratio of the RM to CaO was 5:5, the UCS was optimal. However, when the curing times were 14 and 28 days, the UCS of stabilized soil first increased and then decreased as the content of the RM in the mixture decreased. At this time, when the mass ratio of RM to CaO was 8:2, the maximum UCS was obtained. This might be because the RM activity was not fully activated when the MgO (or CaO) content was low. As the content of MgO (or CaO) increased, its activity was fully activated to reach the maximum UCS value. However, when the optimum content of MgO (or CaO) was exceeded, the MgO (or CaO) would be dispersed in the soil in its natural state, which would weaken the connection between the soil and lead to a reduction in UCS [39,40]. ASTM [41] advised that the UCS of admixtures for improving the engineering properties of fine-grained soil should not be less than 345 kPa. For the RM–CaO stabilized soil, only the 7-day UCS for RL-5 (RM:CaO = 5:5) met this requirement, only the 14-day UCS for RL-5 and RL-2 (RM:CaO = 5:5 and 8:2) met this requirement. Additionally, the 28-day UCS of RL-2, RL-3, RL-4, and RL-5 (RM:CaO = 8:2, 7:3, 6:4, and 5:5) met this requirement. Conversely, the UCS of the RM–MgO stabilized soil met this requirement at all curing times. Overall, the UCS of the RM–CaO stabilized soil was lower than that of the RM–MgO stabilized soil.

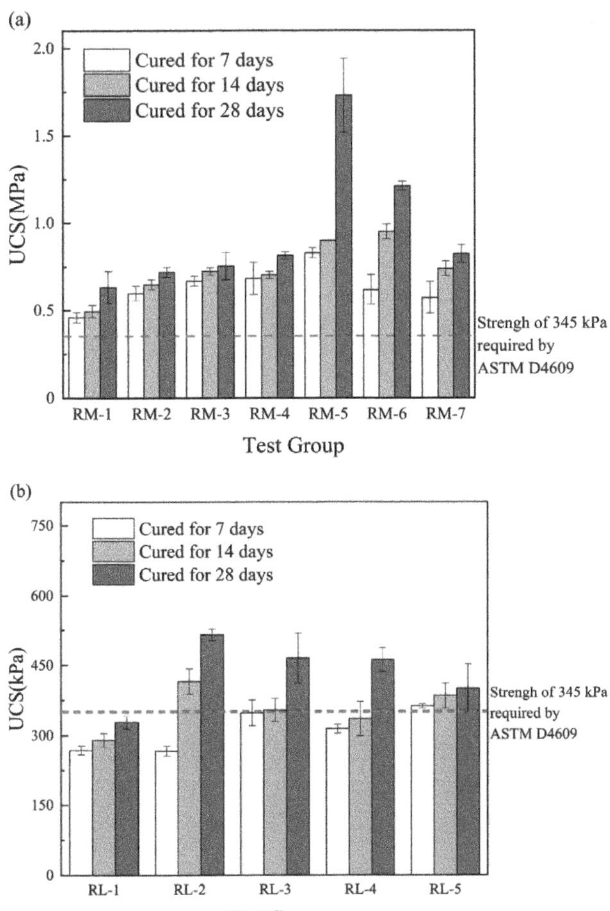

Figure 5. (**a**) UCS of RM–MgO stabilized soil (without NaOH); (**b**) RM–CaO stabilized soil's UCS (without NaOH) [41].

3.2.2. UCS of RM–MgO (or RM–CaO) Stabilized Soil with NaOH

As shown in Figure 6a, the UCS of the RM–MgO stabilized soil increased after adding NaOH. With the decrease in RM content, the UCS showed a trend of increasing and then decreasing, and the maximum 7-day UCS was obtained when the RM content was 30%, while the maximum 14-day and 28-day UCS were obtained when the RM content was 20%. As seen in Figure 6b, for the RM–CaO stabilized soil, the UCS of the soil with NaOH was much higher than that of the stabilized soil without NaOH. As the content of RM continued to decrease, the UCS of the stabilized soil showed a trend of first increasing and then decreasing. When the mass ratio of RM to CaO was 4:6, the maximum UCS of the stabilized soil was obtained. Although RM–MgO (or RM–CaO) could accelerate the UCS development of stabilized soils, excessive residual MgO (or CaO) negatively affected the UCS of the stabilized soils. The optimal ratios for the stabilized soil with NaOH (RM:MgO = 2:8, RM:CaO = 4:6) require more MgO (or CaO) than the optimal ratios for the stabilized soil without NaOH (RM:MgO = 5:5, RM:CaO = 8:2), probably because the addition of NaOH requires more MgO (or CaO) to participate in the hydration reaction, which continues to increase in UCS. Equation (3) was used to calculate the rate of the UCS enhancement of the soil stabilized with RM–MgO (or RM–CaO). The UCS enhancement rate of the RM–MgO stabilized soil ranged from 8.09% to 66.67%, while the UCS enhancement rate of the RM–CaO stabilized soil ranged from 204.6% to 346.6%. According to Cristelo et al. [31], the addition of NaOH accelerated the dissolution of silica and alumina, thereby producing more hydration products to increase the UCS of the RM–MgO (or RM–CaO) stabilized soil. The UCS of all specimens generally met the requirements of ASTM [41] specifications.

$$\Delta S = (S_1 - S_0)/S_0 \times 100\% \quad (3)$$

where ΔS—the rate of the UCS enhancement of the stabilized soil, S_1—the UCS of the stabilized soil with NaOH, and S_0—the UCS of the stabilized soil without NaOH.

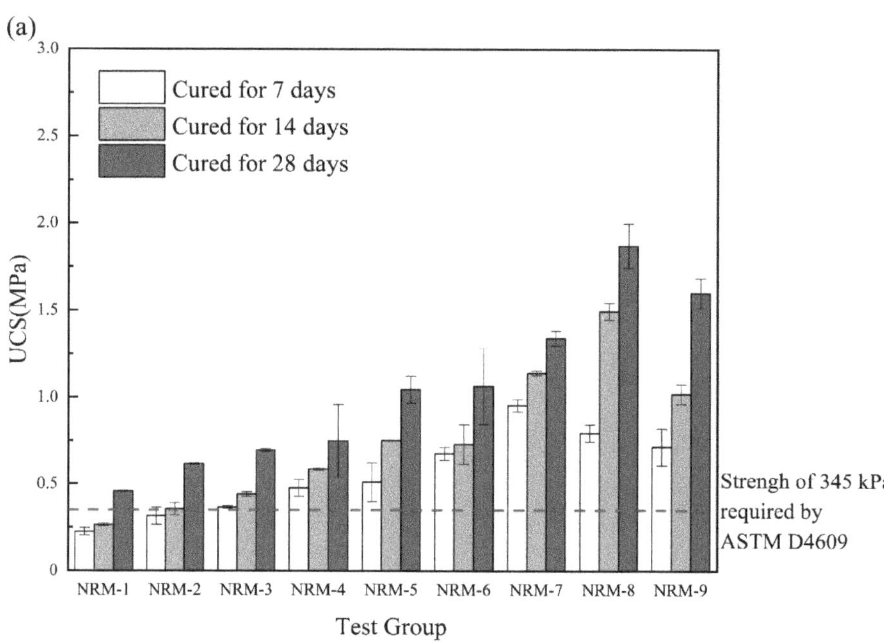

Figure 6. *Cont.*

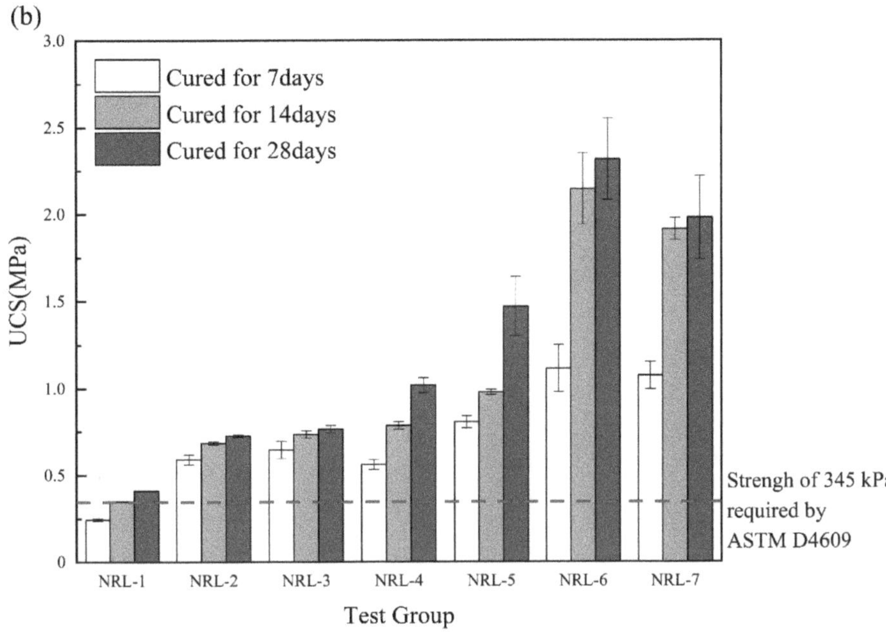

Figure 6. (**a**) UCS of RM–MgO stabilized soil (with NaOH); (**b**) RM–CaO stabilized soil's UCS (with NaOH) [41].

3.3. *Physical, Mechanical, and Microstructural Charateristics of Stabilized Soil under Freeze–Thaw Cycles*

The specimens with the optimal ratios (NRM-8 and NRL-6) determined in Section 3.2 were subjected to freeze–thaw cycles after being cured for 28 days. Their mass and volume were monitored. UCS, NMR, XRD, and SEM tests were performed on the NRM-8 and NRL-6 specimens, and then the performance was compared with pure MgO stabilized soil (NRM-0) and pure CaO stabilized soil (NRL-0).

3.3.1. The Influence of Freeze–Thaw Cycles on Mass Change

Figure 7 shows the mass loss rate of the stabilized soil with an increasing number of cycles. The mass loss rate of the RM–MgO stabilized soil decreased as the cycle number grew, while that of the RM–CaO stabilized soil increased. For the RM–MgO stabilized soil, as the cycle number grew from zero to three, the mass loss rate of NRM-8 and NRM-0 reached their maximum values of 1.52% and 1.77%, respectively. As the cycle number grew to six, the mass loss rate of NRM-8 started to decrease to 1.34%, while that of NRM-0 decreased to 1.70%. As the cycle number progressed from three to ten, the mass loss rate of NRM-8 decreased more than that of NRM-0. For the RM–CaO stabilized soil, when the cycle number progressed from zero to three, the mass loss rate increased to 1.11% for NRL-6 and 1.40% for NRL-0. As the cycle number progressed to six, the mass loss rate of NRL-6 continued to increase to 1.37% and that of NRL-0 increased to 1.87%, while the increase in the mass loss rate of NRL-6 was less than that of NRM-0. When the cycle number progressed from six to ten, the mass loss rate of NRL-6 continued to increase less than that of NRL-0. The mass loss rate of the RM–MgO stabilized soil (or RM–CaO stabilized soil) was always lower than that of the pure MgO stabilized soil (or the pure CaO stabilized soil) for the same cycle number. This indicated that for both stabilized soils, the addition of RM–MgO (or RM–CaO) can effectively prevent the mass loss caused by freeze–thaw cycles. As the cycle number increased from zero to ten, the mass loss of NRM-8 first increased and then decreased, while the mass loss rate of NRL-6 kept increasing, so the RM–MgO stabilized soil had better resistance to mass loss. The reason for this might be that the expansion stress caused by freezing led to internal cracks in the specimen. Then, more external water was transferred into the internal voids and cracks, thus decreasing the mass loss rate of the specimen. When the mass loss of the specimen due to spalling was higher than the mass gained by water absorption, the mass loss rate of the specimen increased [42,43].

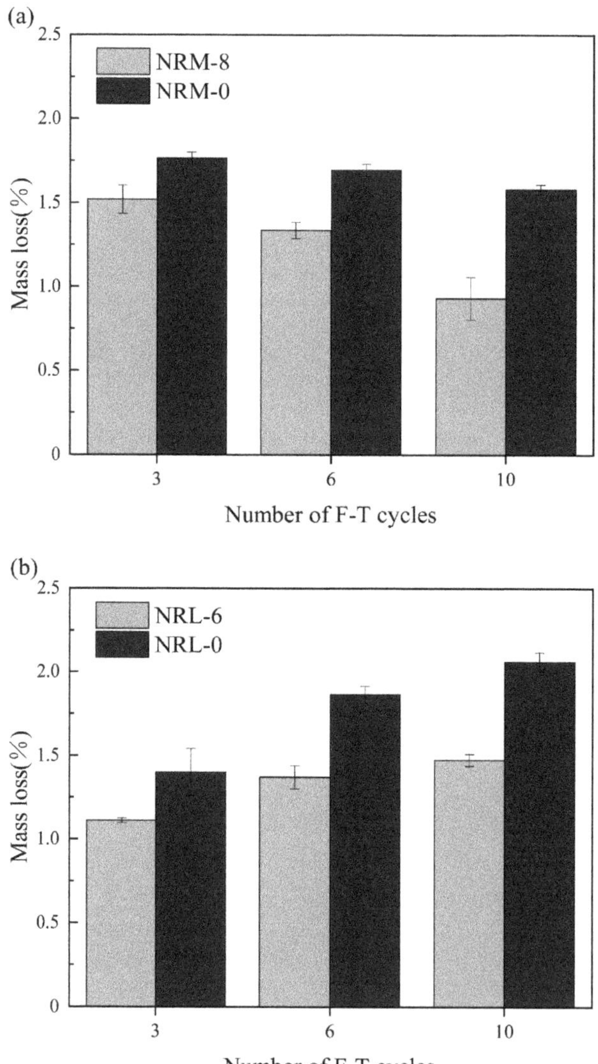

Figure 7. Mass loss rate during freeze–thaw cycles: (**a**) NRM-8, NRM-0; (**b**) NRL-6, NRL-0.

3.3.2. The Influence of Freeze–Thaw Cycles on Volume Change

Figure 8 shows the volume expansion rate of the stabilized soil with the number of cycles. The volume expansion rate of the RM–MgO stabilized soil increased slightly with the increase in cycle number. Conversely, the volume expansion rate of the RM–CaO stabilized soil showed an increasing trend with the cycle number. As the cycle number rose from zero to three in the RM–MgO stabilized soil, the volume expansion of NRM-8 increased to 0.10% and that of NRM-0 rose to 1.29%. This indicated that the volume expansion of freeze–thaw cycles was suppressed by the addition of RM–MgO. From cycles 3 to 10, the volume expansion of NRM-8 increased to 0.17% and the volume expansion of NRM-0 slightly decreased to 1.10%. Furthermore, NRM-8 showed much lower volume expansion compared to NRM-0. For the RM–CaO stabilized soil, when the cycle number was increased from zero to three, the volume expansion rate started to increase to 0.58% for NRL-6 and to 0.66% for NRL-0. As the cycle number rose to 10, the volume expansion rate of NRL-6 increased to 1.24% and that of NRL-0 increased to 1.52%. Additionally, the volume expansion rate of NRL-6 was always lower than that of NRM-0. The volume change rate of the RM–MgO stabilized soil (or the RM–CaO stabilized soil) was always lower than that of the pure MgO (or pure CaO) stabilized soil for the same cycle number. This indicated that for both stabilized soils, the addition of RM–MgO (or RM–CaO) could effectively prevent the volume expansion caused by freeze–thaw cycles. As the cycle number increased from zero to ten, the volume expansion rate of NRM-8 was much lower than that of NRL-6, indicating that the RM–MgO stabilized soil resisted the volume expansion better. According to the results of Konrad [44], the stabilized soil expanded because the temperature difference between the surface and the interior caused water to migrate from the inside to the outside. Some of the pores enlarged by freezing and melting water could not return to their original state, resulting in expansion.

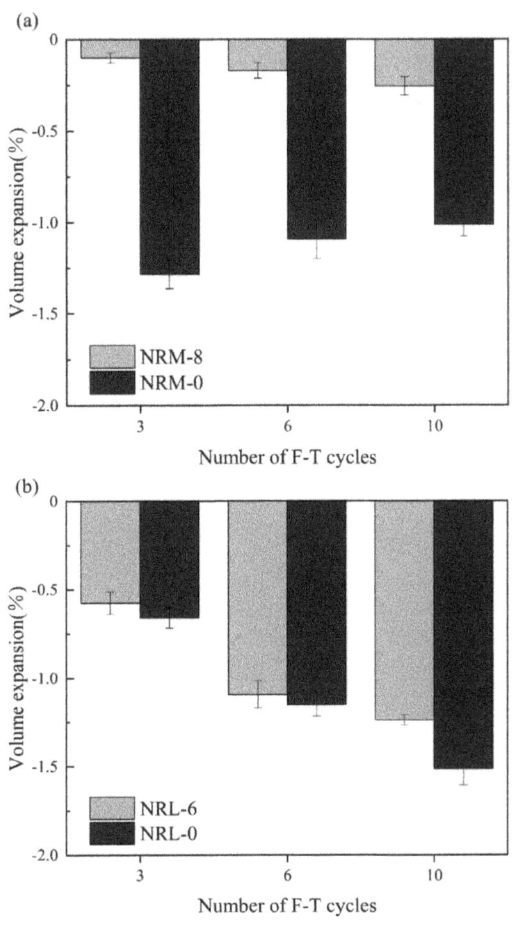

Figure 8. Volume expansion rate during freeze–thaw cycles: (**a**) NRM-8, NRM-0; (**b**) NRL-6, NRL-0.

3.3.3. The Influence of Freeze–Thaw Cycles on UCS

Figure 9 shows the UCS of the stabilized soil during freeze–thaw cycles. The UCS of the stabilized soil showed a decreasing trend as the cycle number increased. For the RM–MgO stabilized soil, when the cycle number increased from zero to three, the UCS of NRM-8 decreased from 1.87 MPa to 1.67 MPa, while the UCS of NRM-0 decreased from 1.69 MPa to 0.84 MPa. The UCS of NRM-8 decreased significantly more than that of NRM-0 (the UCS decrease was the ratio of the difference between the UCS after the freeze–thaw cycles and the UCS without freeze–thaw cycles). As the cycle number increased from three to ten, the UCS values of both NRM-8 and NRM-0 decreased. However, the UCS of NRM-8 was consistently higher than that of NRM-0. For the RM–CaO stabilized soil, when the cycle number increased from zero to ten, the UCS of NRL-6 slightly decreased, but that of NRL-0 first increased and then decreased. Yang et al. [30] found that freeze–thaw cycles destroyed the soil's structure and reduced its strength. Nevertheless, they also triggered the migration of water, which stimulated the hydration of the binders. Consequently, the soil structure was strengthened again, showing first increased and then decreased UCS values. As the cycle number increased from zero to ten, the UCS of NRL-6 (or NRM-8) was always higher than that of NRL-0 (or NRM-0).

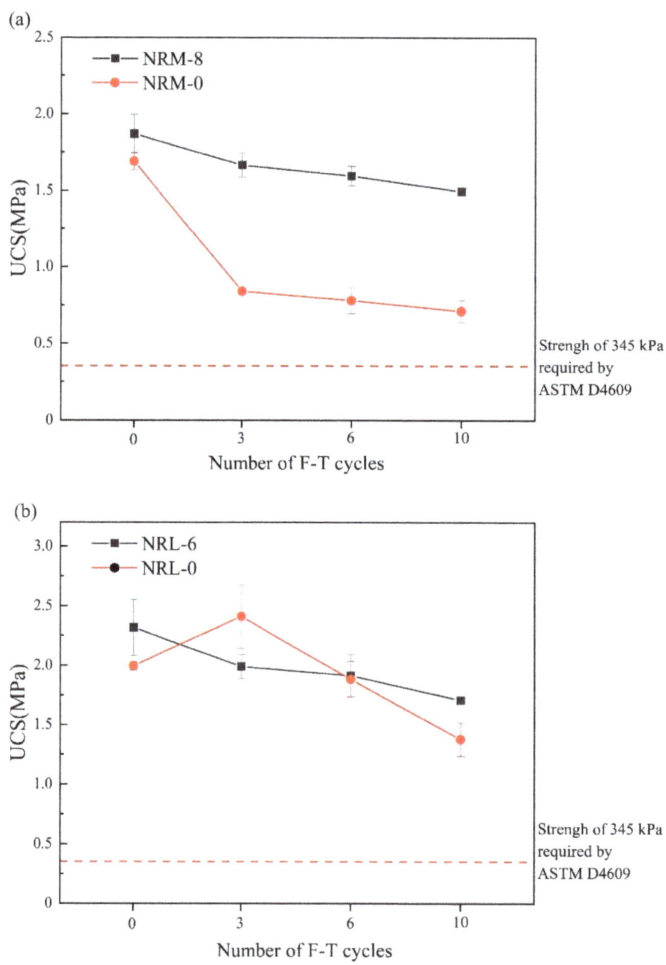

Figure 9. Change in UCS during freeze–thaw cycles: (**a**) NRM-8, NRM-0; (**b**) NRL-6, NRL-0 [41].

The UCS of the stabilized soil generally decreased with the increase in the cycle number, which indicated that freeze–thaw erosion still had a deteriorating effect on the UCS of the soil. The UCS of the soils decreased because the freezing of the water made them swell, and repeated incidents of swelling made the soils looser and weaker [45]. In addition, the UCS of the RM–MgO stabilized soil (or the RM–CaO stabilized soil) was significantly higher than that of the pure MgO stabilized soil (or the pure CaO stabilized soil) after 10 freeze–thaw cycles. The test results showed that for soils, the addition of RM–MgO (or RM–CaO) was effective in mitigating the loss of UCS caused by freeze–thaw erosion. Nevertheless, the UCS of the RM–MgO and RM–CaO stabilized soils during the freeze–thaw cycles still met ASTM [41] strength standards.

3.3.4. NMR

To explore the influence of freeze–thaw cycles on the pore structure of the stabilized soil, NMR tests were used. In Figure 10, the main peaks of the specimens were mainly distributed between 0.1 ms and 10 ms and between 10 ms and 1000 ms according to transverse relaxation times. The transverse relaxation time was positively correlated with pore size. The two main peaks corresponding to pores were defined as small pores and large pores, respectively, and the peak area (i.e., the closed area formed by the T2 curve and the X-axis) represented the pore volume [46].

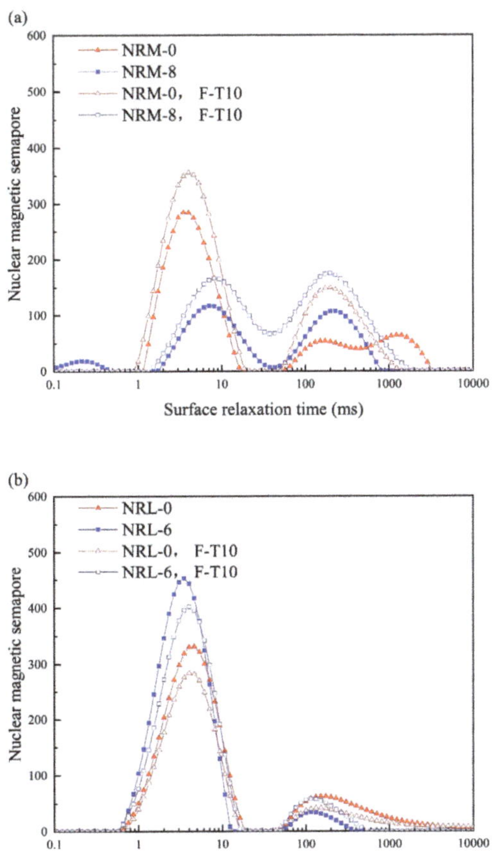

Figure 10. NMR before and after freeze–thaw cycles: (**a**) NRM-8, NRM-0 and (**b**) NRL-6, NRL-0. F-T10: after 10 freeze–thaw cycles.

Figure 10 shows the T2 distribution curves of NRM-0, NRL-0, NRM-8, and NRL-6. From Figure 8a, the peak areas of both the small and large pores in NRM-8 and NRM-0 were increased after 10 freeze–thaw cycles. Figure 8b showed that the peak area of small pores in NRL-6 and NRL-0 decreased after 10 freeze–thaw cycles, while the peak area of large pores increased slightly. Despite the different trends in the peak areas of the large and small pores in the NRL-6, the overall peak areas were basically unchanged after 10 freeze–thaw cycles. This indicated that the freeze–thaw cycles caused the pore size of the small pores in NRL-6 to increase and eventually transformed the small pores into large pores. He et al. [47] observed this phenomenon and found that freeze–thaw cycles transformed small pores in the soil into large pores. It was found that freeze–thaw cycles resulted in an increased pore volume in the stabilized soil of RM–MgO (or RM–CaO), and the total pore volume of NRM-8 (or NRL-6) was more than that of NRM-0 (or NRL-0), which might also account for the difference in UCS. The appearance of ice crystals in the specimen after freezing increased the diameter of the pores in the specimen, and after melting, the pore diameter could not be restored to its original state, which led to an increase in pore volume [47,48].

3.3.5. XRD

Figure 11 provides the XRD results of the freeze–thaw cycle samples. After stabilization with RM–MgO (or RM–CaO), new diffraction peaks appeared in the samples compared to the original soil, indicating that new mineral components were generated in the soil after treatment. In the RM–MgO solidification system, the main hydration product generated was magnesium silicate hydrate (MSH), with diffraction peaks at 18° and 26°. In the RM–CaO solidification system, the main hydration product inside the soil was calcium silicate hydrate (CSH), whose diffraction peaks occurred at 26° and 28°. The production of MSH and CSH might be responsible for the increase in UCS [49]. The crystalline phases detected in the RM–CaO stabilized soil before and after the freeze–thaw cycles were similar, and so were those in the RM–MgO soil. Taking NRM-8 as an example, before the freeze–thaw cycles, MSH was detected at 18° and 26°. After the freeze–thaw cycles, the position and intensity of these diffraction peaks hardly changed. For the NRL-0 sample, before the freeze–thaw cycles, its CSH diffraction peaks were mainly observed at 28°. After 10 cycles, the intensity of this diffraction peak decreased, but a new CSH characteristic peak was detected at 26°. The above results indicated that the freeze–thaw cycles did not decrease the formation of hydration products (CSH or MSH), so the decrease in UCS of the RM–MgO (or RM–CaO) stabilized soil was mainly caused by physical damage.

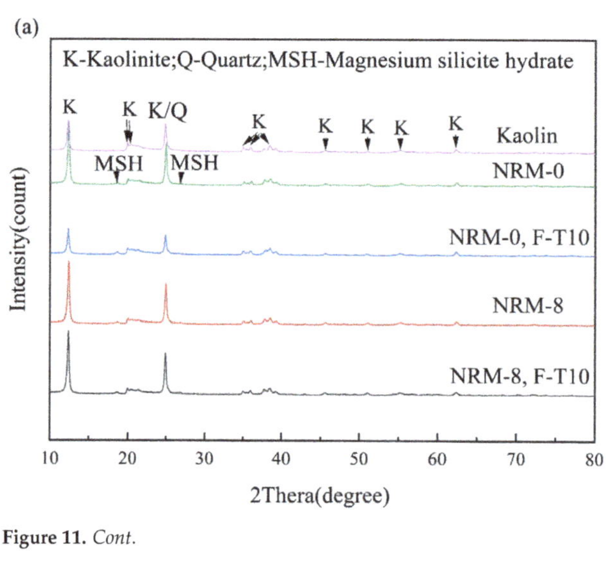

Figure 11. *Cont.*

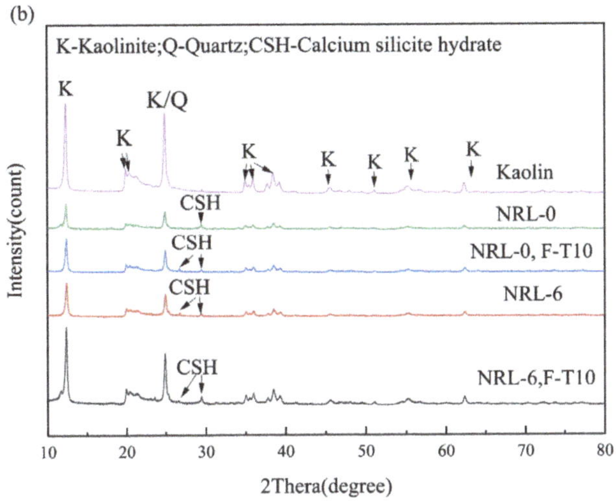

Figure 11. XRD before and after freeze–thaw cycles: (**a**) NRM-8, NRM-0 and (**b**) NRL-6, NRL-0. F-T10: after 10 freeze–thaw cycles.

3.3.6. SEM

A representative image (magnified 1000 times) was selected for each sample in SEM tests, and the images are shown in Figures 12 and 13. After 28 days of curing without freeze–thaw cycles, the soil particles of the RM–MgO stabilized soil (or the RM–CaO stabilized soil) made contact with each other, and only a few small pores were present. After 10 freeze–thaw cycles, fine cracks developed inside the RM–MgO stabilized soil, while the pore size for the RM–CaO stabilized soil increased significantly and its soil structure became looser. This indicated that freeze–thaw cycles reduced the particle density and bonding force in the RM–MgO (or RM–CaO) stabilized soil, resulting in damage to the soil structure and causing the UCS to decrease on a macroscopic level. The reason for this might be that repeated freeze–thaw cycles caused stress inside the soil, which caused a volume expansion in the samples and increased the pore volume in the soil [47]. As the pores increased in size and number, the distance between the pores decreased. Some pores connected or merged with each other, forming cracks.

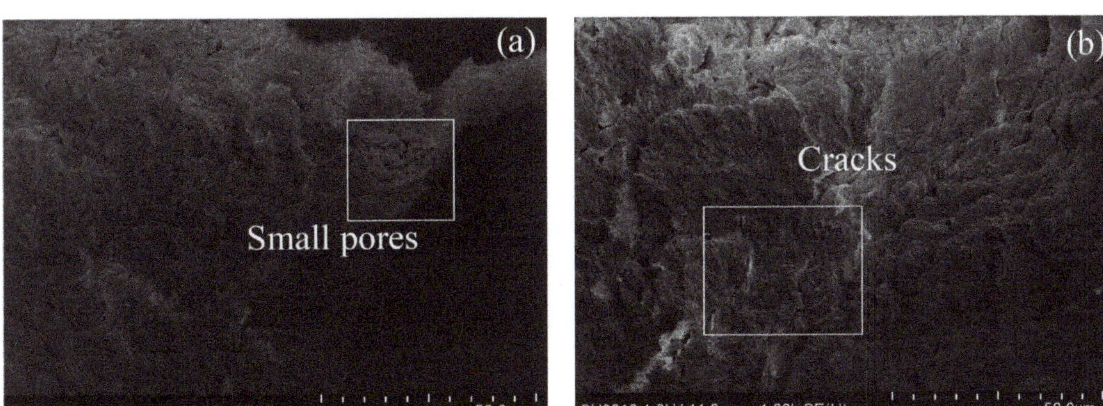

Figure 12. *Cont.*

Figure 12. SEM of RM–MgO stabilized soil: (**a**) NRM-8, without freeze–thaw cycles; (**b**) NRM-8, after 10 freeze–thaw cycles; (**c**) NRM-0, without freeze–thaw cycles; (**d**) NRM-0, after 10 freeze–thaw cycles.

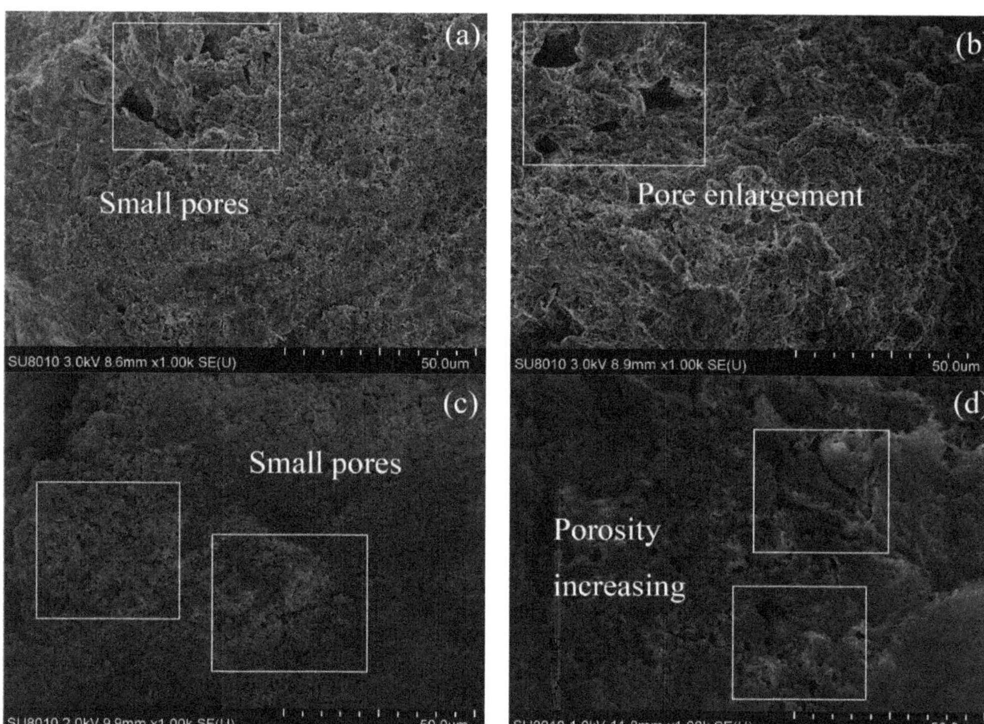

Figure 13. SEM of RM–CaO stabilized soil: (**a**) NRL-6, without freeze–thaw cycles; (**b**) NRL-6, after 10 freeze–thaw cycles; (**c**) NRL-0, without freeze–thaw cycles; (**d**) NRL-0, after 10 freeze–thaw cycles.

4. Stabilization Mechanism

For the RM–MgO stabilized soil, when the RM, MgO, and water were mixed in the soil, the MgO in contact with the water produced a hydration reaction to form $Mg(OH)_2$, as shown in Equation (4). As shown in Equations (5) and (6), some of the $Mg(OH)_2$ dissociated in the water to form Mg^{2+} and OH^-, which promoted the dissolution of reactive alumina and reactive silica in the RM. The dissolved silica and alumina were combined with Mg^{2+} in the soil to form MSH and other gelling substances. Similarly, for the RM–CaO stabilized soil, when the RM, CaO, and water were mixed in the soil, as in Equation (7), CaO generated $Ca(OH)_2$ when it came into contact with the water. Part of the $Ca(OH)_2$ would dissociate the Ca^{2+} and OH^- in the water, as shown in Equations (8) and (9), and the active alumina and active silicon oxide in the RM started to dissolve and polymerize with the free Ca^{2+} in the soil to generate gelling substances, such as CSH. MSH and CSH filled the pores in the soil. The connections between the soil particles became tighter, thus improving the densification of the soil. The UCS of the RM–MgO and RM–CaO stabilized soil was improved. As can be seen in Figures 2 and 3, with the addition of NaOH, more MgO (or CaO) was consumed for the hydration reaction. This is because that the addition of NaOH provided an alkaline environment, which could facilitate hydration reactions to form CSH [30,32]. Additionally, the cementing effect of CSH was stronger than that of MSH [50], which led to a difference in the enhancement between the RM–MgO and RM–CaO stabilized soil. The reaction process is as follows:

$$\mathrm{MgO + H_2O \rightarrow Mg(OH)_2} \tag{4}$$

$$\mathrm{Mg(OH)_2 \rightarrow Mg^{2+} + 2OH^-} \tag{5}$$

$$\mathrm{Mg(OH)_2 + SiO_2 + nH_2O \rightarrow MgO{\cdot}SiO_2{\cdot}(n+1)H_2O} \tag{6}$$

$$\mathrm{CaO + H_2O \rightarrow Ca(OH)_2} \tag{7}$$

$$\mathrm{Ca(OH)_2 \rightarrow Ca^{2+} + 2OH^-} \tag{8}$$

$$\mathrm{Ca(OH)_2 + SiO_2 + nH_2O \rightarrow CaO{\cdot}SiO_2{\cdot}(n+1)H_2O} \tag{9}$$

5. Conclusions

(1) Both MgO and CaO had a good activation effect on the RM. The UCS reached a maximum value of 9.65 MPa when the mass ratio of RM to MgO was 3:4 and a maximum value of 7.48 MPa when the mass ratio of RM to CaO was 3:2.

(2) The UCS of the RM–MgO (or RM–CaO) stabilized soil first increased and then decreased with the decrease in RM content. The UCS reached a maximum value of 1.73 MPa when the mass ratio of the RM–MgO stabilized soil (without NaOH) was 5:5 and a maximum value of 514.95 kPa when the mass ratio of the RM–CaO stabilized soil (without NaOH) was 8:2. Compared with the RM–MgO stabilized soil after adding NaOH, the enhancement of the UCS of the RM–CaO stabilized soil by NaOH was much greater. The optimum ratios for the RM–MgO stabilized soil (with NaOH) and the RM–CaO stabilized soil (with NaOH) were 2:8 (1.87 MPa) and 4:6 (2.32 MPa), respectively. The stabilized soil's UCS values are, in increasing order, as follows: the RM–CaO stabilized soil (without NaOH); RM–MgO stabilized soil (without NaOH); RM–MgO stabilized soil (with NaOH); and RM–CaO stabilized soil (with NaOH).

(3) The ability of the stabilized soil to resist freeze–thaw erosion was significantly improved by adding the RM. With the addition of the NaOH solution, the mass loss rate, volume expansion rate, and UCS reduction in the RM–MgO (or RM–CaO) stabilized soil after the freeze–thaw cycles were significantly lower than those of the pure CaO or pure MgO stabilized soil. The RM–MgO stabilized soil showed better resistance to mass loss and volume expansion than the RM–CaO stabilized soil.

(4) The main reason for the elevated UCS of the RM–MgO (or RM–CaO) stabilized soil was the formation of cementitious materials, such as CSH and MSH. Freeze–thaw cycles did not weaken the formation of its CSH and MSH but caused the pore volume of the specimen to become larger and cracks to form inside the specimen. The decrease in the UCS of the specimen was mainly due to these physical damages.

The test method for the determination of the California bearing ratio, immediate bearing index, and linear swelling was not considered in this study. If red mud is utilized in road base materials, these tests should be considered. So, in our next study, we will consider these tests.

Author Contributions: Conceptualization, L.L.; Data curation, W.L., K.H. and F.C.; Formal analysis, W.L., L.L. and Y.C.; Investigation, L.L.; Methodology, F.C.; Resources, F.C. and K.Y.; Software, Y.C. and K.Y.; Validation, K.H.; Writing—original draft, K.H.; Writing—review and editing, W.L. All authors have read and agreed to the published version of the manuscript.

Funding: The authors would like to acknowledge the following grants: Open Project Funding of Key Laboratory of Intelligent Health Perception and Ecological Restoration of Rivers and Lakes, Ministry of Education, Hubei University of Technology (HGKFZP008), the Joint Funds of the Natural Science Foundation of Hubei Province (2022CFD130), the Key Research and Development Program of Hubei Province (No. 2021BGD015), and the Key Research and Development Program of Hubei Province (2023BCB116).

Institutional Review Board Statement: Not applicable.

Informed Consent Statement: Not applicable.

Data Availability Statement: All data, models, and code generated or used during this study appear in the submitted article.

Conflicts of Interest: The authors declare no conflicts of interest.

References

1. Yuan, B.; Yuan, S.; Straub, C.; Chen, W. Activation of binary binder containing fly ash and portland cement using red mud as alkali source and its application in controlled low-strength materials. *J. Mater. Civ. Eng.* **2020**, *32*, 04019356. [CrossRef]
2. Liu, R.-X.; Poon, C.-S. Effects of red mud on properties of self-compacting mortar. *J. Cleaner Prod.* **2016**, *135*, 1170–1178. [CrossRef]
3. Qaidi, S.M.; Tayeh, B.A.; Ahmed, H.U.; Emad, W. A review of the sustainable utilisation of red mud and fly ash for the production of geopolymer composites. *Constr. Build. Mater.* **2022**, *350*, 128892. [CrossRef]
4. Li, W.; Yan, X.; Niu, Z.; Zhu, X. Selective recovery of vanadium from red mud by leaching with using oxalic acid and sodium sulfite. *J. Environ. Chem. Eng.* **2021**, *9*, 105669. [CrossRef]
5. Archambo, M.; Kawatra, S. Red mud: Fundamentals and new avenues for utilization. *Miner. Process. Extr. Metall. Rev.* **2021**, *42*, 427–450. [CrossRef]
6. Liu, W.; Chen, X.; Li, W.; Yu, Y.; Yan, K. Environmental assessment, management and utilization of red mud in China. *J. Clean. Prod.* **2014**, *84*, 606–610. [CrossRef]
7. Li, S.; Zhang, R.; Feng, R.; Hu, B.; Wang, G.; Yu, H. Feasibility of recycling Bayer process red mud for the safety backfill mining of layered soft bauxite under coal seams. *Minerals* **2021**, *11*, 722. [CrossRef]
8. Mi, H.; Yi, L.; Wu, Q.; Xia, J.; Zhang, B. A review of comprehensive utilization of red mud. *Waste Manag. Res.* **2022**, *40*, 1594–1607. [CrossRef]
9. Chavan, S.; Salokhe, S.; Nadagauda, P.; Patil, S.; Mane, K. An investigational study on properties of concrete produced with industrial waste red mud. *Mater. Today Proc.* **2021**, *42*, 733–738. [CrossRef]
10. Kavak, A.; Baykal, G. Long-term behavior of lime-stabilized kaolinite clay. *Environ. Earth Sci.* **2012**, *66*, 1943–1955. [CrossRef]
11. Li, W.; Chen, Y.; Li, R.; Xiao, H.; Li, L.; Yang, J. A new method for carbonating magnesia and its possible application to soil stabilization and CO_2 capture. *Constr. Build. Mater.* **2023**, *383*, 131263. [CrossRef]
12. Alam, S.; Das, S.K.; Rao, B.H. Strength and durability characteristic of alkali activated GGBS stabilized red mud as geo-material. *Constr. Build. Mater.* **2019**, *211*, 932–942. [CrossRef]
13. Sahile, K. Utilization of red mud-fly ash reinforced with cement in road construction applications. *Adv. Mater. Sci. Eng.* **2021**, *2021*, 3728652. [CrossRef]
14. Mukiza, E.; Zhang, L.; Liu, X.; Zhang, N. Utilization of red mud in road base and subgrade materials: A review. *Resour. Conserv. Recycl.* **2019**, *141*, 187–199. [CrossRef]
15. Li, R.; Zhang, T.; Liu, Y.; Lv, G.; Xie, L. Calcification–carbonation method for red mud processing. *J. Hazard. Mater.* **2016**, *316*, 94–101. [CrossRef]
16. Fei, J.-C.; Min, X.-B.; Wang, Z.-X.; Pang, Z.-H.; Liang, Y.-J.; Ke, Y. Health and ecological risk assessment of heavy metals pollution in an antimony mining region: A case study from South China. *Environ. Sci. Pollut. Res.* **2017**, *24*, 27573–27586. [CrossRef]
17. Liu, D.-G.; Min, X.-B.; Ke, Y.; Chai, L.-Y.; Liang, Y.-J.; Li, Y.-C.; Yao, L.-W.; Wang, Z.-B. Co-treatment of flotation waste, neutralization sludge, and arsenic-containing gypsum sludge from copper smelting: Solidification/stabilization of arsenic and heavy metals with minimal cement clinker. *Environ. Sci. Pollut. Res.* **2018**, *25*, 7600–7607. [CrossRef]
18. Vigneshwaran, S.; Uthayakumar, M.; Arumugaprabu, V. Development and sustainability of industrial waste-based red mud hybrid composites. *J. Cleaner Prod.* **2019**, *230*, 862–868. [CrossRef]
19. Li, Y.-C.; Min, X.-B.; Ke, Y.; Chai, L.-Y.; Shi, M.-Q.; Tang, C.-J.; Wang, Q.-W.; Liang, Y.-J.; Lei, J.; Liu, D.-G. Utilization of red mud and Pb/Zn smelter waste for the synthesis of a red mud-based cementitious material. *J. Hazard. Mater.* **2018**, *344*, 343–349. [CrossRef]

20. Zhang, J.; Li, S.; Li, Z.; Liu, C.; Gao, Y. Feasibility study of red mud for geopolymer preparation: Effect of particle size fraction. *J. Mater. Cycles Waste* **2020**, *22*, 1328–1338. [CrossRef]
21. Sreelekshmi, S.; Chandran, R.S.; Varghese, S.R.; Raju, T.; Ramaswamy, K. Rheological and strength behaviour of red mud-slag geopolymer composites. *IOP Conf. Ser. Mater. Sci. Eng.* **2020**, *989*, 012005. [CrossRef]
22. Feigl, V.; Anton, A.; Uzigner, N.; Gruiz, K. Red mud as a chemical stabilizer for soil contaminated with toxic metals. *Water Air Soil Pollut.* **2012**, *223*, 1237–1247. [CrossRef]
23. Aswathy, M.; Salini, U.; Gayathri, V. Utility of lime and red mud in clay soil stabilization. *Geotech. Charact. Geoenviron. Eng.* **2019**, *1*, 19–26. [CrossRef]
24. Sridevi, G.; Sahoo, S.; Sen, S. Stabilization of expansive soil with red mud and lime. *Ground Improv. Tech. Geosynth.* **2019**, *2*, 259–268. [CrossRef]
25. Ma, Q.; Duan, W.; Liu, X.; Fang, P.; Chen, R.; Wang, T.; Hao, Z. Engineering Performance Evaluation of Recycled Red Mud Stabilized Loessial Silt as a Sustainable Subgrade Material. *Materials* **2022**, *15*, 3391. [CrossRef]
26. Suo, C.; Yao, X.; Song, Z.; Dong, X. Mechanical and leaching characteristics of red mud residue solidified/stabilized high Cu (II)-contaminated soil. *Environ. Earth Sci.* **2022**, *81*, 9. [CrossRef]
27. Hu, W.; Nie, Q.; Huang, B.; Su, A.; Du, Y.; Shu, X.; He, Q. Mechanical property and microstructure characteristics of geopolymer stabilized aggregate base. *Constr. Build. Mater.* **2018**, *191*, 1120–1127. [CrossRef]
28. Guo, S.; Lu, Z.; Liu, G.; Zhuang, B.; Fan, Y.; Zheng, Z. Influence of Freeze-Thaw Cycles on Mechanical Response of Levee Pavement. *Adv. Civ. Eng.* **2021**, *2021*, 6639303. [CrossRef]
29. Zhou, B.; Sanchez, M.; Shastri, A.; Lee, J. Mechanical Behavior of Soil Subjected to Freezing-Thawing Cycles. *PanAm Unsaturated Soils* **2017**, *2017*, 300–307. [CrossRef]
30. Yang, Z.; Li, X.; Li, D.; Wang, Y.; Liu, X. Effects of long-term repeated freeze-thaw cycles on the engineering properties of compound solidified/stabilized Pb-contaminated soil: Deterioration characteristics and mechanisms. *Int. J. Environ. Res. Public Health* **2020**, *17*, 1798. [CrossRef]
31. Cristelo, N.; Glendinning, S.; Fernandes, L.; Pinto, A.T. Effect of calcium content on soil stabilisation with alkaline activation. *Constr. Build. Mater.* **2012**, *29*, 167–174. [CrossRef]
32. Pavithra, P.; Srinivasula Reddy, M.; Dinakar, P.; Hanumantha Rao, B.; Satpathy, B.; Mohanty, A. Effect of the Na_2SiO_3/NaOH ratio and NaOH molarity on the synthesis of fly ash-based geopolymer mortar. *Geo-Chicago* **2016**, *2016*, 336–344. [CrossRef]
33. *ASTM D698-12e2*; Standard Test Methods for Laboratory Compaction Characteristics of Soil Using Standard Effort (12,400 ft-lbf/ft^3 (600 kN-m/m^3)). ASTM: West Conshohocken, PA, USA, 2012.
34. *ASTM D1633*; Standard Method for Compressive Strength of Molded Soil-Cement Cylinders. ASTM: West Conshohocken, PA, USA, 2017.
35. *ASTM D2166/D2166M-16*; Standard Test Method for Unconfined Compressive Strength of Cohesive Soil. ASTM: West Conshohocken, PA, USA, 2016.
36. Chen, R.; Cai, G.; Dong, X.; Mi, D.; Puppala, A.J.; Duan, W. Mechanical properties and micro-mechanism of loess roadbed filling using by-product red mud as a partial alternative. *Constr. Build. Mater.* **2019**, *216*, 188–201. [CrossRef]
37. Zhongping, Y.; Yao, W.; Xuyong, L.; Shupei, R.; Hui, X.; Jiazhuo, C. The effect of long-term freeze-thaw cycles on the stabilization of lead in compound solidified/stabilized lead-contaminated soil. *Environ. Sci. Pollut. Res.* **2021**, *28*, 37413–37423. [CrossRef]
38. Nie, Q.; Hu, W.; Ai, T.; Huang, B.; Shu, X.; He, Q. Strength properties of geopolymers derived from original and desulfurized red mud cured at ambient temperature. *Constr. Build. Mater.* **2016**, *125*, 905–911. [CrossRef]
39. Yi, Y.; Gu, L.; Liu, S.; Jin, F. Magnesia reactivity on activating efficacy for ground granulated blastfurnace slag for soft clay stabilisation. *Appl. Clay Sci.* **2016**, *126*, 57–62. [CrossRef]
40. Millogo, Y.; Hajjaji, M.; Ouedraogo, R. Microstructure and physical properties of lime-clayey adobe bricks. *Constr. Build. Mater.* **2008**, *22*, 2386–2392. [CrossRef]
41. *ASTM D4609*; Standard Guide for Evaluating the Effectiveness of Admixture for Soil Stabilization. ASTM: West Conshohocken, PA, USA, 2008.
42. Zhang, W.; Pi, Y.; Kong, W.; Zhang, Y.; Wu, P.; Zeng, W.; Yang, F. Influence of damage degree on the degradation of concrete under freezing-thawing cycles. *Constr. Build. Mater.* **2020**, *260*, 119903. [CrossRef]
43. Lu, Y.; Liu, S.; Zhang, Y.; Li, Z.; Xu, L. Freeze-thaw performance of a cement-treated expansive soil. *Cold Reg. Sci. Technol.* **2020**, *170*, 102926. [CrossRef]
44. Konrad, J. Physical processes during freeze-thaw cycles in clayey silts. *Cold Reg. Sci. Technol.* **1989**, *16*, 291–303. [CrossRef]
45. Li, G.; Ma, W.; Zhao, S.; Mao, Y.; Mu, Y. Effect of freeze-thaw cycles on mechanical behavior of compacted fine-grained soil. *J. Cold Reg. Eng.* **2012**, *2012*, 72–81. [CrossRef]
46. Dong, J.; Lyu, H.; Xu, G.; He, C. NMR-based study on soil pore structures affected by drying–wetting cycles. *Arab. J. Sci. Eng.* **2020**, *45*, 4161–4169. [CrossRef]
47. He, J.; Zhang, L.; Zhang, C. Durability Analysis of Sludge Solidified with Soda Residue Subjected to Dry-Wet and Freeze-Thaw Cycles. *Adv. Civ. Eng.* **2021**, *2021*, 9960416. [CrossRef]
48. Zheng, Y.; Ma, W.; Mu, Y.; Bing, H. Analysis of soil structures and the mechanisms under the Action of freezing and thawing cycles. *J. Cold Reg. Eng.* **2015**, *2015*, 25–33. [CrossRef]

49. Li, W.; Li, R.; Chen, Y.; Xiao, H. Comparison of Two Sulfate-Bearing Soils Stabilized with Reactive Magnesia-Activated Ground Granulated Blast Furnace Slag: Swelling, Strength, and Mechanism. *Buildings* **2023**, *13*, 230. [CrossRef]
50. Hekal, E.E.; Kishar, E.; Mostafa, H. Magnesium sulfate attack on hardened blended cement pastes under different circumstances. *Cement Concrete Res.* **2002**, *32*, 1421–1427. [CrossRef]

materials

MDPI

Article

Experimental Study on Effects of CO_2 Curing Conditions on Mechanical Properties of Cement Paste Containing CO_2 Reactive Hardening Calcium Silicate Cement

Young-Jin Kim, Sang-Rak Sim and Dong-Woo Ryu *

Department of Architectural Engineering, Daejin University, Pocheon-si 11159, Gyeonggi-do, Republic of Korea; rladud316@gmail.com (Y.-J.K.); simsr@daejin.ac.kr (S.-R.S.)
* Correspondence: dwryu@daejin.ac.kr; Tel.: +82-31-539-1949

Abstract: Human survival is threatened by the rapid climate change due to global warming caused by the increase in CO_2 emissions since the Second Industrial Revolution. This study developed a secondary cement product production technology by replacing cement, a conventional binder, with calcium silicate cement (CSC), i.e., CO_2 reactive hardening cement, to reduce CO_2 emissions and utilize CO_2 from the cement industry, which emits CO_2 in large quantities. Results showed that the carbonation depth, compressive strength increase rate, and CO_2 sequestration rate increased as the CSC content increased, suggesting that CSC can be applied as a secondary cement product.

Keywords: CO_2 curing; mineral carbonation; calcium silicate cement; CO_2 sequestration; decalcification

1. Introduction

Humanity is threatened by global warming due to the increase in CO_2 emissions since the Second Industrial Revolution. Based on the "Global Warming 1.5 °C Special Report" from the Intergovernmental Panel on Climate Change, the average global temperature increased by approximately 1.09 °C after industrialization as of 2018, and the increase is expected to reach 1.5 °C between 2030 and 2052 [1]. This prompted the international community to establish the goal of 2050 carbon neutrality. To this end, carbon emissions should be effectively reduced, or high carbon emission reduction rates should be allocated to the energy conversion, waste, transport, and construction sectors, which generate high carbon emissions in the industrial sector [2].

In particular, the cement industry, which accounts for the majority of CO_2 emissions from the mineral industry, is classified as a high-carbon-emitting industry [3]. Carbon emissions are increasing annually on a global scale. The 10 major countries contributing to carbon emissions are listed in Table 1 [4]. Carbon capture utilization (CCU) based on the carbonation reaction of cementitious materials has been presented as a key technology for CO_2 reduction in the cement industry. In general, conventional concrete carbonation reactions have been recognized as a degradation factor that corrodes steel reinforcement because of the loss in alkalinity inside concrete caused by the carbonation of $Ca(OH)_2$, which is a hydration product. However, studies have reported that proper carbonation reactions can improve strength and durability through the densification of the microstructure inside concrete [5]. Concrete is the second most consumed resource worldwide annually, and the implementation of efficient carbonation technology is expected to significantly reduce CO_2 emissions from the cement industry [6,7].

To reduce CO_2 emissions and utilize CO_2 during the production and curing processes of cement, CO_2 reactive hardening calcium silicate cement (CSC) has been extensively investigated. CSC primarily consists of rankinite (C_3S_2) and wollastonite (CS). It is non-hydraulic, as it does not react with water; however, it hardens through reactions with CO_2, thereby developing strength [8]. The carbonation curing mechanism of CSC entails the penetration of CO_2 into capillary pores during the initial stages of the reaction. This leads to the dissolution and ionization of water within the pores, resulting in the formation of H^+, HCO^{3-}, and $CO_3{}^{2-}$ ions. Specifically, the generation of significant quantities of H^+ decreases the pH of the pore water and triggers the dissolution of Ca^{2+} from the C_3S_2 and CS phases. Then, H_4SiO_4 is polymerized to form a SiO_2 gel with a high degree of polymerization. During the subsequent process of dissolution and polymerization, a slow depletion of H^+ ions occurs, leading to the restoration of the pH of the pore solution. This creates an environment that is conducive to the formation of $CaCO_3$. The formation of the SiO_2 gel and $CaCO_3$ around the C_3S_2 and CS phases while curing due to the above-mentioned reaction is the carbonation curing mechanism of CSC (see Figure 1) [9].

Table 1. Carbon emissions per capita in major countries [4].

Country	CO_2 Emissions per Capita (tCO_2)	CO_2 Emissions ($MtCO_2$)
China	8.0	11,472.37
United States of America	15	5007
India	1.9	2710
Russian Federation	12	1756
Japan	8.6	1067
Iran	8.5	749
Germany	8.1	675
Saudi Arabia	19	672
Indonesia	2.3	619
South Korea	12	616

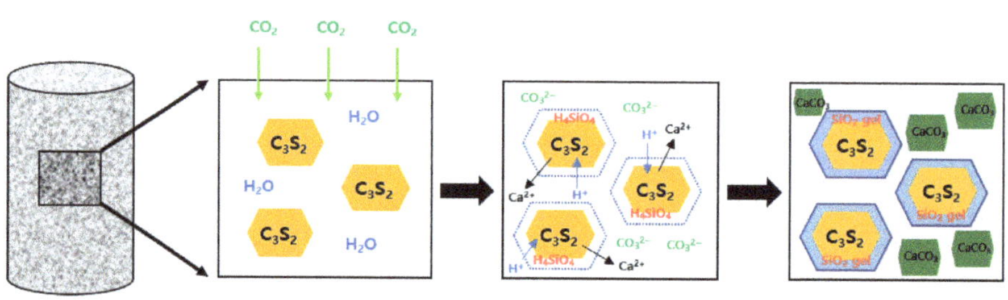

Figure 1. CSC curing mechanism.

CSC production can reduce CO_2 emissions by up to 70% compared with ordinary Portland cement (OPC) production owing to the reduced consumption of limestone and the reduced firing temperature (1250 °C). In addition, CO_2 reduction due to the additional CO_2 sequestration resulting from the carbonation curing process of secondary cement products can be expected [10,11].

Similarly, Bao Lu et al. [12] formulated a clinker that mostly comprises C_3S_2 and γ-C_2S mineral phases to realize a cement clinker with low carbon content. They subjected it to carbonation curing and observed a direct correlation between the amount of CO_2 fixation and the increase in compressive strength. Ashraf et al. [13] found that the highest amount of $CaCO_3$ produced during the maximum carbonization of CSC is approximately 40% (by weight of paste).

In this study, a paste was prepared with different CSC contents and CO_2 curing conditions, and basic mechanical properties were examined for the production of CSC-based secondary wet cement products for realizing CCU. CSC is non-hydraulic and develops strength through carbonation reactions. Because demolding is essential for the application of homogeneous carbonation curing to the entire surface of the paste containing CSC, blast-furnace slag cement (BSC) was added as a binder instead of OPC for minimum strength development. To realize CSC-based secondary cement products, strength should be developed for efficient demolding in the typical secondary cement product production process. Moreover, the optimal carbonation curing conditions and optimal CSC content that enables the development of strength equal to or higher than the required strength for conventional secondary cement products should be derived.

2. Experimental

2.1. Experimental Factors and Levels

The experimental factors and levels are presented in Table 2. The 0, 50, and 100 wt% BSC were replaced with CSC for cement paste. The corresponding specimens were labeled as P-CSC0, P-CSC50, and P-CSC100, respectively. Paste specimens with dimensions of $50 \times 50 \times 50$ mm^3 were prepared for the experiment.

Table 2. Experimental factors and levels.

<table>
<tr><th>Factors</th><th colspan="2">Title 2</th><th>Levels</th><th>Notes</th></tr>
<tr><td>Specimens</td><td colspan="2">W/B
CSC (%)</td><td>0.36
0, 50, 10</td><td>Paste</td></tr>
<tr><td rowspan="3">Curing</td><td colspan="2">Pre-curing</td><td>Steam curing</td><td>KS L 4004</td></tr>
<tr><td rowspan="2">Carbonation Curing</td><td>CO_2 20%</td><td>4, 8, 12, 24 h</td><td></td></tr>
<tr><td>CO_2 99%</td><td>4, 8, 12, 24 h</td><td></td></tr>
<tr><td></td><td colspan="2">Mechanical properties</td><td>Compressive strength</td><td>KS F 5105</td></tr>
<tr><td></td><td colspan="2">Quantitative analysis</td><td>Carbonation depth,
XRD, TG-DTA,
FT-IR, SEM</td><td></td></tr>
</table>

2.2. Materials

In this study, BSC and CSC, which are commonly used for the production of secondary cement products and suitable for KS L 5210, from the company H (Seoul, Republic of Korea) were used. The chemical compositions of the materials were analyzed using X-ray fluorescence. The results are presented in Table 3. The primary mineral facies of CSC were analyzed using X-ray diffraction (XRD, Rigaku D, max 2200, Tokyo, Japan), as shown in Figure 2. A carbonation curing period of less than 24 h was applied considering the secondary product production process. The cement paste was analyzed, excluding the influence of the aggregate, to identify the hardening characteristics of cement containing CSC based on the carbonation reaction products.

Table 3. Chemical composition of binder (obtained by X-ray fluorescence spectroscopy (Rigaku D, ZSX PrimuslV, Tokyo, Japan)).

	Chemical Composition (wt%)				
	CaO	SiO_2	Al_2O_3	Fe_2O_3	MgO
BSC	52.64	19.86	7.41	2.21	4.2
CSC	48.28	44	0.45	0.67	0.46

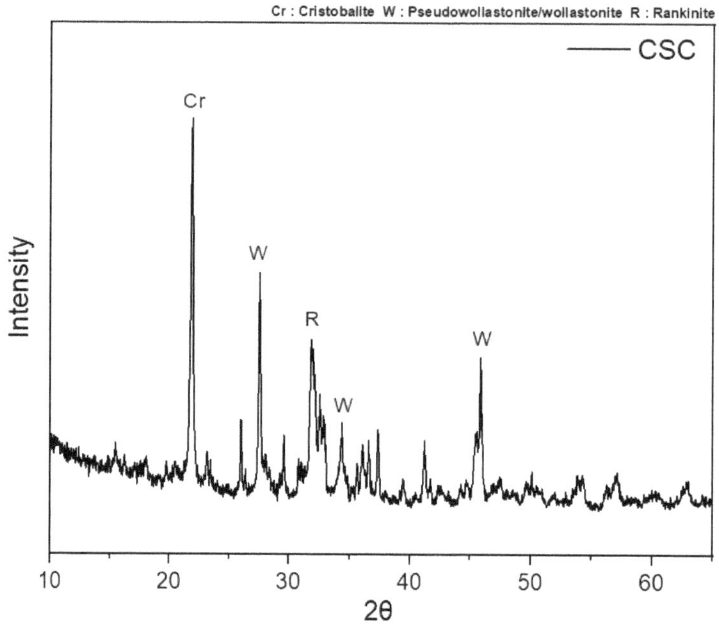

Figure 2. Main mineral facies of CSC.

2.3. Curing Method

Regarding the curing of the secondary wet cement products, after steam curing was performed as pre-curing before demolding, carbonation curing was applied uniformly to all the surfaces, excluding the bottom surface, of each specimen within 24 h after demolding.

Steam curing was performed following KS L 4002 (empty concrete block). The accumulated temperature was set to 500 °C·h by applying a pre-curing temperature of 20 °C, a maximum temperature of 65 °C, and a heating rate of 20 °C/h (Figure 3). For carbonation curing, CO_2 concentrations of 20% and 99% and curing periods of 4, 8, 12, and 24 h were applied.

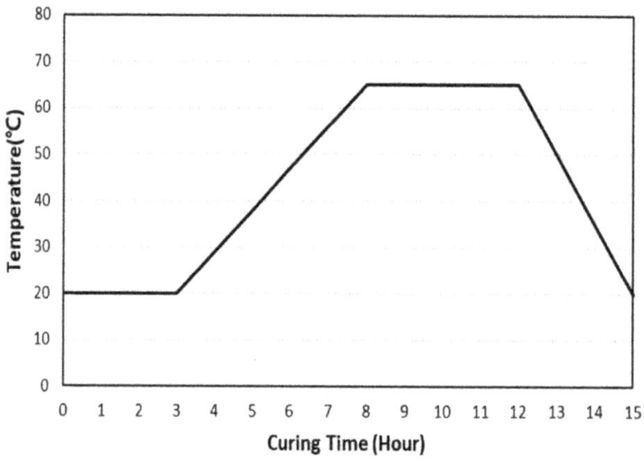

Figure 3. Steam curing method.

2.4. Test Methods

2.4.1. Compressive Strength Evaluation

To evaluate the compressive strength based on the carbonation curing conditions, a test was conducted following KS L 5105 (compressive strength test method for hydraulic cement mortar). Specimens with dimensions of $50 \times 50 \times 50$ mm^3 were prepared. The compressive strength was measured for three samples upon completion of carbonation curing (4, 8, 12, and 24 h) at CO_2 concentrations of 20% and 99%, and the average value was used.

2.4.2. Carbonation Depth Evaluation

To evaluate the carbonation degree of paste containing CSC, 1% phenolphthalein solution was sprayed on the surface of each carbonation-cured specimen. The carbonation depth was measured at four left and right positions at equal intervals in areas that were not discolored, and the average value was calculated, as shown in Figure 4.

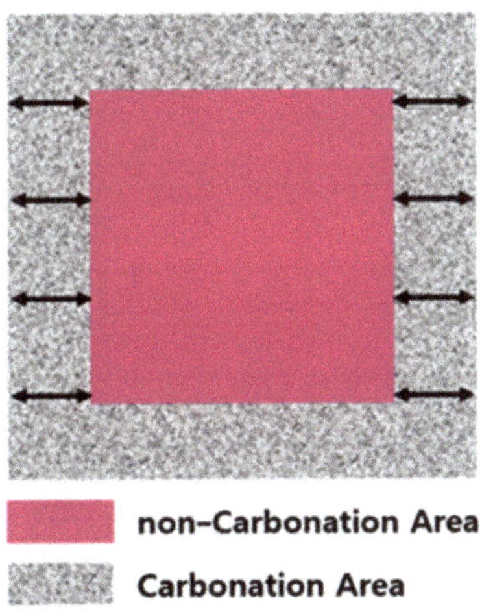

Figure 4. Carbonation depth measurement.

2.5. Quantitative Analysis

For quantitative analysis, 5 mm thick cubic samples were collected at depths of 5, 10, and 15 mm from the surface of each specimen using a microsampling cutter upon completion of carbonation curing. The collected samples were immersed in acetone solution for 2 d to stop hydration. They were crushed to a size of less than 75 μm, and the crushed powder was stored in a vacuum desiccator to maintain completely dry conditions.

2.5.1. XRD Analysis

XRD analysis was conducted by collecting samples at depths of 5, 10, and 15 mm from the surface of each specimen. Measurements were performed using a tube current of 30 mA and a voltage of 40 kV. The XRD analysis was conducted in the scanning angle range of 5–65°.

2.5.2. Thermogravimetric Analysis (TGA)

Thermogravimetry/differential thermal analysis (TG-DTA, STA 2500, Netzsch, Selb, Germany) was conducted to analyze the carbonation rate. Each processed powder was placed in a TGA sample holder, and the weight reduction rate was measured in the temperature range of 25–1000 °C at a heating rate of 10 °C/min in a high-purity nitrogen (99.999%) environment. For the main weight reduction of carbonated cement paste containing CSC, the weight reduction rate caused by the pyrolysis of the reaction products of BSC and CSC was measured.

2.5.3. Fourier Transform Infrared Spectroscopy (FT-IR)

In an FT-IR (ALPHA2, Bruker, Ettlingen, Germany) analysis, scanning was performed in the wavenumber range of 400–4000 cm^{-1} with a resolution of 2 cm^{-1} in the attenuated total internal reflection mode for the samples collected near a depth of 5 mm from the surface.

Typical carbonation reactions of cement generate $CaCO_3$ and silica gel by inducing the carbonation of $Ca(OH)_2$ and C-S-H, which are hydration products. The carbonation reaction of $Ca(OH)_2$ can promote the strength improvement effect by filling pores inside the hardened body, but the carbonation of C-S-H, which is responsible for strength, may reduce the strength through the contraction caused by decalcification [14–17]. To identify the carbonation degree of C-S-H generated by BSC, which is the binder of cement paste containing CSC, carbonation characteristics based on the carbonation period were analyzed using FT-IR spectroscopy, as this method can measure C-O bonds and Si-O chain structures.

3. Results

3.1. Carbonation Depth Analysis Results

The carbonation depth measurement results are presented in Figures 5 and 6. When carbonation curing was performed at a CO_2 concentration of 20%, the carbonation depth tended to increase as the CSC content increased. When carbonation curing was performed for 24 h, a carbonation depth of 12.3 mm was observed for the P-CSC100 specimen. However, almost no carbonation occurred in the P-CSC0 (0 mm) and P-CSC50 (3.63 mm) mixes. For the P-CSC0 mix, no carbonation occurred because the penetration of CO_2 was obstructed by the dense pore structure, as a high compressive strength of 55 MPa was developed by performing steam curing as pre-curing [18].

In addition, the carbonation depth rapidly increased in certain sections when carbonation curing was performed. The carbonation depth rapidly increased after 12 h of curing for the P-CSC50 mix and 8 h of curing for the P-CSC100 mix. This is because the diffusion of CO_2 was reduced at the beginning of carbonation curing by the pores in the specimen saturated with moisture [19], as carbonation curing was performed immediately upon completion of steam curing under a 100% relative humidity condition as pre-curing. However, the carbonation depth was increased by the increased diffusion rate of CO_2 along with the evaporation of moisture.

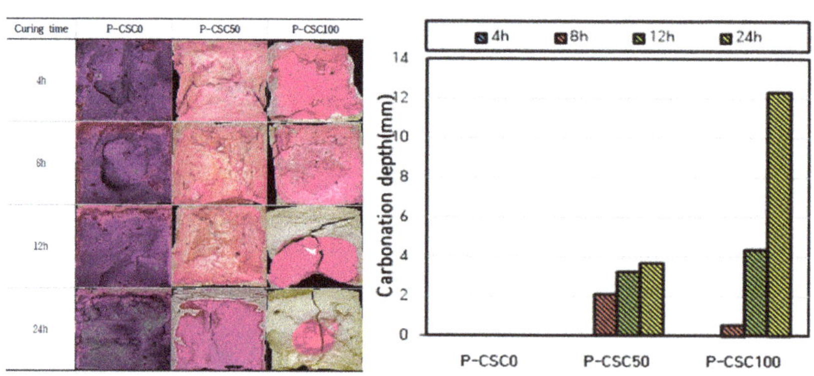

Figure 5. CO_2 concentration 20% curing carbonation depth measurement result.

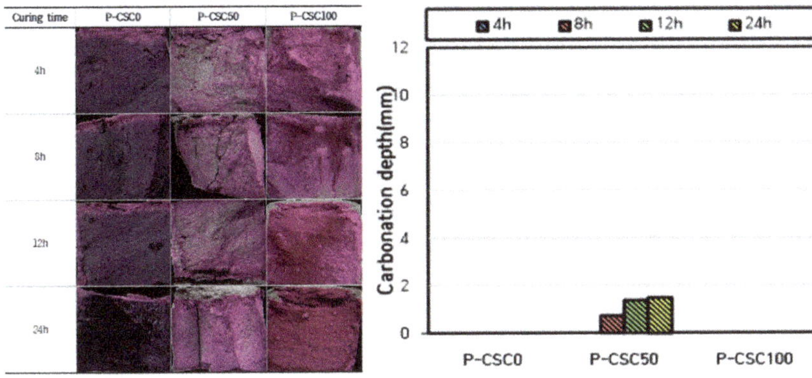

Figure 6. CO_2 concentration 99% curing carbonation depth measurement result.

When the carbonation depth was measured for the specimens subjected to carbonation curing at a CO_2 concentration of 99%, the P-CSC0 mix exhibited the same carbonation depth as that cured at a CO_2 concentration of 20%. The highest carbonation depth was observed for the P-CSC50 mix, but it was as low as 0–2 mm. Moreover, the depth of carbonation could not be determined for the P-CSC100 mix. This is because the diffusion of CO_2 decreased as micropores were filled with $CaCO_3$, which was generated by the rapid carbonation of the specimen surface at the beginning of the reaction due to high-concentration CO_2 diffusion [20].

3.2. Results of XRD Analysis and TG-DTA

Figure 7 shows the XRD patterns of the specimens collected at depths of 5, 10, and 15 mm subjected to carbonation curing at a CO_2 concentration of 20%. Table 4 presents the TG-DTA results at positions where the carbonation depth was measured by spraying the phenolphthalein solution.

Table 4. TG-DTA results.

Title 1		**CO_2 20%**		**CO_2 99%**	
		400–500 °C	**600–800 °C**	**400–500 °C**	**600–800 °C**
CSC0	4 h—5 mm	1.48	1.55	1.36	2.76
	8 h—5 mm	1.63	2.36	1.53	3.05
	12 h—5 mm	1.23	2.76	1.2	2.3
	24 h—5 mm	1.54	2.04	1.23	2.36
CSC50	4 h—5 mm	1.04	3.64	1.45	3.32
	8 h—5 mm	1.24	4.23	1.68	2.72
	12 h—5 mm	0.63	5.31	1.06	3.04
	24 h—5 mm	0.95	7.2	1.23	3.02
CSC100	4 h—5 mm	1.17	1.78	0.88	2.57
	8 h—5 mm	0	0.84	0.53	4.13
	12 h—5 mm	0	11.39	1.55	1.26
	24 h—5 mm	0.16	12.04	1.6	1.36
	24 h—10 mm	0.17	11.05	1.2	1.02
	24 h—15 mm	0.13	9.04	1.16	0.97

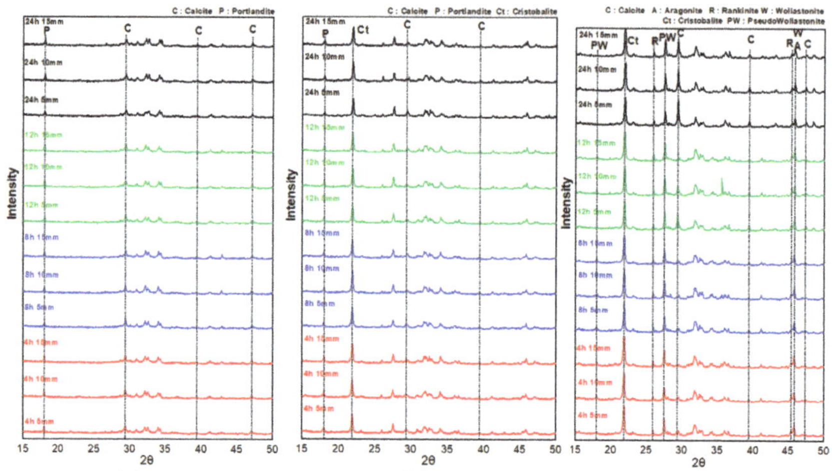

(a) (b) (c)

Figure 7. CO_2 concentration 20% curing XRD pattern: (**a**) P-CSC0, (**b**) P-CSC50, and (**c**) P-CSC100.

In the XRD analysis results for the CO_2 concentration of 20%, the P-CSC0 mix exhibited $Ca(OH)_2$ and calcite peaks. The $Ca(OH)_2$ peak intensity decreased and the calcite peak intensity tended to slightly increase toward the surface. However, no significant peak change with the curing period was observed. Additionally, the TG-DTA measurement results indicated that the weight reduction rate tended to slightly increase as the carbonation curing period increased in the range of 600–800 °C, which is the pyrolysis range of $CaCO_3$.

For the P-CSC50 and P-CSC100 mixes containing CSC, the peaks of pseudowollastonite, wollastonite, rankinite, and cristobalite, which are minerals that constitute CSC, and those of calcite and aragonite, which are carbonate minerals, were observed [13,21]. The XRD analysis results for the P-CSC50 mix indicated that the intensities of the $Ca(OH)_2$ and calcite peaks were reduced and increased, respectively, after 8 h of carbonation curing. In the TG-DTA results for the samples of the 8, 12, and 24 h cured specimens collected near a depth of 5 mm from the surface, weight reduction rates of 4.23, 5.31, and 7.2 wt% were observed in the temperature range of 600–800 °C, respectively, confirming that the amount of $CaCO_3$ generated increased as the carbonation curing period increased, in agreement with the XRD analysis results.

The XRD analysis results showed that, for the P-CSC100 mix, no significant formation of carbonate minerals occurred until 8 h of carbonation curing. However, the intensity of the calcite peak tended to rapidly increase after 12 h. For the specimens subjected to carbonation curing for 24 h, an increase in the calcite peak intensity for the specimen collected at a depth of 15 mm was confirmed. When TG-DTA was conducted using samples collected at a depth of 5 mm from the surface of the specimens subjected to carbonation curing for 4 and 8 h, weight reduction rates of 1.78 and 0.84 wt% were observed in the temperature range of 600–800 °C, respectively, indicating no carbonation. A high weight reduction rate of 11.39 wt% was observed for the 5 mm samples of the specimens subjected to carbonation curing for 12 h, and weight reduction rates of 12.04 to 9.82 wt% were observed for the 5–15 mm samples of the specimens subjected to carbonation curing for 24 h, confirming the occurrence of carbonation.

Figure 8 shows the XRD analysis results by depth for the P-CSC0, P-CSC50, and P-CSC100 paste specimens subjected to carbonation curing (4, 8, 12, and 24 h) at a CO_2 concentration of 99%. Table 4 presents the weight reduction rate measurement results for the temperature ranges of 400–550 °C and 600–800 °C obtained via TG-DTA.

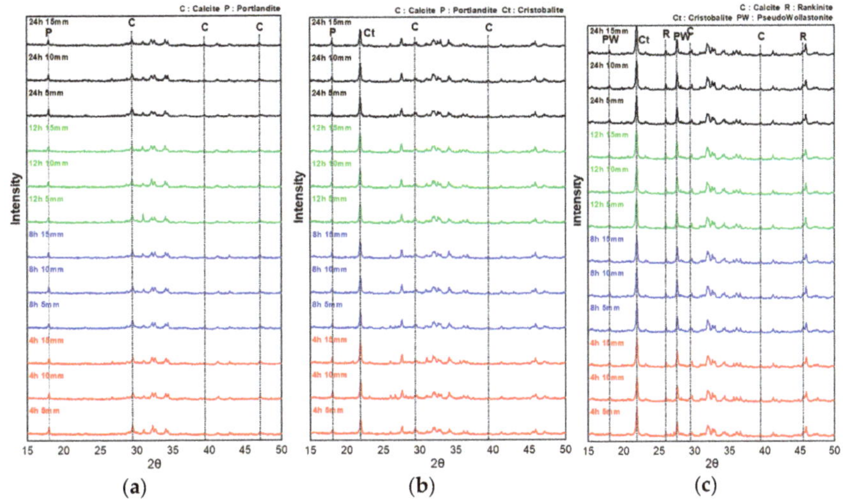

Figure 8. CO_2 concentration of 99% curing XRD pattern: (**a**) P-CSC0, (**b**) P-CSC50, and (**c**) P-CSC100.

In the XRD analysis results, peaks of mineral facies similar to those of the specimens of the same mixes carbonation-cured at a CO_2 concentration of 20% were observed. In contrast to the specimens cured at a CO_2 concentration of 20%, for which the intensities of the peaks of carbonate minerals increased with the carbonation curing period, all the specimens cured at a CO_2 concentration of 99% exhibited slight increases in the intensities of the peaks of carbonate minerals only at a depth of 5 mm from the surface when carbonation curing was performed for 8 h. Despite the increase in the carbonation curing period, no further changes in the intensities of the peaks of carbonate minerals were observed.

Additionally, the TG-DTA results showed that the weight reduction rate of $CaCO_3$ increased slightly only at a depth of 5 mm from the surface after 8 h of carbonation curing in the 600–800 °C range, which is the pyrolysis temperature range of $CaCO_3$. Despite the increase in the carbonation curing period, no carbonation was confirmed, as no change in the weight reduction rate was observed.

Thus, as discussed in Section 3.1, the quantitative XRD analysis and TG-DTA indirectly showed that the CO_2 diffusion rate was reduced by the pores inside the hardened body saturated with moisture at the beginning of the carbonation curing (4 h) and that the diffusion rate of CO_2 decreased as micropores were filled with $CaCO_3$, which was generated by the rapid carbonation of the specimen surface due to high-concentration CO_2 diffusion, after 8 h of carbonation curing.

3.3. FT-IR Analysis Results

Figure 9 shows the FT-IR analysis results for P-CSC0 subjected to carbonation curing at a CO_2 concentration of 20%. A peak at 3645 cm^{-1} corresponding to $Ca(OH)_2$, a cement hydration product, was observed for the non-carbonated sample of the P-CSC0 mix [22]. However, the intensity of the $Ca(OH)_2$ peak decreased after 12 h of carbonation curing and was not observed at 24 h of carbonation curing. The specimens subjected to carbonation curing exhibited rapid increases in the intensities of the peaks at 874 and 1420 cm^{-1} after 4 h of curing. These peaks suggested that $CaCO_3$ was generated by C-O stretching [23]. The C-O stretching peaks indicated the generation of $CaCO_3$ caused by the carbonation of $Ca(OH)_2$. The carbonation was natural, as peaks at 874 and 1420 cm^{-1} were also observed for non-carbonated specimens, although they were weak compared with those of the carbonated specimens.

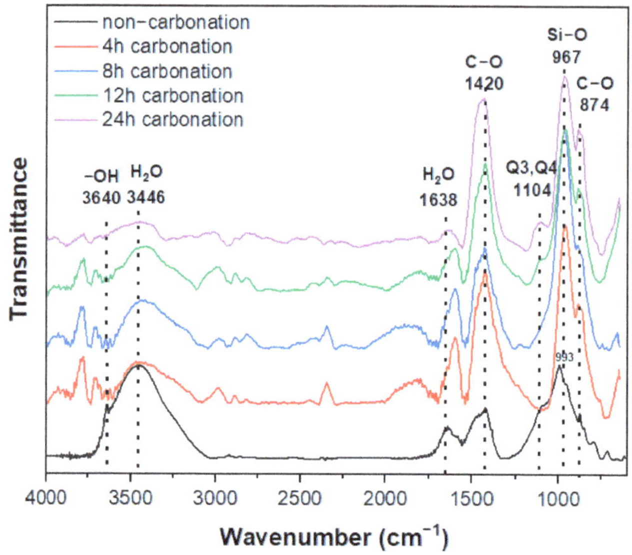

Figure 9. FT-IR analysis results of 20% concentration CO_2-cured CSC0.

The peaks at 993 and 967 cm^{-1} for the non-carbonated and carbonated specimens corresponded to the Si-O stretching vibration of the Q^2 site [22]. These peaks are observed for hydrated cement paste containing C-S-H. Their positions depend on the Ca/Si ratio, and their intensities are increased by the decalcification of C-S-H during the carbonation process [24]. When C-S-H is completely carbonated, the peak of Q^2 moves to 1020 and 1140 cm^{-1}, corresponding to the peaks of Q^3 and Q^4, respectively (silica gel). The P-CSC0 specimen exhibited a peak at 1104 cm^{-1} after 12 h of carbonation curing, confirming the decalcification of C-S-H.

3.4. Compressive Strength Analysis Results

Figure 10 shows the compressive strength measurement results for the paste containing CSC. The compressive strengths measured immediately upon the completion of steam curing were 55.75, 18.34, and 1.37 MPa for CSC contents of 0%, 50%, and 100%, respectively, indicating that the strength decreased as the CSC content increased.

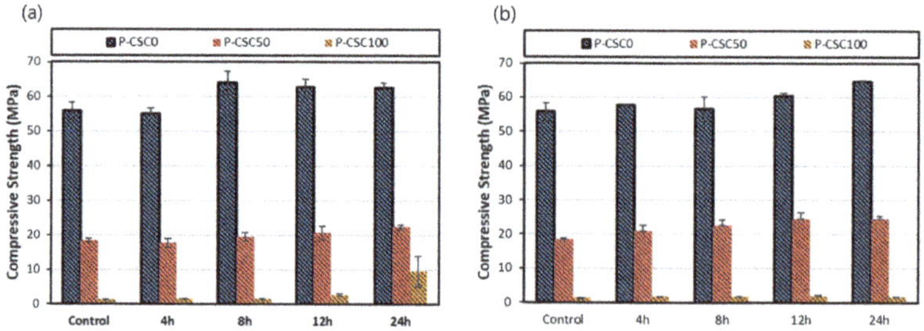

Figure 10. Compressive strengths results: (**a**) CO_2 20% and (**b**) CO_2 99%.

For the P-CSC0 mix, no significant change in compressive strength after 4 h of carbonation curing at a CO_2 concentration of 20% was observed. The compressive strength was increased by approximately 9 MPa after 8 h and was reduced by approximately 1.2 MPa after 12 h compared with 8 h. Thereafter, no strength changes were observed. Notably, the compressive strength increased after 8 h of carbonation curing owing to the densification of the structure caused by the generation of $CaCO_3$, as confirmed by the TG-DTA results. The strength was reduced after 12 h of carbonation curing because of the decalcification of C-S-H caused by an increase in the intensity of the 1104 cm^{-1} peak corresponding to silica gel (Q3 and Q4), as shown in Figure 6. Meanwhile, at a CO_2 concentration of 99%, the compressive strength increased after 4 h of carbonation curing, and the generation of $CaCO_3$ was confirmed by the TG-DTA results. The strength slightly decreased and then increased after 8 h of carbonation curing. This occurrence may be similar to the result for the 20% CO_2 concentration.

The compressive strength of the P-CSC50 mix after carbonation curing was measured to be within the range of 17.62–22.5 MPa at a CO_2 concentration of 20% and 20.89–24.56 MPa at a CO_2 concentration of 99%. When the CO_2 concentration was 20%, the compressive strength was 2 MPa lower on average than when it was 99%, despite the increase in carbonation depth. This difference is because mechanical properties deteriorated when sufficient CO_2 diffusion into a deep part of concrete occurred compared with the case where carbonation occurred only in the surface layer because of the increase in the carbonation area of C-S-H (decalcification) [25]. At a CO_2 concentration of 99%, however, the formation of a dense $CaCO_3$ layer on the surface suppressed the diffusion of CO_2 into the inside, suppressed the internal CSC carbonation reaction, and reduced the size of the C-S-H decalcification section, thereby increasing the compressive strength.

The P-CSC100 mix exhibited a compressive strength of approximately 1.3 MPa immediately after steam curing, confirming that CSC had no hydration reaction. When it was subjected to carbonation curing at a CO_2 concentration of 20%, its compressive strength gradually increased after 12 h and rapidly increased to approximately 9.5 MPa after 24 h. However, it exhibited no compressive strength development until 24 h of carbonation curing at a CO_2 concentration of 99%. When the weight reduction rate of $CaCO_3$ measured via TG-DTA was compared with the compressive strength based on the carbonation curing period, the weight reduction rate was confirmed after 12 h when strength development began and increased after approximately 24 h at a CO_2 concentration of 20%. At a CO_2 concentration of 99%, the weight reduction rate of $CaCO_3$ was higher than that at a CO_2 concentration of 20% at the beginning of the reaction (after 4 h to 8 h) but did not increase despite the increase in the carbonation curing period. This accounts for the absence of strength improvement.

In addition, SEM analysis was conducted on the P-CSC50 mix with a 20% CO_2 concentration to determine the cause of the increase in compressive strength based on the carbonation curing period. Figure 11 shows a distinct disparity in the microstructure between the 4 and 24 h carbonation curing specimens. The 4 h curing specimen exhibited the formation of large pores, and the 24 h curing specimen exhibited a dense pore structure. However, the large pores observed in the 4 h specimen were not apparent.

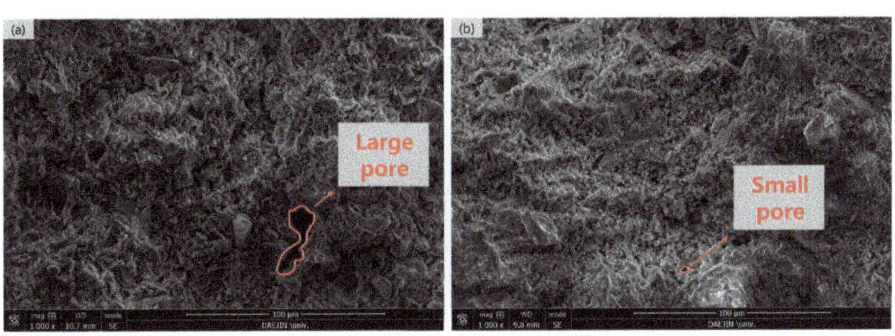

Figure 11. SEM images of 20% P-CSC50 specimens with CO_2 concentration based on reaction time: (**a**) Carbonation curing time 4 h and (**b**) carbonation curing time 24 h.

4. Discussion

4.1. Effect of Moisture on Carbonation Curing

Moisture is essential for the carbonation reaction of typical concrete as CO_2 and Ca^{2+} ions are dissolved in water for the reaction. However, the presence of a considerable amount of water in the concrete pores reduces the diffusion rate of CO_2 to the order of 1/10,000 of that in gas, slowing the carbonation reaction [26,27].

Moradllo et al. [19] investigated the effect of water saturation on the carbonation of CSC mortars and found that carbonation did not occur even when carbonation curing was performed if the internal water saturation of the mortar was 100%. They reported that the carbonation rate increased approximately four times as the water saturation decreased from 80% to 10% [19].

When the carbonation depth was examined based on the above-described carbonation curing conditions, the carbonation depth rapidly increased under certain carbonation curing period conditions. This indicated that the diffusion of CO_2 was hindered at the beginning of the carbonation curing by pores in the specimen saturated with moisture, as the carbonation curing was performed immediately upon the completion of steam curing under the 100% relative humidity condition as pre-curing. However, the carbonation depth was increased by the increased diffusion rate of CO_2 along with the evaporation of moisture. Thus, a drying process is necessary to increase the carbonation rate at the beginning of the carbonation reaction.

4.2. Changes in Mechanical Properties during Carbonation Curing

The typical carbonation reactions of BSC generate $CaCO_3$ and silica gel by inducing the carbonation of $Ca(OH)_2$ and C-S-H, which are hydration products. The carbonation reaction of $Ca(OH)_2$ can promote the strength improvement effect by filling the pores inside the hardened body, but the carbonation of C-S-H, which is responsible for strength, may reduce the strength through the contraction caused by decalcification. Therefore, in this study, before the evaluation of mechanical properties after carbonation curing, we assumed that CSC would be hardened by the carbonation reactions of CS and C_3S_2 and that the mechanical properties of BSC would be changed by the carbonation reactions of the hydration products generated by steam curing. In the compressive strength analysis results based on this, the strength improvement effect was offset by the decalcification of C-S-H, which is a BSC-based hydration product, in sections where the compressive strength decreased despite the increase in the amount of $CaCO_3$ due to carbonation curing. Therefore, the proper carbonation reaction conditions should be derived considering both the changes in mechanical properties caused by the carbonation reaction of the hydration products of BSC and the mechanical property improvement due to the carbonation reaction products of CSC.

5. Conclusions

In this study, the basic mechanical properties and CO_2 sequestration rate of paste were evaluated based on the CSC content and carbonation curing conditions to develop secondary cement products containing CSC for CO_2 reduction. The optimal CSC content and carbonation curing condition were determined.

The carbonation depth results for cement paste with different CSC contents indicated that the carbonation depth tended to increase as the CSC content increased. Moreover, the carbonation depth rapidly increased in certain sections. This appears to be because the diffusion rate of CO_2 was reduced at the beginning of carbonation curing by the pores in the specimen saturated with moisture, as the carbonation curing was performed immediately upon the completion of steam curing, or the carbonation depth was increased by the increased diffusion rate of CO_2 along with the evaporation of moisture. The carbonation depth was higher for the 20% CO_2 concentration curing condition than for the 99% CO_2 concentration curing condition. This appears to be because the diffusion rate of CO_2 decreased as micropores were filled with $CaCO_3$, which was rapidly generated on the specimen surface by the high-concentration CO_2 at the beginning of the carbonation reaction.

The compressive strength results for paste containing CSC indicated that the compressive strength tended to increase after carbonation curing. However, for the P-CSC0 mix, strength degradation sections were observed. FT-IR analysis results indicated that these sections were caused by the decalcification of C-S-H, which is a BSC-based hydration product. Accordingly, for the P-CSC50 mix, CSC was expected to increase the strength through carbonation, but the strength improvement effect was offset by the decalcification of C-S-H. For the P-CSC100 mix, no strength improvement at a CO_2 concentration of 99% was observed, in contrast to the case of a CO_2 concentration of 20%.

This study conducted a basic investigation to develop secondary products of cement utilizing CSC to advance the CCU technology in the cement industry. Furthermore, the application of CSC is a technological approach that can effectively decrease CO_2 emissions throughout the cement production process while utilizing CO_2. It is expected to serve as a direct solution for CO_2 reduction and contribute to the advancement of CCU technology in the cement industry.

Author Contributions: Original draft manuscript, Y.-J.K.; Investigation, writing—review and editing, S.-R.S.; Supervision, writing—review and editing, D.-W.R. All authors have read and agreed to the published version of the manuscript.

Funding: This research was conducted with the support of the Ministry of Trade, Industry, and Energy's National Research and Development Program (Project Number: RS-2023-00155521).

Institutional Review Board Statement: Not applicable.

Informed Consent Statement: Not applicable.

Data Availability Statement: The data presented in this study are available upon request from the corresponding author.

Conflicts of Interest: The authors declare no conflict of interest.

References

1. IPCC. *Global Warming of 1.5 °C*; World Meteorological Organization: Geneva, Switzerland, 2018.
2. Ministry of Environment. *2023 Nationally Determined Contribution*; Ministry of Environment: Sejong, Republic of Korea, 2021.
3. Ministry of Environment. *National Greenhouse Gas Inventory Report of Korea*; Ministry of Environment: Sejong, Republic of Korea, 2022.
4. Xian, X.; Zhang, D.; Lin, H.; Shao, Y. Ambient pressure carbonation curing of reinforced concrete for CO_2 utilization and corrosion resistance. *J. CO_2 Util.* **2022**, *56*, 101861. [CrossRef]
5. Friedlingstein, P.; O'Sullivan, M.; Jones, M.W.; Andrew, R.M.; Gregor, L.; Hauck, J.; Le Quéré, C.; Luijkx, I.T.; Olsen, A.; Peters, G.P.; et al. Global Carbon Budget 2022. *Earth Syst. Sci. Data Discuss.* **2022**, *14*, 4811–4900. [CrossRef]
6. UN Environment; Scrivener, K.L.; John, V.M.; Gartner, E.M. Eco-efficient cements: Potential economically viable solutions for a low-CO_2 cement-based materials industry. *Cem. Concr. Res.* **2018**, *114*, 2–26. [CrossRef]
7. World Business Council for Sustainable Development. *Cement Technology Roadmap 2009: Carbon Emissions Reductions up to 2050*; World Business Council for Sustainable Development: Geneva, Switzerland, 2009.
8. Wang, X.; Guo, M.; Ling, T. Review on CO_2 curing of non-hydraulic calcium silicates cements: Mechanism, carbonation and performance. *Cem. Concr. Compos.* **2022**, *133*, 104641. [CrossRef]
9. Hwa, S.S. A Study on the Manufacturing of Carbon Dioxide Absorption Hardening Calcium Silicate Based Eco-Friendly Cement. Master's Thesis, Korea National University of Transportation, Chungju, Republic of Korea, 2021.
10. Nicholas, D.; Vincent, M.; Soda, S.; Jason, B.; Moro, F. Environmental Impact of carbonated calcium silicate cement-based concrete. *Constr. Mater.* **2018**, *172*, 1–9.
11. Vincent, M.; Sada, S.; Andrew, D. Properties of Solidia Cement and Concrete. In Proceedings of the 1st International Conference on Innovation in Low Carbon Cement & Concrete Technology, London, UK, 24–26 June 2019.
12. Lu, B.; Shi, C.; Hou, G. Strength and microstructure of CO_2 cured low-calcium clinker. *Constr. Build. Mater.* **2018**, *188*, 417–423. [CrossRef]
13. Ashraf, W.; Olek, J. Elucidating the accelerated carbonation products of calcium silicate using multi-technique approach. *J. CO_2 Util.* **2018**, *23*, 61–74. [CrossRef]
14. Li, L.; Liu, Q.; Dao, V.; Wu, M. Dimensional change of cement paste subjected to carbonation in CO_2 sequestration and utilization context: A critical review on the mechanisms. *J. CO_2 Util.* **2023**, *70*, 102444.

15. Manns, W.; Wesche, K. Variation in Strength of Mortars Made of Different Cements due to Carbonation. *Cem. Assoc. Jpn.* **1968**, *3*, 385–393.
16. Matsusato, H.; Ogawa, K.; Funato, M.; Sato, T. Studies on the Carbonation of Hydrated Cement and its Effect on Microstructure and Strength. *Natl. Counc. Cem. Build. Mater.* **1992**, *5*, 363–369.
17. Xiao, J.; Li, J.; Zhu, B.; Fan, Z. Experimental study on strength and ductility of carbonated concrete elements. *Constr. Build. Mater.* **2002**, *16*, 187–192. [CrossRef]
18. Kwon, Y.R.; Kim, H.S.; Lee, C.Y.; Cheng, H.M.; Ahn, T.S. Carbonation and Strength of High Strength Binary Concrete Used Ground Granulated Blast Furnace Slag. *J. Korea Concr. Insitiute* **2015**, *20*, 561–564.
19. Moradllo, M.K.; Ghantous, R.M.W.; Quinn, S.; Aktan, V.; Reese, S.; Weiss, W.J. Quantifying drying and carbonation in calcium silicate-cement systems using neutron radiography. *ACI Mater. J.* **2022**, *119*, 231–242.
20. Cui, H.; Tang, W.; Liu, W.; Dong, Z.; Xing, F. Experimental study on effects of CO_2 concentrations of concrete carbonation and diffusion mechanisms. *Constr. Build. Mater.* **2015**, *93*, 522–527. [CrossRef]
21. Ashraf, W.; Olek, J.; Sahu, S. Phase evolution and strength development during carbonation of low-lime calcium silicate cement (CSC). *Constr. Build. Mater.* **2019**, *210*, 473–482. [CrossRef]
22. Jiang, L.; Wu, Q.; Huo, Z.; Zhu, Z.; Wu, F.; Lu, B. An approach to improve compressive strength of cement paste at low temperature by carbonation curing. *Constr. Build. Mater.* **2023**, *365*, 130128. [CrossRef]
23. Pan, X.; Shi, C.; Farzadnia, N.; Hu, X.; Zheng, J. Properties and microstructure of CO_2 surface treated cement mortars with subsequent lime-saturated water curing. *Cem. Concr. Compos.* **2019**, *99*, 89–99. [CrossRef]
24. Ylmén, R.; Wadsö, L.; Panas, I. Insights into early hydration of Portland limestone cement from infrared spectroscopy and isothermal calorimetry. *Cem. Concr. Res.* **2010**, *40*, 1541–1546.
25. Dong, B.; Qiu, Q.; Xiang, J.; Huang, C.; Xing, F.; Han, N. Study on the carbonation behavior of cement mortar by electrochemical impedance spectroscopy. *Materials* **2014**, *7*, 218–231. [CrossRef] [PubMed]
26. Mizokami, Y.; Ichiro, T. Cross-point of water and carbon in the leaf-CO_2 diffusion conductance. *Jpn. J. Ecol.* **2016**, *66*, 477–487.
27. Genty, B.; Meyer, S.; Piel, C.; Badeck, F.; Liozon, R. CO_2 diffusion inside leaf mesophyll of ligneous plants. In *Photosynthesis: Mechanisms and Effects*; Garab, G., Ed.; Kluwer Academic Publishers: Dordrecht, The Netherland, 1998; Volume 1, pp. 3961–3966.

materials

Article

Effects of Fiber and Surface Treatment on Airport Pavement Concrete against Freeze–Thawing and Salt Freezing

Lei Xu [1], Yong Lai [2], Daoxun Ma [3], Junjie Wang [1,*], Molan Li [1], Le Li [1], Zhibin Gao [4], Yan Liu [2], Pukang He [1] and Yi Zhang [1]

[1] Department of Civil Engineering, Tsinghua University, Beijing 100084, China
[2] Beijing Super-Creative Technology Co., Ltd., Beijing 100621, China
[3] China Airport Construction Group Corporation of CACC, Beijing 100621, China
[4] Capital Airport Group Co., Ltd., Beijing 100621, China
* Correspondence: junjiewang@tsinghua.edu.cn

Abstract: Airport pavement concrete often suffers from freeze–thawing damage in high latitude and cold areas. In addition, the use of aircraft deicer makes the airport pavement concrete suffer from salt-freezing damage. To improve the durability of airport pavement concrete, modified polyester synthetic fiber (FC), cellulose fiber (CF), and basalt fiber (BF) reinforced concrete were prepared in this paper. The mechanical strength, pore structure, and frost resistance (freeze–thawing and salt freezing) of fiber-reinforced concrete were investigated. The effects of the combined action of fiber (fiber type and content) and surface treatment methods (spraying silane and impregnating silane) on the frost resistance of concrete were investigated. The results show that the flexural strength of concrete is positively correlated with the elastic modulus of fiber, but has little effect on the compressive strength. Fiber can reduce mass loss and dynamic modulus loss of concrete subjected to frost damage. FC more effectively improved the frost resistance of concrete than CF. After 30 cycles of salt freezing, the spalling amount of concrete sprayed or soaked with silane was decreased by 65.5% and 55.5%, respectively. Adding fiber and impregnating silane reduced the spalled concrete by up to 70.5%. Spraying silane treatment is better than impregnating silane treatment in enhancing the frost resistance of concrete because a better silane condensation reaction is achieved with spraying silane.

Keywords: fiber-reinforced concrete; freeze–thawing; salt freezing; airport pavement; silane; spraying and immersion

1. Introduction

As one of the most basic and important infrastructures of the airport, airport pavement has a significant impact on flight safety and airport operation efficiency. In addition to geometric factors such as length, width, and slope, airport pavement must also meet the requirements of strength, stiffness, flatness, and durability. Cement concrete pavement is the main structural form of airport pavement, which is affected repeatedly by various environmental climates. For example, the pavement concrete of some airports in high-latitude cold areas is damaged by freezing and thawing all year round, resulting in performance degradation, shedding, and failure [1–5].

In addition to the common failure form of gradual cracking, denudation, and structural failure under the action of repeated freeze–thaw cycles, airport pavement is also damaged by surface denudation caused by freezing in the presence of deicing salt (glycol aircraft deicing fluid). Some concrete with good water-resistant freeze–thaw damage tests will also suffer serious salt freeze–thaw erosion after being sprinkled with deicing salt for several winters, and the pavement aggregate will be exposed, which will seriously affect its use [6,7]. The typical damage of salt freezing gradually extends from the surface to the inside. The surface mortar is peeled off and the aggregate is exposed, but the concrete under the denudation layer is basically hard and intact, and layered denudation traces can be seen on the damaged section.

To improve the flexural strength and frost resistance of pavement materials, fibers are often added to concrete to improve the brittleness and durability of concrete and prevent damage caused by various factors [8–11]. Fiber has a significant effect on the mechanical properties of concrete under freeze–thaw conditions, and the improvement effect of different types and contents of fiber is different [12–14]. Many other types of composite concrete and applications are also proposed to improve concrete durability [15,16].

In addition to the durability of fiber-reinforced concrete, since the 1980s, the United States has used organosilane materials to spray and soak the concrete structures of roads and bridges to increase the durability of structures. Many durability technical standards also take surface silane impregnation as an effective surface protection measure [17–19]. Wang et al. and Jiang et al. systematically studied the effects of silane brushing times, concrete curing age, construction temperature, mix proportion, and other conditions on the silane impregnation effect [20,21].

In summary, at present, research on freeze–thaw damage of fiber-reinforced concrete and silane impregnation protection has been carried out, but research on the protection of concrete against freeze–thaw damage by different fiber types, dosages, and different surface treatment methods is still limited. This study attempts to fill this gap through the mechanical test, freeze–thaw test, and salt-freezing test, combined with SEM, XRD, and MIP measurements.

2. Experimental Program

2.1. Raw Materials

Cement: The cement is P.O 42.5 ordinary Portland cement (conforming to Chinese national standard GB175-2007) with a specific surface area of 335 m^2/kg [22]. The initial and final setting times were 180 min and 235 min, the flexural and compressive strengths were 4.9 MPa and 22.5 MPa at 3 days, and the flexural and compressive strengths were 8.6 MPa and 48.3 MPa at 28 days, respectively. All these data are from the cement plant (Fushun Aosaier Technology Co., Ltd., Fushun, China).

Coarse Aggregates: The coarse aggregates are subdivided into large (16–31.5 mm) and small (4.75–16 mm) gravel and both of them are well graded; the properties are shown in Table 1 and sieving curves are shown in Figure 1a.

Table 1. Basic properties of coarse aggregates.

Characteristic	Large Gravel 16–31.5 mm	Small Gravel 4.75–16 mm
Apparent density	2.716	2.731
Water absorption rate	0.63	0.60
Mud content (%)	0.9	0.8
Crushing value (%)	/	17.9

Note: The mud content refers to the content of particles less than 0.075 mm.

Fine Aggregates: The fine aggregates were natural sand (0–4.75 mm), whose apparent density was 2.655 g/cm^3, fineness modulus was 2.3, sand mud content was 2.9%, and its sieving curve is shown in Figure 1b.

Fibers: Three types of fiber were used in this study (provided by Beijing Zhongqi Zhuochuang Technology Development Co., Ltd., Beijing, China). Ultra-high strength modified polyester synthetic fiber (FC) is a new material specially developed for airport road surface concrete. Cellulose fiber (CF) is a sheet-like monomer of fiber monofilament. Basalt fiber (BF) is a continuous fiber drawn from natural basalt. The morphology of the three types of fibers is shown in Figure 2, and their physical and mechanical properties are shown in Table 2. The results of the fiber characteristic test are provided by the corresponding manufacturers.

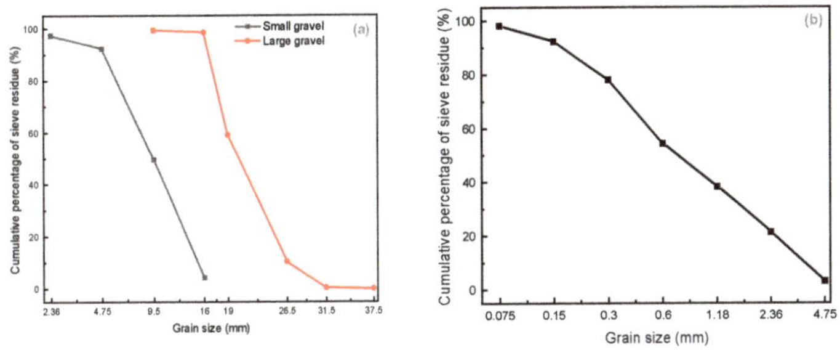

Figure 1. Coarse aggregate (**a**) and fine aggregate (**b**) sieving curve.

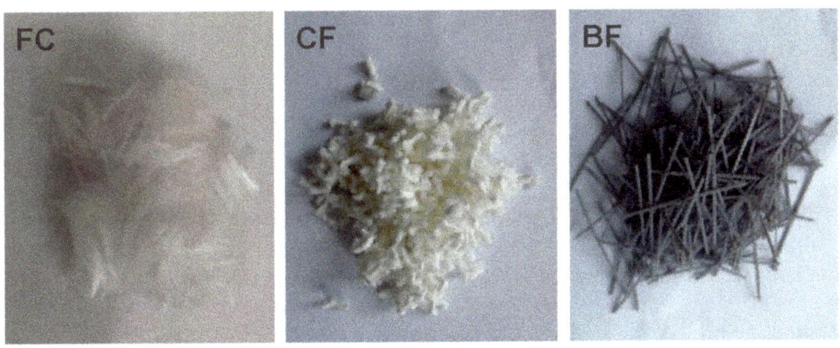

Figure 2. Morphology of the three types of fibers.

Table 2. Fiber physical and mechanical properties index.

Fiber Type	Diameter μm	Tensile Strength/MPa	Density g/cm^3	Modulus of Elasticity/GPa	Elongation %	Length mm
FC	40	1252	1.31	11.7	21	18
CF	18	960	1.10	9	3.5	18
BF	150	1930	2.65	80	3	20

Additives and other chemical reagents: The superplasticizer is an air-entraining polycarboxylic acid water-reducing agent, which is in accordance with the provisions of GB8076-2008 Concrete Admixtures [23]. The glycol deicing liquid used for the salt-freezing test is FCY-1A-SAE Type I aircraft deicing anti-icing liquid, which is a light yellow transparent liquid with a specific gravity of 1.1120 ± 0.015 at 20 °C. Silane for impregnation and spraying is a compound of a variety of long-chain silanes. All water used in this experiment was tap water.

2.2. Mix Design and Specimen Preparation

To ensure the uniformity of the concrete mixture, concrete was mixed with a forced mixer. Test samples were manufactured in the laboratory. The mixing procedure of fiber concrete used in the test is as follows: first pour stones, sand, cement, and fiber into the mixer successively, mixing for 120 s, then pour the mixed liquid of water reducer and water into the mixer, mixing for another 120 s, and cast the test block after the fiber is evenly dispersed. The test samples were filled at one time when forming, and then placed on the vibration table for 10 s. After vibration, the surface was troweled, and the samples were water cured (20 °C) after demoulding at 24 h.

The concrete mix consists of eight mix groups according to different fiber content and fiber types, whether a water-reducing agent is used or not, and different sand ratios. Their mixed proportion and relevant composition information are shown in Table 3 below. There are 8 recipes in total, labeled G1 to G8. The water to cement ratio was kept constant at 0.4. The sand ratio is the percentage of the sand mass in the total mass of sand and stone in the concrete. In this experiment, the sand ratio was controlled to be 0.3. Air entraining and water-reducing superplasticizer with a solid content of 2% (relative to the mass of cement), i.e., 6.6 kg/m^3, was used in the concrete mixes except for G7.

Table 3. Experimental mix design.

Group	Cement (kg/m^3)	Sand (kg/m^3)	Large Gravel (kg/m^3)	Small Gravel (kg/m^3)	Water (kg/m^3)	Sand Ratio	Fiber Content (kg/m^3)	Fiber Type	Superplasticizer (kg/m^3)
G1	330	611.40	784.63	641.97	132.00	0.30	0	0	6.6
G2	330	611.40	784.63	641.97	132.00	0.30	1.0	FC	6.6
G3	330	611.40	784.63	641.97	132.00	0.30	1.4	FC	6.6
G4	330	611.40	784.63	641.97	132.00	0.30	1.8	FC	6.6
G5	330	611.40	784.63	641.97	132.00	0.30	1.4	CF	6.6
G6	330	611.40	784.63	641.97	132.00	0.30	1.4	BF	6.6
G7	330	611.40	784.63	641.97	132.00	0.30	1.4	FC	0
G8	330	570.64	807.05	660.31	132.00	0.28	1.4	FC	6.6

2.3. Mechanical Tests

After mixing, prismatic specimens of 150 × 150 × 550 mm^3 were molded and compacted according to the GBT 50081-2019 [24]. After 24 h of being protected with a plastic film, the specimens were demoulded and placed in a wet chamber where they were water cured until 28 days. Three specimens in each mixed group were tested for flexural strength according to GBT 50081-2019 [24]. The two halves resulting from the flexural tests were then tested for compressive strength according to the same standard.

2.4. Freeze–Thaw Cycle Test

The freeze–thaw cycle test was carried out according to a quick freezing method in GB 50082-2009 [25], which simulated frost damage to the airport pavement in cold regions. The size of the sample was 100 × 100 × 400 mm prismatic specimen. At the age of 28 days, the specimen soaked in water was taken out and directly used for the test. After different freeze–thaw cycles, the loss of mass and dynamic modulus were measured.

2.5. Salt-Freezing Test

The salt-freezing test was carried out according to MH 5006-2015 [26], and the experiment was based on the single-side salt-freezing test in the freezing–thawing damage test [27,28]. The test specimen for the salt freezing were cylindrical, and the size was Φ150 mm × 100 mm. Except for the exposure surface, epoxy resin was applied on the other surfaces of the specimen after 28 days of curing. The salt-freezing solution was prepared with 97% distilled water and 3% ethylene glycol. The working temperature was −20 °C to 20 °C. The measurement, weighing, and appearance inspection of the test piece was carried out instantly, and the piece to be tested was covered with wet cloth. The single-sided freeze–thaw test box automatically carried out the freeze–thaw cycle according to the predefined freeze–thaw cycle program. The one typical cycle in the freeze–thaw cycle program is shown in Figure 3 below.

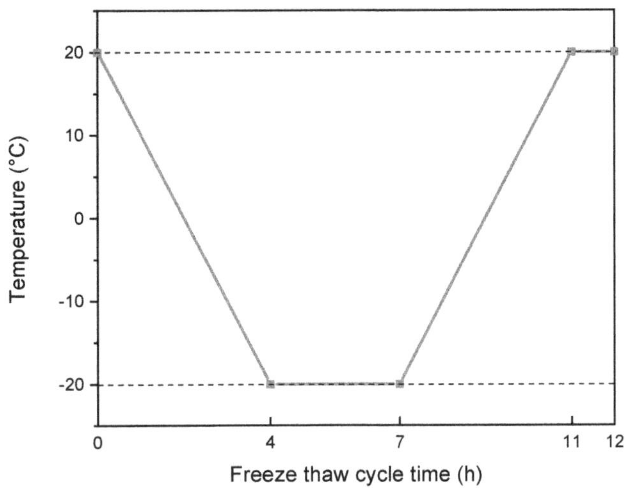

Figure 3. One typical cycle in a freeze–thaw cycle program.

2.6. Mercury Intrusion Porosimetry (MIP)

After the concrete samples were cured to the specified age, they were crushed with a small hammer. The selected pastes, particles about the size of soybeans, were soaked in sufficient absolute ethanol for 24 h to stop hydration, and then placed in a vacuum drying oven at 40 °C for vacuum drying for 3 days. After that, the pore structure was tested by the Autopore IV9510 mercury intrusion instrument (McMurray Teck Instrument Co., Ltd., Atlanta, GA, USA). A selection of 3–5 g samples were taken for each test, the pressure range was 0.0036–414 MPa, and the measured minimum pore diameter was about 3 nm.

2.7. X-ray Diffraction (XRD)

XRD patterns were collected by using a Bruker D8 powder diffractometer (Bruker, Germany) with a CuKα radiation source (λ = 1.5418 Å) at 40 kV and 40 mA to identify the variation in the crystalline phases. Scanning angles (2θ) were from 5° to 70° at a rate of 0.6 s/step and 0.02° per step. The samples for XRD were dried at 40 °C for 2 days and then ground to powder ($\leq$75 μm).

2.8. Scanning Electron Microscopy (SEM)

Field emission environmental scanning microscope (FEI QUANTA 200 FEG, USA) and a working voltage of 15.0 kV was used for the SEM test to observe the micromorphology of the samples under high vacuum mode. The sample preparation method is as follows: use the cutting machine to cut the large test piece into small pieces, and then crush it before observations. The fresh broken surface, which was the target surface, was coated with carbon before being placed into the measurement chamber.

3. Results

3.1. Compressive and Flexural Strength of Concrete

Figure 4 shows the compressive and flexural strength results of different concrete mixes. From the results of G3 and G7, the effect of the superplasticizer on the strength can be observed. G3 and G8 groups have different sand ratios. Moreover, the effect of fiber variety was investigated (G3, G5, and G6) and can be compared with G1 without fiber. In the mixes with different FC fiber contents (0 kg/m^3, 1.0 kg/m^3, 1.4 kg/m^3, and 1.8 kg/m^3), the effect of fiber content on the strength of concrete can be discussed. The compressive strength and flexural strength of the eight groups, G1–G8, at 28 days are shown in Figure 4a,b.

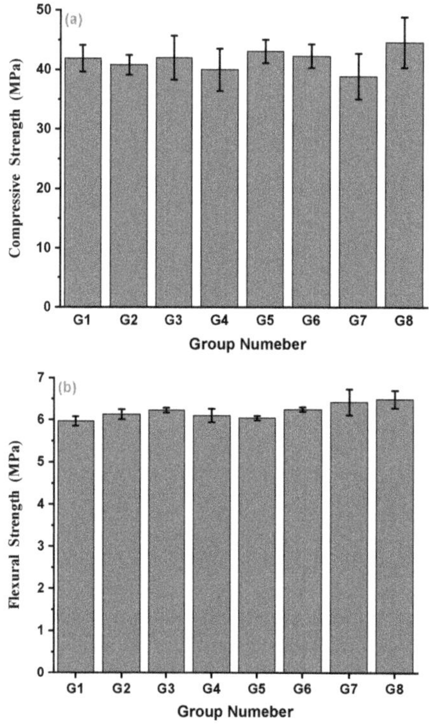

Figure 4. Compressive strength (**a**) and flexural strength (**b**) at 28 days.

From the experimental results, the addition of superplasticizer (G3) increases the 28 d compressive strength of concrete but decreases the 28 d flexural strength of concrete. Compared with G3 and G8, the sand ratio is reduced from 0.3 to 0.28, and the compressive and flexural strength of concrete at 28 days is increased by 6.19% and 8.7%, respectively. The compression failure of concrete may include the failure of the aggregate itself in addition to the bonding interface between the cement paste itself and the aggregate. Then, a larger sand ratio could reduce the amount of coarse aggregate and the ability of aggregate in concrete to bear its damage, resulting in a reduction in compressive strength.

When the fiber type changes but the fiber content is relatively low, the 28 d compressive strength of fiber concrete changes little, and the effect of improving the compressive strength of the fiber concrete is not obvious. Among the three fibers, CF (G5) showed the highest compressive strength. Compared with G1 (without fiber), the flexural strength of G3 (FC), G5 (CF), and G6 (BF) at 28 days increased by 4.36%, 1.17%, and 4.64%, respectively. After adding fiber, the 28 d flexural strength of the concrete is improved. The flexural strength of the high modulus fiber (FC) concrete G3 was the highest, the flexural strength of the medium modulus fiber (BF) concrete G6 was the middle one, and the flexural strength of the low modulus fiber (CF) concrete G5 was the lowest. The flexural strength of fiber-reinforced concrete was positively correlated with the elastic modulus of fiber. The 28 d flexural strength of all specimens was greater than 5.75 MPa, which meets the requirements of flexural strength of airport pavement concrete [23].

The change of FC content from G1–G4 is 0 kg/m^3, 1.0 kg/m^3, 1.4 kg/m^3, and 1.8 kg/m^3, respectively. From the experimental results, the 28 d compressive strength of concrete does not change significantly and fluctuates within a reasonable range, among which the 28 d compressive strength of G3 is greater. The flexural strength of the three groups with 1.0 kg/m^3, 1.4 kg/m^3, and 1.8 kg/m^3 fibers increased by 2.65%, 4.36%, and 2.18%, respectively.

3.2. Frost Resistance of Fiber Reinforced Concrete under Freeze–Thawing

From the above mechanical results, three groups G1, G3, and G5 were selected for freeze–thaw cycle experiments. The mass loss and relative dynamic elastic modulus under different freeze–thaw cycles are shown in Figures 5 and 6, respectively.

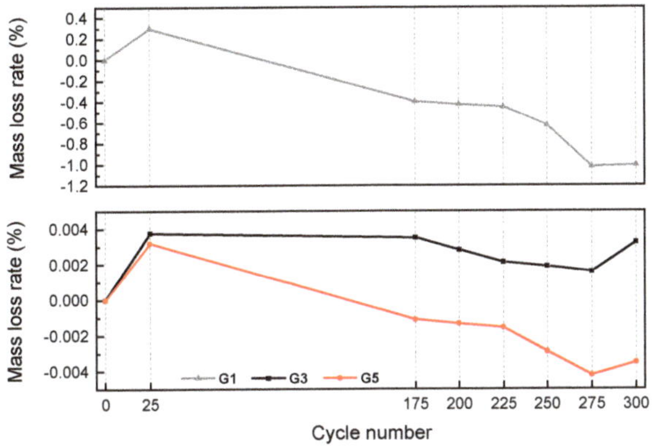

Figure 5. Mass loss after different freeze–thaw cycles (note: the plus and minus directions of the Y axis represent the increase and decrease in mass, respectively).

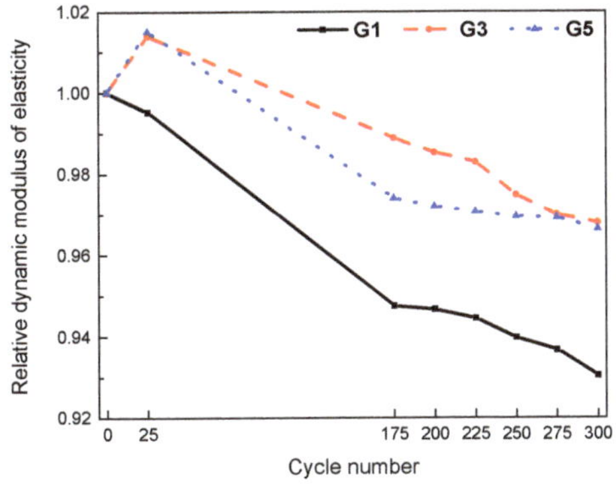

Figure 6. Relative dynamic elastic modulus loss after different freeze–thaw cycles.

After 200 freeze–thaw cycles, the surface of the reference concrete (G1) specimen showed local peeling and dropping. The spalling phenomenon on the surface of fiber-reinforced concrete was not obvious. Compared with the FC concrete, the surface holes of the CF concrete were more obvious. After 300 freeze–thaw cycles, the surface of the reference concrete (G1) specimen showed flake falling, the number of new holes increased, and the denudation was obvious. Individual new holes appeared in the FC concrete. The original holes on the surface of the CF-fiber-reinforced concrete specimens increased, but there was still less spalling than the reference concrete specimens. The different trend from cycles 275 to 300 could be explained by the fact that after 275 freeze–thaw cycles, the mass peels off completely. Therefore, 275 to 300 cycles may cause the freeze–thaw water absorption icing mass to be greater than the peeling mass, resulting in a mass increase.

In G1 without fiber, the mass loss of 300 freeze–thaw cycles was about 1%, and there was almost no mass loss in G3 and G5. It is obvious that the use of fiber can improve the frost resistance of concrete. Among them, compared with CF, FC is more effective in improving the spalling mass loss of concrete caused by frost resistance.

Similar to the experimental results of mass loss rate, the relative dynamic elastic modulus of the two groups with fibers was greater than that of the control group without fiber. After 300 freeze–thaw cycles, the relative dynamic elastic modulus of the control group (G1) was reduced to 0.93, while the relative dynamic elastic modulus of G3 and G5 was reduced to 0.968 and 0.966, respectively. The improvement increment of the relative dynamic elastic modulus was 4.03% and 3.86%, respectively. Compared with the concrete with CF, the frost resistance of concrete with FC is better.

3.3. *Effect of Surface Treatment Methods on the Salt Freezing of Fiber-Reinforced Concrete*

Since G3 had the highest resistance to freeze–thawing cycles, this mix and the control mix G1 were explored further to investigate the effect of surface treatment methods on the salt-freezing performance. Different surface treatment methods were considered and the results are shown in Figure 7. Troweling means that the formed specimen was troweled; roughening means that the surface of the formed specimen was roughened, a basic requirement of the airport pavement (Figure 8); spraying silane means that the forming surface of the concrete specimen was sprayed with 300 g/m^2 silane; and silane impregnation means that the formed surface of the concrete specimen was soaked in the silane (5 mm) for one day.

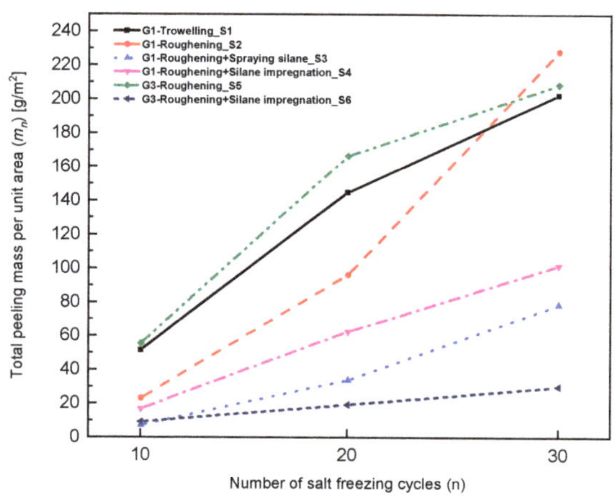

Figure 7. Variation of total spalling mass per unit area with the number of cycles.

According to the requirements of high friction coefficient, the airport pavement needs to be roughened [26]. In addition, to resist the highly cold environment, silane protection is often used in practical projects [29]. In this paper, four surface treatments are carried out for G1 without fiber, namely, troweling, roughening, spraying silane, and silane impression. Two surface treatments are carried out for G3 mixed with FC, roughened only, and roughening + silane impregnated. The results of 30 salt-freezing cycles on the above six groups (S1–S6) are shown in Figure 7.

It is obvious from the experimental results that no matter how the surface is treated and whether fibers are added or not, the total peeling mass of the three groups impregnated or sprayed with silane is less than that of the three groups not sprayed or soaked with silane. This shows that the use of silane can significantly improve the salt-frost resistance of concrete. Compared with S2, the spalling quality of concrete after 30 times of salt freezing was reduced by 65.5% and 55.5%, respectively, in S3 (silane is sprayed) and S4 (silane is soaked). Compared with S5 with S2, the spalling quality of concrete after 30 times of salt freezing was reduced by 8.54%. Compared with S6 with S4, the spalling quality of concrete after 30 times of salt freezing was reduced by 70.5%. It is obvious that fiber can increase the salt-freezing resistance of concrete, and this effect is more significant in specimens with impregnating silane.

Figure 8. Roughening treatment.

3.4. Pore Structure Characteristics of Fiber-Reinforced Concrete

Figures 9 and 10 show the porosity, pore volume, and pore distribution in all mixes. Cement-based materials are porous materials and the pore characteristics of materials have a decisive impact on their properties. An MIP test was carried out on eight groups of concrete, and important parameters such as total porosity, absolute void volume, and pore distribution can be obtained, as shown in Figures 9a,b and 10 below. The effects of superplasticizer, sand ratio, fiber content, and fiber types on porosity were discussed from different aspects in the following parts:

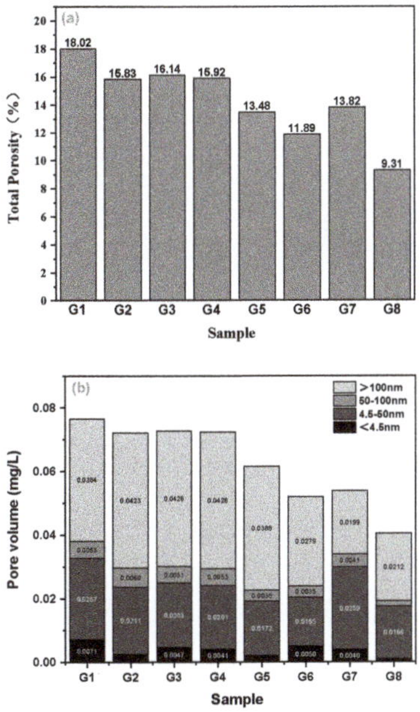

Figure 9. Total porosity (**a**) and absolute pore volume (**b**) of concrete.

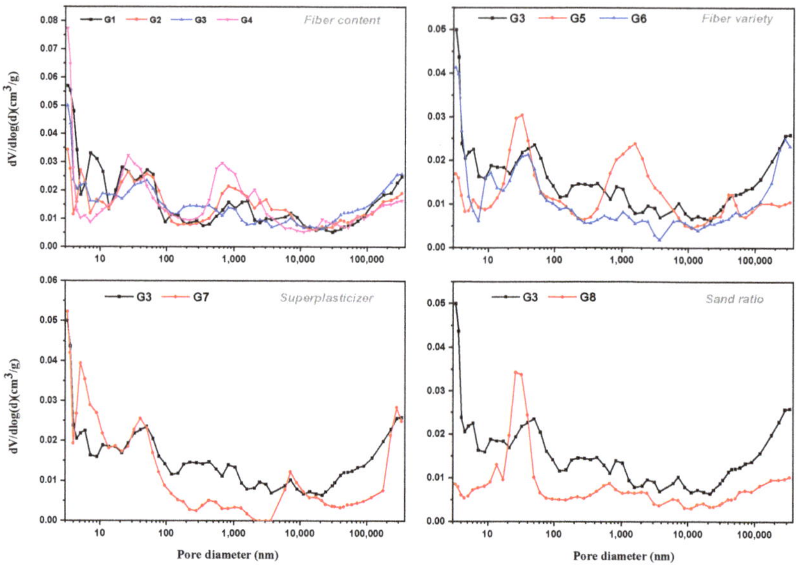

Figure 10. Size distribution of mercury injection voids in concrete.

3.4.1. Effect of Superplasticizer and Sand Ratio

Compared with G1, the use of air entraining superplasticizer increases the total porosity (G3 and G7), and also increases the pore volume of <4.5 nm, 50–100 nm, and >100 nm except for the pore diameter of 4.5–50 nm. The characteristic pore size—the pore size corresponding to the highest peak value of pore size distribution—of the group without superplasticizer is smaller, which may be more favorable for the concrete to show better mechanical, but not frost, resistance.

From Figure 9, it seems that there is no direct relationship between the sand ratio and the pores of concrete. The sand rate often affects the durability of concrete after hardening by affecting the workability of a concrete mixture. The appropriate sand ratio is that the concrete mixture has good workability and is easy to vibrate and compact. To a certain extent, it can improve the compactness and impermeability of concrete, improve the resistance to the destructive effect of external corrosive media, and effectively reduce the degree of erosion and delay the development speed of erosion. For concrete in a freeze–thaw environment, demonstration of good impermeability depends on the ability of the concrete to resist freeze–thaw damage. If the sand ratio is too large or too small, the fluidity of the concrete mixture will be reduced. On one hand, the concrete is not easy to mix evenly and the homogeneity is poor; on the other hand, during the pouring process, the concrete was not easy to be vibrated and compacted. After pouring, honeycombs and cavities easily formed inside the concrete mix and inside the formwork, and pitted surfaces easily formed on the surface, increasing the porosity inside the hardened concrete. In general, in this experiment, the sand ratio of 0.28 seems to be more favorable than that of 0.3: the porosity of concrete with a sand ratio of 0.28 was lower, and the total porosity was reduced by about 42.3% compared with that of concrete with a sand ratio of 0.3. The absolute pore volume of each size of concrete with a sand ratio of 0.28 was also low, and the characteristic pore size was also less than that of concrete with a sand ratio of 0.3.

3.4.2. Effect of Fiber Contents and Types

When comparing G2, G3, and G4 groups with the G1 group without fiber, it was observed that the addition of fiber into the cement matrix improved the compactness of the matrix. This can be explained by the fact that these micro fibers penetrated into the pores of the pastes and thus reduced the porosity. The porosity of the three groups with fiber was less than that of the G1 group without fiber, and the reduction range was more than 10%. In terms of the fiber types, concrete with basalt fibers showed a lower porosity, which might not be a good option for frost resistance. BF is a typical silicate fiber, which is close to the density of cement concrete. It is easier to disperse and evenly distribute in the concrete during mixing, overcoming the shortcomings of uneven distribution such as agglomeration of other fibers. FC fiber has a large surface area, which restricts the micro-cracks so that they are not connected, but its density is small, the diameter of monofilament is small, and there is a thickening effect, which is not conducive to the vibration compaction of concrete. Therefore, the porosity of FC concrete was the largest among the three fibers, which could be beneficial to resist frost damage.

3.5. XRD Results

The results of eight groups of XRD for paste sampling in concrete are shown in Figure 11 below. Although the sample was taken to knock out the clean paste as much as possible, it will inevitably be doped with quartz sand. In addition, the main products of cement hydration, calcium hydroxide (portlandite), and AFt can be seen in XRD. This shows that when the type of fiber, the amount of fiber, the use of superplasticizer, and the size of sand ratio changes, the main hydration products in the paste do not obviously change. The hydration products portlandite and AFt did not change greatly and these products can also be seen in the electron microscope results in Figure 12.

In addition, for G8 with a low sand ratio and G7 without a water reducing agent, the XRD results show that the peak value of calcium hydroxide, the hydration product of cement, is lower than that of their respective control groups. It was determined that the use of a low sand ratio without a water reducing agent affects the workability of concrete, and then affects the hydration degree of concrete. The G1 group was not mixed with fiber and the amount of calcium hydroxide was also lower than that of the corresponding control group. It can be considered that the use of fiber in concrete will not affect the hydration of cement, but may promote the hydration degree. According to the consistency of their calcium hydroxide content, it appears that the kind of fiber and the amount of fiber used in G2-G6 fiber-reinforced concrete will not significantly affect the degree of hydration.

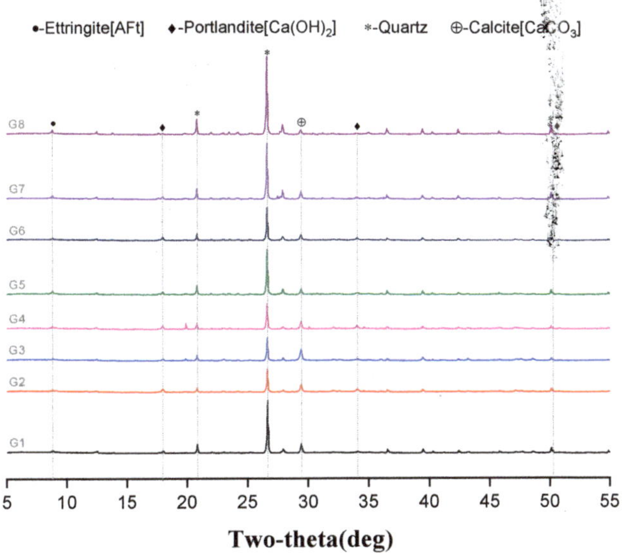

Figure 11. XRD pattern of 8 groups of concrete.

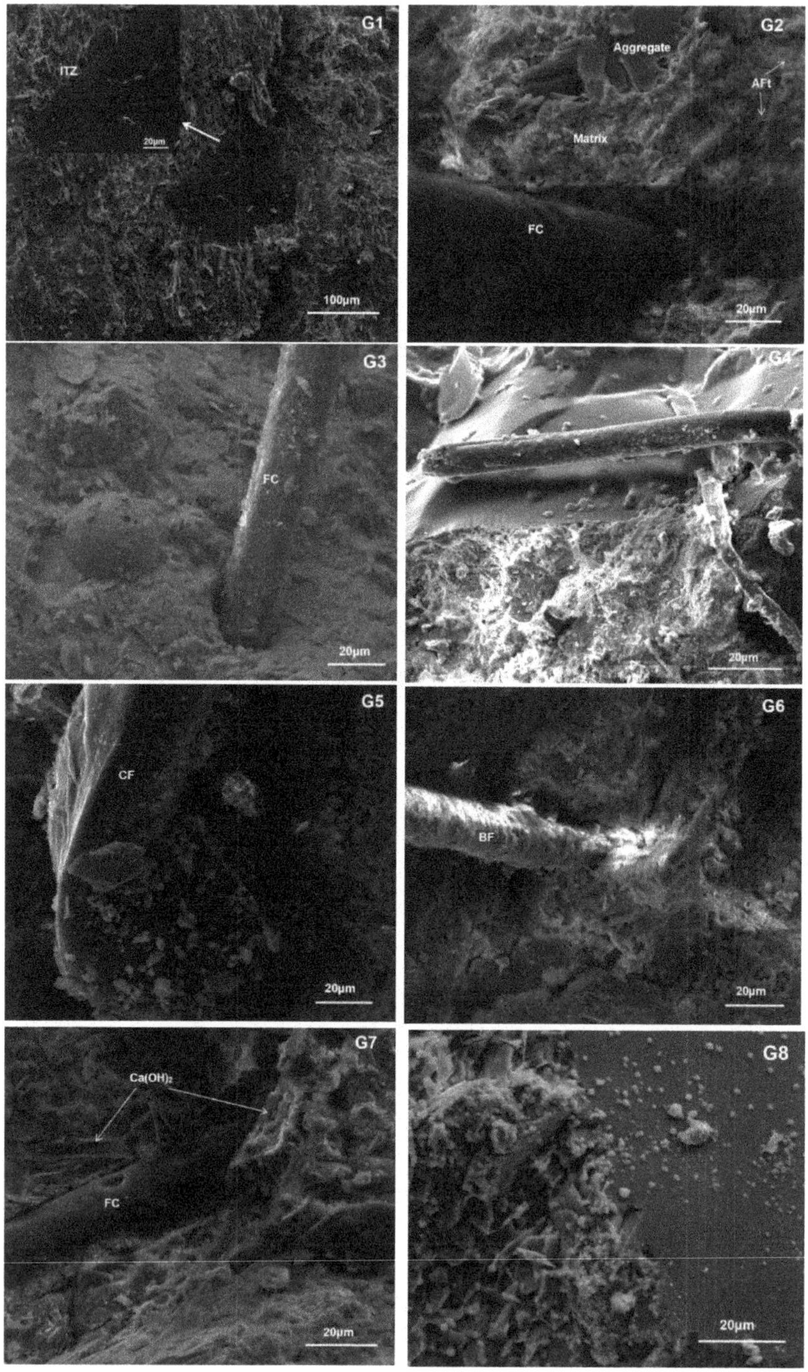

Figure 12. Electron microscope pictures of 8 groups of concrete.

3.6. Microstructure of Interface Transition Zone

The test results show that after adding fiber, the frost resistance of the concrete is greatly improved. In order to study the modification mechanism of fiber in the concrete matrix, SEM was used to analyze the microstate of the interfacial transition zone (ITZ) of concrete specimens and the interfacial structure characteristics of the fiber-cement paste and the improvement effects were analyzed [30–32]. The results are shown in Figure 12. It can be clearly seen that the microscopic morphology of FC (G3), CF (G5), and BF (G6) fibers under the electron microscope is different. The surface of the BF fiber is the roughest, and interacts with the surrounding mortar, which is consistent with its highest flexural strength. By adding an appropriate amount of fiber into the concrete, the fiber is randomly distributed in three dimensions and completely bonded with the concrete. When the concrete cracks appear, the existence of fiber can provide a bridge relay and increase the bonding force inside the concrete, so as to reduce the expansion of large voids and cracks in the process of concrete hardening [33,34]. More importantly, it reduces cracks, changes pore structure, reduces pore connectivity, effectively reduces water transmission channels, makes the damage caused by freezing and thawing relatively independent, and effectively controls the occurrence of freezing and thawing, so as to enhance the frost resistance [4]. In addition, the fibers are overlapped with each other and a fiber frame can be formed. The external bending force is greatly absorbed by the fiber frame in the matrix, and fibers are added to effectively reduce the occurrence of local damage, so as to enhance the flexural strength and frost resistance.

4. Discussion

4.1. Influence of Fiber Types and Contents on Concrete Strength

The compressive and flexural strength of concrete are the basic mechanical indexes of airport pavement concrete. From the results in Figure 4, it can be seen that the addition of fiber has little effect on the compressive strength, but the flexural strength has been significantly improved. In addition, the reinforcing effect of different fibers on flexural strength is different, showing a positive correlation trend with an elastic modulus of fiber. The fiber with a large elastic modulus has a greater reinforcing effect on the flexural strength of concrete, which may be why the crack resistance effect of high elastic modulus fiber is more obvious [35–41]. At the same time, the content of fiber has a great influence on the strength of concrete. According to the fiber reinforcement theory, the increase in fiber content up to an optimum content can lead to an increase in the number of fibers per unit volume and the increase in fiber spacing. However, it must be based on concrete compaction [39,42–45]. If the fiber content is too high (shown as >1.4 kg/m^3 in this study), it will lead to insufficient mixing and vibration of concrete, resulting in unconsolidated concrete and reduced strength. Therefore, the optimal fiber content recommended in this study is 1.4 kg/m^3.

4.2. Effect of Fiber Types on Frost Resistance of Concrete

Frost resistance is another important technical index of airport pavement concrete. By comparing the test of the relative dynamic elastic modulus and mass loss of concrete mixed with CF and FC fiber and concrete without fiber, it can be concluded that adding fiber can improve the frost resistance of concrete, due to the creation of a mechanical bite between fiber and substrate. The stress suffered in concrete can be borne by the tensile and shear resistance of the fibers, so as to improve the strength of the interface between the fiber and the substrate, which is beneficial to alleviate the freeze–thaw temperature stresses in the concrete and improve the frost resistance [36]. From the perspective of fiber types, the chemical composition of CF fiber belongs to polysaccharide, and the surface is more hydrophobic than the modified FC fiber. Therefore, the thickness of the water film interface is larger, and the microstructure results show that the interface strength is weak (Figure 12 G4). From this point of view, the improvement effect of CF on the frost resistance of concrete is not as good as FC, which agrees with the freeze–thawing results. It can be proved that freeze–thaw cycles impair the fiber/matrix interfacial bond, which is attributed to be the main cause of the degradation of durability [37]. Therefore, the interfacial bonding between fiber and matrix is a performance that should be paid special attention to in airport pavement concrete. The FC fiber used in this experiment realized a high tensile strength (900–1300 MPa), high modulus (8000–13,000 MPa), high toughness, low elongation (10–25%), high alkali resistance (96%), and ultra-large aspect ratio (>300) required by major and special concrete projects, which can fully achieve the bridging function of cement and other mixed materials and effectively prevent concrete cracking [38]. During construction mixing, FC fiber can be evenly distributed in concrete, establish multi-dimensional structure, and be applied in airport runways and other projects, which can improve the durability of airport pavement, realize the excellent performance of freeze–thaw resistance, salt alkali resistance, and corrosion resistance [46,47], and effectively ensure the safety of flight takeoff and landing.

4.3. *Effect of Surface Treatments on Salt Freezing of Concrete*

The results of salt-freezing indicated that silane spraying (S3) is more efficient than silane impregnation (S4). Compared with no silane treatment, the mass loss after 30 salt-freezing cycles in concrete can be reduced by 65.5% by spraying silane. Silane is a small molecular substance, and liquid silane has low viscosity and can penetrate to a certain depth through the open pores on the concrete surface. Under the action of water in the concrete, silane monomer molecules undergo a hydrolysis reaction to produce hydroxysilane [29].

$$\text{R-Si-(OR}')_3 + n\text{H}_2\text{O} \rightarrow \text{R-Si-(OR}')_{3-n}(\text{OH})_n + n\text{R}'\text{OH} \quad (1)$$

where: $n \leq 3$ indicates the degree of silane hydrolysis. The hydrolysate hydroxysilane has high activity and the condensation reaction can continue in hydroxysilane. The equation is as follows:

$$2\text{R-Si-(OH)}_3 \rightarrow \text{R-(OH)}_2\text{SiOSi(OH)}_2\text{-R} + \text{H}_2\text{O} \quad (2)$$

The above condensation reaction can occur between hydroxysilane and between hydroxysilane and hydration products on the concrete surface. The former makes silane form a network structure of dimer or even polymer on the concrete surface, and the latter makes silane establish a chemical connection with the concrete surface [48]. Through the two-step reactions (1) and (2), silane forms a stable polymerization network on the surface of concrete pores, to which alkyl R gives hydrophobicity to the surface of concrete pores.

The condensation reaction will cause a dehydration process. A relatively dry concrete (pore) surface is more conducive to reaction (2), and reaction (2) is easier to carry out.

5. Conclusions

In this paper, three types of fiber (FC, CF, and BF) reinforced concrete were prepared. The physical properties, and freeze–thawing and salt-freezing performance were investigated, and the microstructure and pore distribution of different fiber-reinforced concretes were characterized by SEM and MIP. Some favorable measures to improve the frost resistance of pavement concrete are put forward. The main conclusions are as follows:

(1) All fibers improve concrete flexural strength but have little effect on compressive strength. Under the same mass fraction (1.4 kg/m^3), the flexural strength of FC and BF concrete increased by 4.36% and 4.64% compared with that of ordinary concrete, but CF concrete increased by only 1.17%. The flexural strength of fiber-reinforced concrete is positively correlated with the elastic modulus of fiber.

(2) The use of fiber can improve the frost resistance of concrete, characterized by less mass loss and dynamic modulus loss. After 300 freeze–thaw cycles, the relative dynamic elastic modulus of the control group (G1) was reduced to 0.93, while the relative dynamic elastic modulus of G3 and G5 with fibers was reduced to 0.968 and 0.966, respectively, increasing the relative dynamic elastic modulus by 4.03% and 3.86%, respectively. Compared with CF, FC is more effective in improving concrete frost resistance.

(3) Regardless of fiber types, the total peeling amount of concrete impregnated or sprayed with silane was less than that of concrete not sprayed or impregnated with silane. The use of silane can significantly improve the salt-frost resistance of concrete. After 30 cycles of salt freezing, the spalling amount of concrete sprayed or soaked with silane was decreased by 65.5% and 55.5%, respectively. Furthermore, adding fiber and impregnating silane reduced the spalled concrete by up to 70.5%.

(4) It is recommended to add FC with a content of 1.4 kg/m^3 in the airport pavement concrete, and apply silane spraying on the exposure surface to protect the concrete. Spraying silane treatment was better than impregnating silane treatment in enhancing the frost resistance of concrete because the dry concrete surface is more conducive to silane condensation reaction in the spraying treatment. Proper surface treatment is more important than the effect of fiber when the concrete is subjected to salt freezing.

Author Contributions: Conceptualization, J.W.; Data curation, L.X., Y.L. (Yong Lai), D.M., J.W. and M.L.; Formal analysis, L.X., Y.L. (Yong Lai), D.M., J.W., M.L., L.L., Z.G., Y.L. (Yan Liu), P.H. and Y.Z.; Funding acquisition, J.W.; Investigation, L.X., Y.L. (Yong Lai), D.M. and M.L.; Methodology, L.X., Y.L. (Yong Lai), D.M., J.W., M.L., L.L., Z.G. and Y.L. (Yan Liu); Project administration, J.W.; Software, J.W.; Supervision, J.W.; Visualization, L.X., Y.L. (Yong Lai), D.M., M.L., L.L. and Z.G.; Writing—original draft, L.X., J.W., L.L., Z.G. and Y.L. (Yan Liu); Writing—review & editing, L.X., J.W., L.L., Z.G., Y.L. (Yan Liu), P.H. and Y.Z. All authors have read and agreed to the published version of the manuscript.

Funding: This research was funded by the National Natural Science Foundation of China (Nos. 52008232 and 52038004).

Institutional Review Board Statement: Not applicable.

Informed Consent Statement: Not applicable.

Data Availability Statement: The data presented in this study are available on request from the corresponding author.

Acknowledgments: The authors wish to acknowledge the support provided from Beijing Super-Creative Technology Co., Ltd.

Conflicts of Interest: The authors declare no conflict of interest.

References

1. Li, W.; Wu, Y.; Di, L. Study on freeze-thaw test of airport pavement concrete in cold area. *Subgrade Eng.* **2011**, *2*, 123–125.
2. Green, M.; Bisby, L.; Beaudoin, Y.; Labossière, P. Effect of freeze-thaw cycles on the bond durability between fibre rein. *Can. J. Civ. Eng.* **2000**, *27*, 949–959. [CrossRef]
3. Kankam, C. Impact resistance of palm kernel fiber-reinforced concrete pavement slab. *J. Ferrocem.* **1999**, *29*, 279–286.
4. Cui, Y.; Chen, Y.; Cen, G.; Peng, G. Comparative Study on the Effect of organic and inorganic fiber on the anti-wheel impact performance of airport pavement concrete under freeze-thaw environment. *Constr. Build. Mater.* **2019**, *211*, 284–297. [CrossRef]
5. Chen, Y.; Cen, G.; Cui, Y. Comparative study on the effect of synthetic fiber on the preparation and durability of airport pavement concrete. *Constr. Build. Mater.* **2018**, *184*, 34–44. [CrossRef]
6. Farnam, Y.; Todak, H.; Spragg, R.; Weiss, J. Electrical response of mortar with different degrees of saturation and deicing salt solutions during freezing and thawing. *Cem. Concr. Comp.* **2015**, *59*, 49–59. [CrossRef]
7. Deja, J. Freezing and deicing salt resistance of blast furnace slag concretes. *Cem. Concr. Comp.* **2003**, *25*, 357–361. [CrossRef]
8. Paluri, Y.; Mogili, S.; Mudavath, H.; Pancharathi, R. A study on the influence of steel fibers on the performance of Fine Reclaimed Asphalt Pavement (FRAP) in pavement quality concrete. *Mater. Today Proc.* **2020**, *32*, 657–662. [CrossRef]
9. Chen, Y.; Cen, G.; Cui, Y. Comparative analysis on the anti-wheel impact performance of steel fiber and reticular polypropylene synthetic fiber reinforced airport pavement concrete under elevated temperature aging environment. *Constr. Build. Mater.* **2018**, *192*, 818–835. [CrossRef]
10. Sassani, A.; Arabzadeh, A.; Ceylan, H.; Kim, S.; Sadati, S.; Gopalakrishnan, K.; Taylor, P.; Abdualla, H. Carbon fiber-based electrically conductive concrete for salt-free deicing of pavements. *J. Clean. Prod.* **2018**, *203*, 799–809. [CrossRef]
11. Par, K.; Paulino, G.; Roesler, J. Cohesive fracture model for functionally graded fiber reinforced concrete. *Cem. Concr. Res.* **2010**, *40*, 956–965.
12. Wu, H.; Zhao, J.; Wang, Z. Study on micro-structure and durability of fiber concrete. *Res. J. Appl. Sci. Eng. Technol.* **2013**, *5*, 659–664. [CrossRef]
13. Liu, D.; Xu, W.; Cheng, Z.; Zhou, Z.; Cai, X.; Zhao, B. Improvement test on frost resistance of concrete and engineering application of test fruitage. *Environ. Earth. Sci.* **2013**, *69*, 161–170.

14. Miao, C.; Mu, R.; Qian, T.; Wei, S. Effect of sulfate solution on the frost resistance of concrete with and without steel fiber reinforcement. *Cem. Concr. Res.* **2002**, *32*, 31–34. [CrossRef]
15. Joshani, M.; Koloor, S.S.R.; Abdullah, R. Damage mechanics model for fracture process of steel-concrete composite slabs. *Appl. Mech. Mater.* **2012**, *165*, 339–345. [CrossRef]
16. Shokravi, H.; Mohammadyan-Yasouj, S.E.; Koloor, S.S.R.; Petrů, M.; Heidarrezaei, M. Effect of Alumina Additives on Mechanical and Fresh Properties of Self-Compacting Concrete: A Review. *Processes* **2021**, *9*, 554. [CrossRef]
17. *BS EN 1504-2004*; Products and Systems for the Protection and Repair of Concrete Structures-Definitions, Requirements, Quality Control and Evaluation of Conformity. British Standards Institution: London, UK, 2004.
18. *JTS 153-2015*; Standard for Durability Design of Port and Waterway Engineering Structure. China Communication Press Co., Ltd.: Beijing, China, 2015.
19. *JTJ 275-2000*; Corrosion Prevention Technical Specification for Concrete Structures of Marine Harbour Engineering. China Communication Press Co., Ltd.: Beijing, China, 2000.
20. Wang, Q.; Luan, H.; Fan, Y. Chloride resistance of concrete with isobutyl-triethoxy-silane. *J. Guangxi Univ. (Nat. Sci. Ed.)* **2015**, *40*, 876–882.
21. Jiang, X.; Qin, M.; Xiao, J.; Zhang, Y.; Xu, W. Influence of silane impregnation conditions on anti-corrosion effect of concrete. *Railw. Eng. Chin.* **2013**, *07*, 146–149.
22. *GB 175-2007*; Common Portland Cement. China Quality Standards Publishing & Media Co., Ltd.: Beijing, China, 2007.
23. *GB 8076-2008*; Concrete Admixtures. China Quality Standards Publishing & Media Co., Ltd.: Beijing, China, 2008.
24. *GB 50081-2019*; Standard for Test Methods of Concrete Physical and Mechanical Properties. China Quality Standards Publishing & Media Co., Ltd.: Beijing, China, 2019.
25. *GB 50082-2009*; Standard for Test Methods of long-Term Performance and Durability of Ordinary Concrete. China Quality Standards Publishing & Media Co., Ltd.: Beijing, China, 2009.
26. *MH 5006-2015*; Specifications for Construction of Aerodrome Cement Concrete Pavement. Civil Aviation Administration of China: Beijing, China, 2015.
27. Paul, S.C.; Babafemi, A.J. Performance of strain hardening cement-based composite (SHCC) under various exposure conditions. *Cogent Eng.* **2017**, *4*, 1345608. [CrossRef]
28. Meng, C.; Li, W.; Cai, L.; Shi, L.; Jiang, C. Experimental research on durability of high-performance synthetic fibers reinforced concrete: Resistance to sulfate attack and freezing-thawing. *Constr. Build. Mater.* **2020**, *262*, 120055. [CrossRef]
29. Li, K.; Jing, W.; Yang, R. Review on silane impregnation of concrete surface and its long-term hydrophobic performance. *J. Chin. Ceram Soc.* **2019**, *47*, 1181–1190.
30. Wang, J.; Xu, L.; Li, M.; He, H.; Wang, Y.; Xiang, D.; Lin, S.; Zhong, Y.; Zhao, H. Effect of pre-carbonation on the properties of cement paste subjected to high temperatures. *J. Build. Eng.* **2022**, *51*, 104337. [CrossRef]
31. Xu, L.; Wang, J.; Li, K.; Lin, S.; Li, M.; Hao, T.; Ling, Z.; Xiang, D.; Wang, T. A systematic review of factors affecting properties of thermal-activated recycled cement. *Resour. Conserv. Recy.* **2022**, *185*, 106432. [CrossRef]
32. Wang, J.; Wang, Y.; Yu, J.; Xu, L.; Li, M.; Cheng, J.; Li, Z. Effects of sodium sulfate and potassium sulfate on the properties of calcium sulfoaluminate (CSA) cement based grouting materials. *Constr. Build. Mater.* **2022**, *353*, 129045. [CrossRef]
33. Zhang, J.; Stang, H.; Li, V. Crack bridging model for fiber reinforced concrete under fatigue tension. *Int. J. Fatigue.* **2001**, *23*, 655–670. [CrossRef]
34. Zhang, J.; Stang, H. Application of stress crack width relationship in predicting the flexural behavior of fiber reinforced concrete. *Cem. Concr. Res.* **1998**, *28*, 439–452. [CrossRef]
35. Camps, G.; Turatsinze, A.; Sellier, A.; Escadeillas, G.; Bourbon, X. Steel-fibre-reinforcement and hydration coupled effects on concrete tensile behaviour. *Eng. Fract. Mech.* **2008**, *75*, 5207–5216. [CrossRef]
36. Mohammadi, Y.; Singh, S.; Kaushik, S. Properties of steel fibrous concrete containing mixed fibres in fresh and hardened state. *Constr. Build. Mater.* **2008**, *22*, 956–965. [CrossRef]
37. Dong, F.; Wang, H.; Yu, J.; Liu, K.; Guo, Z.; Duan, X.; Qiong, X. Effect of freeze-thaw cycling on mechanical properties of polyethylene fiber and steel fiber reinforced concrete. *Constr. Build. Mater.* **2021**, *295*, 123427. [CrossRef]
38. Wang, Y.; Zhang, S.; Niu, D.; Fu, Q. Quantitative evaluation of the characteristics of air voids and their relationship with the permeability and salt freeze-thaw resistance of hybrid steel-polypropylene fiber-reinforced concrete composites. *Cem. Concr. Comp.* **2022**, *125*, 104292. [CrossRef]
39. Jones, P.; Austin, S.; Robins, P. Predicting the flexural load-deflection response of steel fibre reinforced concrete from strain, crack-width, fibre pull-out and distribution data. *Mater. Struct.* **2008**, *41*, 449–463. [CrossRef]
40. Małek, M.; Jackowski, M.; Łasica, W.; Kadela, M. Characteristics of Recycled Polypropylene Fibers as an Addition to Concrete Fabrication Based on Portland Cement. *Materials* **2020**, *13*, 1827. [CrossRef] [PubMed]
41. Małek, M.; Jackowski, M.; Łasica, W.; Kadela, M.; Wachowski, M. Mechanical and Material Properties of Mortar Reinforced with Glass Fiber: An Experimental Study. *Materials* **2021**, *14*, 698. [CrossRef]
42. Wang, Y.; Lu, H.; Wang, J.; He, F. Effects of highly crystalized nano C-S-H particles on performances of Portland cement paste and its mechanism. *Crystals* **2020**, *10*, 816. [CrossRef]
43. Hang, H.; Wang, Y.; Wang, J. Effects of aggregate micro fines (AMF), aluminum sulfate and polypropylene fiber (PPF) on properties of machine-made sand concrete. *Appl. Sci.* **2019**, *9*, 2250.

44. Wang, J.; Xie, J.; He, J.; Sun, M.; Zhao, J. Combined use of silica fume and steel fibre to improve fracture properties of recycled aggregate concrete exposed to elevated temperature. *J. Mater. Cycles Waste Manag.* **2020**, *22*, 62–877. [CrossRef]
45. Xie, J.; Zhao, J.; Wang, J.; Huang, P.; Liu, J. Investigation of the high-temperature resistance of sludge ceramsite concrete with recycled fine aggregates and GGBS and its application in hollow blocks. *J. Build. Eng.* **2021**, *34*, 101954. [CrossRef]
46. Wang, J. Steady-state chloride diffusion coefficient and chloride migration coefficient of cracks in concrete. *J. Mater. Civ. Eng.* **2017**, *29*, 0417117. [CrossRef]
47. Wang, J.; Liu, E. The relationship between steady-state chloride diffusion and migration coefficients in cementitious materials. *Mag. Concr. Res.* **2020**, *72*, 1016–1026. [CrossRef]
48. Herb, H.; Gerdes, A.; Brenner-Weiß, G. Characterization of silane-based hydrophobic admixtures in concrete using TOF-MS. *Cem. Concr. Res.* **2015**, *70*, 77–82. [CrossRef]

materials

MDPI

Article

Laboratory Study and Field Validation of the Performance of Salt-Storage Asphalt Mixtures

Yangsen Cao [1], Xinzhou Li [1], Zhuangzhuang Liu [1,2,*], Jiarong Li [1], Fan Zhang [1,*] and Baozeng Shan [1]

[1] School of Highway, Chang'an University, Xi'an 710064, China
[2] Key Laboratory for Special Area Highway Engineering of Ministry of Education, Chang'an University, South 2nd Ring Road Middle Section, Xi'an 710064, China
* Correspondence: zzliu@chd.edu.cn (Z.L.); fanzhang@chd.edu.cn (F.Z.)

Abstract: The traditional method of removing ice and snow on roads carries the risk of damaging roads and the environment. In this circumstance, the technology of salt-storage asphalt pavement has gradually attracted attention. However, snow-melting salts may also have an impact on asphalt mixture performance. To explore the effect of snow-melting salts on the mechanical and surface properties of salt-storage asphalt mixtures (SSAM), SSAMs were prepared with styrene–butadiene–styrene (SBS)-modified asphalt and high-elastic asphalt (HEA) as binders and snow-melting salts as fillers. The influence of the type of asphalt binder and the content of snow-melting salt on the performance of the SSAM was preliminarily investigated through laboratory tests. The results show that the high-temperature, low-temperature, and moisture resistance performance of the SBS group SSAM decreased by 9.8–15.1%, 1.6–12.3%, and 6.3–19.4%, respectively, compared with SBS00. The higher the amount of snow-melting salt, the greater the performance drop. The three mechanical properties of the HEA group containing high-elastic agent TPS are 11.3–19.7%, 4.2–12.3%, and 4.8–13.3% higher than that of the SBS group. Even when the content of snow-melting salt is 50% or 75%, the mechanical properties of the HEA group are better than that of SBS00 without snow-melting salt. Snow-melting salt has clear advantages in improving the anti-skid performance but decreases the anti-spalling performance. The surface properties of the HEA group were also better than that of the SBS group. Considering the mechanical properties and surface properties, the comprehensive performance of the HEA group is better than that of the SBS group, and HEA50 has the best comprehensive performance. In addition, the construction performance of the SSAM has also been verified, and the production of SSAM according to the hot mix asphalt can meet the specification requirements.

Keywords: road engineering; salt-storage asphalt mixture; mechanical properties; surface properties; construction performance; high-elastic agents

1. Introduction

Snow and ice on the road affect traffic safety. In China, 15–35% of traffic accidents in winter are related to snow and ice [1,2]. Ice and snow removal has always caught the attention of road practitioners. Currently, the methods of road ice and snow removal mainly include the passive method and active method [3]. Passive snow removal mainly relies on traditional methods such as spreading snow-melting salts, mechanical snow removal, and manual snow removal. These three passive ice and snow removal methods are generally used in combination. Although traffic is guaranteed to a certain extent, there are problems such as untimely snow removal, the large consumption of manpower and material resources, damage to pavements, and pollution of the environment [4]. The active method mainly involves changing the pavement material or structure so that ice and snow can easily be removed from the pavement. With the development of pavement technology, the method of active ice and snow removal has gradually attracted researchers' attention.

Active ice and snow removal technologies include elastic pavement [5,6], thermal pavement [7,8], and salt-storage pavement [9,10]. The elastic pavement technology provides the asphalt pavement with a certain elasticity by replacing the fine aggregate in the asphalt mixture with rubber particles. Under the vehicle load, the non-cooperative deformation of the ice layer and the pavement promotes ice layer rupture to achieve the purpose of deicing [11]. The structure of thermal pavements is changed to allow heating cables [12] or heat pipes [13] to be arranged inside the pavement. The purpose of melting ice and snow is achieved by employing cable heating or thermal fluid in heat pipes. Salt-storage pavement is to reduce the freezing point of water by changing the composition of asphalt mixture materials, thereby inhibiting icing. For example, snow-melting salt replaces mineral powder [14], or salt-storage aggregate replaces traditional aggregate [15]. Among the three active snow removal technologies, elastic pavement technology increases the flexibility of the pavement, its mixing process and rolling process are not yet mature, and there may also be a risk of increasing vehicle fuel consumption [3].

Thermal pavements have a high efficiency for melting ice and snow, but have the disadvantages of a high energy consumption, high cost, and complicated construction and maintenance [3]. Salt-storage pavement technology inherits the construction technology behind current pavements; this technology is favored by researchers and managers due to its low cost and successful snow-melting effect [16].

Salt-storage pavement technology originated in Europe in the 1960s, and similar research was carried out earlier in Switzerland and Germany. Verglimit, an antifreeze developed in Switzerland, reduces the freezing point of pavements to −20 °C [17]. In 1974, Austria laid salt-storage asphalt pavements on Europa Bridge [18]. Japan also introduced salt-storage technology in the 1970s, and in the 1990s Mafilon products were innovated and successfully promoted. The snow removal validity period of Mafilon can reach 5–6 years [19]. China suffered a major snowstorm in early 2008, which not only affected traffic, but also caused huge damage to societal infrastructure. In the summer of 2008, the Lanshang expressway paved the salt-storage asphalt pavement for the first time in China [18].

Some achievements have been made in the research on salt-storage pavement technology. Feng et al. [20] studied the effect of salt and freeze–thaw cycles on asphalt pavement materials, finding that salt can reduce the low-temperature performance of asphalt. When the salt content is lower than 3%, the reduction in the strength of asphalt mixtures is mainly caused by freeze–thaw cycles. When the salt content is higher than 3%, the corrosion of the salt accelerates the destruction of the asphalt mixture. Lysbakken and Lalagüe [21] studied the reliability of the BOBO 20 instrument in detecting the salt content of the pavement by testing the concentration of the salt solution, dry salt, and recrystallized salt content of the pavement. The test results indicate that the instrument is not sensitive to the content of dry salt and recrystallized salt. L. Borst and Brown [22] and Hossain et al. [23] studied the effect of pavement type on the chloride ion release rate and snow melting rate, respectively. In terms of chloride ion release rate, porous asphalt was released the fastest, followed by porous interlocking pavement, and finally porous concrete. In terms of the snow melting rate, asphalt concrete has the fastest snow melting rate, and cement concrete and interlocking concrete have similar snow melting rates. Roseen et al. [24] compared the control level of chloride ions during winter pavement maintenance between a porous asphalt mixture and densely graded asphalt mixture. Under the same ice and snow coverage and anti-skid index, the salt consumption of the porous asphalt mixture is 64–77% less than that of the densely graded asphalt mixture. Derya et al. [25] dissolved formate in a hydrophilic gel medium and then dispersed the gel in SBS polymer-modified asphalt. The anti-condensation ice results show that the modified asphalt containing formate starts to freeze 20 min later than the base asphalt at −14 °C. Over the 67-day test period, 1.07–10.8% of potassium formate was released. Zheng et al. [26] studied the long-term performance of anti-icing pavement and pointed out that snow-melting salt can improve the high-temperature performance of asphalt mixtures and reduce low-temperature performance and moisture resistance. The long-term anti-condensation ice model of pavements was established by indoor salt precipitation tests. In addition to the anti-condensation ice model, İkiz and Galip [27] studied the decay law of salt in porous pavements and cement concrete pavements with time and traffic volume, establishing a residual salt prediction model.

The above research on salt-storage pavement has promoted the development of ice and snow melting technology. The release performance of chloride ions and the performance of melting ice and snow are the functional performance of salt-storage pavement. However, the excellent mechanical properties of salt-storage pavements are fundamental for meeting service conditions. However, studies on the mechanical properties of SSAMs are limited, and some studies found that snow-melting salts can damage some properties of asphalt mixtures, such as low-temperature performance, moisture resistance, etc. [28–30]. Therefore, further fundamental research on the mechanical properties of SSAM needs to be further carried out. As a surface layer or wear layer, the anti-skid performance of the SSAM is also very important. At this stage, snow-melting salts are mostly chlorine-based snow-melting salts. From the perspective of mixture composition, snow-melting salt can replace traditional limestone fillers. However, the chemical properties of chlorine-based materials and alkaline limestone fillers are quite different, which may also affect the spalling resistance of asphalt mixtures, further leading to the loosening of the wear layer. In addition, most of the current research on SSAMs focuses on indoor research, and there are few reports on the verification of its field performance.

To further explore the effect of snow-melting salts on the mechanical properties and surface properties of SSAMs, SSAMs were prepared with SBS-modified asphalt and HEA as adhesives, and slow-release snow-melting salt as a limestone filler substitute. The basic mechanical properties of SSAM were studied using high-temperature performance tests, moisture resistance performance tests, and low-temperature performance tests. The surface properties of SSAMs were evaluated by anti-skid performance tests and anti-spalling performance tests. Finally, in terms of the physical project, the engineering performance and construction quality of the SSAM were tested.

2. Materials and Methods

2.1. Materials and Mix Design

2.1.1. Materials

There are two kinds of asphalt that are typically used in indoor pre-studies: SBS I-D modified asphalt and HEA. HEA is prepared by adding TPS high-elastic agent into SBS I-D modified asphalt. The technical indicators of the two asphalts are shown in Table 1. The coarse aggregate and fine aggregate are limestones from a quarry in Liquan, Xi'an, China. Aggregate technical indicators are shown in Table 2. The snow-melting salt in the SSAM was added in the form of filler, so there are two types of fillers used in the test: conventional limestone filler and snow-melting salt filler. The technical indicators of the two fillers are shown in Table 3. Stone mastic asphalt (SMA) with a good durability is selected for the wear layer, and the technical indicators of lignin fibers are shown in Table 4.

Table 1. Basic indicators of the two asphalts.

Index		Unit	SBS-Modified Asphalt	HEA	Standard	Test Method [31]
Penetration (100 g, 5 s, 25 °C)		0.1 mm	54	49	40–60	T0604-2011
Penetration index		-	0.55	0.68	≥0	T0604-2011
Ductility (5 °C, 5 cm/min)		cm	31	55	≥20	T0605-2011
Softening point		°C	80	94	≥60	T0606-2011
Kinematic viscosity at 135 °C		Pa·s	1.773	1.861	≤3	T0620-2000
Elastic recovery at 25 °C		%	90	98	≥75	T0662-2000
After TFOT	Quality loss	%	−0.213	−0.13	±1.0	T0610-2011
	Penetration ratio at 25 °C	%	70.5	76.6	≥65	T0604-2011
	Ductility (5 °C, 5 cm/min)	cm	16	31	≥15	T0605-2011

Table 2. Basic indicators of aggregates.

Aggregate	Index	Test	Standard	Test Method [32]
Coarse aggregate	Crush value (%)	10	≤26	T0316-2005
	Los Angeles attrition loss (%)	9.4	≤28	T0317-2005
	Firmness (%)	8	≤12	T0314-2000
	Soft stone content (%)	1.2	≤3	T0320-2000
	Needle-like content (%)	8.7	≤15	T0312-2005
Fine aggregate	Apparent specific gravity	2.663	≥2.5	T0328-2005
	Bulk specific gravity	2.731	/	T0330-2005
	Firmness (>0.3 mm) (%)	4	≤12	T0340-2005
	Sand equivalent (%)	70	≥60	T0334-2005

Table 3. Basic indicators of the two fillers.

Filler	Index	Test	Standard	Test Method [32]
Limestone	Moisture content (%)	0.2	≤1.0	T0332-2005
	Apparent specific gravity	2.762	/	T0352-2000
	Bulk specific gravity	2.667	/	T0352-2000
	Hydrophilic coefficient	0.73	<1	T0353-2000
	Plasticity index (%)	2.7	<4	T0354-2000
	Appearance	Qualified	No agglomeration	T0355-2000
Snow-melting salt	Appearance	White powder	No agglomeration	/
	Moisture content (%)	0.2	≤0.5	T0332-2005
	Apparent specific gravity	2.170	/	T0352-2000
	Bulk specific gravity	2.136	/	T0352-2000
	Salt content (%)	56	50 ± 10	/
	pH	8.3	8–8.5	/

Table 4. Technical indicators of lignin fibers.

Index	Ash Content/%	pH	Oil Absorption	Moisture Content/%	Width/mm	Bulk Density/(g/cm^3)
Value	17	7.3	4–6 times the fiber mass	<5	<0.045	0.2–0.4

2.1.2. Mix Design

The gradation type of the wear layer in this study is SMA-10. To ensure that the wear layer is applicable in follow-up research, the suspension compact asphalt mixture AC-16 is selected as the lower bearing layer of SMA-10. The gradation curves of SMA-10 and AC-16 are shown in Figure 1. Snow-melting salt is added to asphalt mixes as fillers. Due to the large difference in density between the snow-melting salt and the limestone filler, the equal volume replacement method was used to replace the limestone filler with snow-melting salt. SMA-10 asphalt mixtures are named in terms of asphalt type and snow-melting salt replacement ratio. There are seven kinds of SMA-10 asphalt mixture, namely SBS00, SBS50, SBS75, SBS100, HEA50, HEA75, and HEA100. The asphalt aggregate ratio of SMA-10 and AC-16 was determined using the Marshall test. The optimal asphalt aggregate ratio was finally determined to be 6.2% for SMA-10 and 4.2% for AC-16.

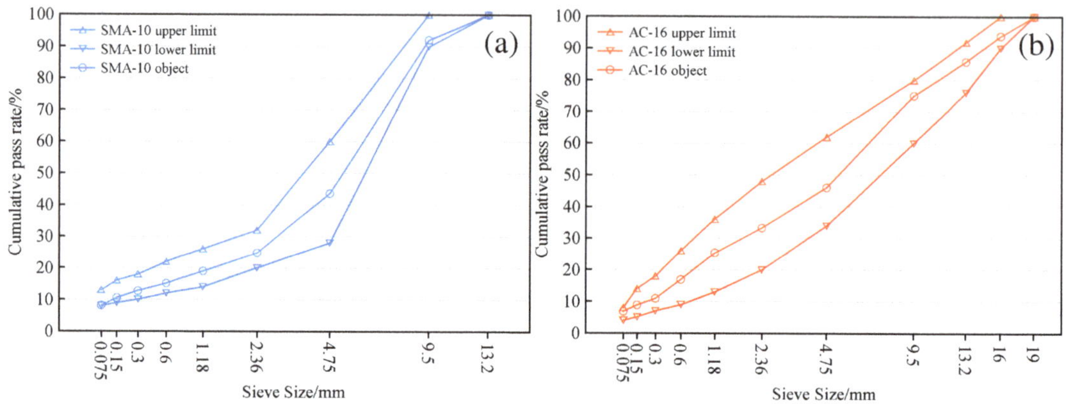

Figure 1. Gradation curves of SMA-10 and AC-16. (**a**) Gradation curve of SMA-10; (**b**) Gradation curve of AC-16.

2.2. Laboratory Test Methods

2.2.1. High-Temperature Performance Test

To truly reflect the anti-rutting performance of SSAM at a high temperature, SMA-10 + AC-16 composite specimens were prepared for rutting tests. The size of the specimen is $300 \times 300 \times 50$ mm^3, of which the thickness of SMA-10 is 20 mm, and the thickness of AC-16 is 30 mm. The specimen was compacted twice: first, the cushion layer AC-16 was formed, and then the SMA-10 was rolled on the cushion layer. The test was carried out according to the specification [31], the test temperature was 60 °C, and the wheel pressure was 0.7 MPa. Rutting at 45 min and 60 min was recorded, and the dynamic stability was calculated accordingly. The average value of three sets of tests was taken as the final dynamic stability to evaluate the high-temperature performance of different salt-storage wear layers.

2.2.2. Moisture Resistance Performance Test

As an alkaline aggregate, limestone filler has a good adhesion to asphalt; therefore, asphalt is a better coating for the aggregate. In the salt-storage wear layer, since the filler is replaced with slow-release snow-melting salt, even if the surface layer of the snow-melting salt is wrapped with lipophilic components, its interaction with asphalt is still not as good as that of the limestone filler, which may affect the moisture resistance performance of salt-storage wear layer.

The immersion Marshall test was used to test the stability of standard Marshall specimens before and after immersion and compare the changes in stability before and after immersion. First, standard Marshall specimens were prepared and divided into two groups: one group was placed in a 60 °C constant temperature water tank for half an hour, and the other group was placed in a 60 °C water tank for 48 h. After that, the stability of each group of specimens was measured, and the residual stability ratio was calculated.

The freeze–thaw splitting test was also needed to divide the mixture specimens into two groups. One group is to measure the splitting strength at room temperature, and the other group requires the specimens to be saturated with water under vacuum conditions, and then the specimens are frozen at −18 °C. After that, the specimens were placed in a constant temperature water tank at 60 °C for 24 h, and the splitting strength of the specimens was finally determined. The moisture resistance performance was evaluated by the ratio of splitting strength of two groups of specimens under freeze–thaw and unfreeze–thaw conditions. The asphalt film on the aggregate surface is more susceptible to the effects of water at low temperatures, so it peeled off from the aggregate surface. Therefore, compared with the water immersion Marshall test, the freeze–thaw splitting test can better reflect the moisture resistance of the salt-storage wear layer.

2.2.3. Low-Temperature Performance Test

Winter is the season when salt-storage asphalt pavements are most required, so these pavements must have a sufficient low-temperature crack resistance. The low-temperature cracking of asphalt pavement mainly occurs in two different ways: material shrinkage cracking and fatigue cracking. In this paper, the low-temperature shrinkage performance of different types of salt-storage wear layers is studied using a low-temperature bending test.

The size of the specimen is 30 × 35 × 250 mm^3. Loading was carried out by a three-point loading method with a span of 200 mm until the specimen broke. The loading rate was 50 mm/min, and the experiments were carried out at −10 °C. Low-temperature performance is usually evaluated by a low-temperature failure strain after the test. In addition to the low-temperature failure strain, the fracture energy density was also used to evaluate the low-temperature crack resistance of the mixture. Strain energy density, as a kind of energy stored in a unit volume, can reflect the fracture ability of a material.

2.2.4. Surface Anti-Skid Performance Test

The friction between the tires and the pavement is an important factor to ensure the safe driving of the vehicle, especially in winter. Therefore, it is necessary to conduct a skid resistance test on the wear layer. This time, the pendulum tribometer BM-III was used to determine the anti-skid properties of different types of salt-storage wear layers. The test specimens used in the test are the same as those used in the rutting test.

2.2.5. Surface Spalling Resistance Test

The spalling resistance of the wear layer largely depends on the adhesion between the asphalt and the aggregate. Conventional tests for measuring adhesion include the boiling method, water immersion method, etc. [31], and these tests are greatly affected by subjective factors. In this study, the Schellenberg binder drainage test and the Cantabro test were used to measure the adhesion of the salt-storage wear layer. The Schellenberg binder drainage test can also be used to judge whether there is an excess of free asphalt in the mixture, and to verify the optimal asphalt aggregate ratio of the mixture. Referring to the specification [31], the Schellenberg binder drainage test was measured using a beaker. A specified amount of asphalt mixture was weighed and placed in a beaker, and then the beaker was kept in an oven at 185 °C for 60 min. Afterward, the asphalt mixture was poured out, the mass of the residual asphalt on the wall of the beaker was weighed, and the asphalt loss was calculated.

During the Cantabro test, the prepared Marshall specimens were placed in a constant temperature water tank at 20 °C for 20 h. Then, the specimen was placed in the Los Angeles abrasion testing machine for 300 revolutions. The mass loss of the specimen after the test is the loss rate of the Cantabro test.

2.3. Field Performance Verification of Salt-Storage Wear Layer

2.3.1. Project Overview

Xi'an is located in the Guanzhong Plain, which is hot in summer and cold in winter. The pavement surface can easily freeze in winter, creating a great hidden danger to traffic safety. Snow removal methods such as sprinkling snow-melting salt and mechanical cleaning will not only corrode pavement structures and damage the surrounding environment, but also consume a lot of manpower and material resources. To eliminate road safety hazards and reduce the resources required for ice and snow removal, on 30 June 2019, an experimental road with a length of 520 m, intended for snow-melting tests, was laid on the Jinghe Bridge of National Highway 310 based on the indoor pre-study. The implementation of the project provides a reference for the construction of pavements that can automatically melt snow in the Guanzhong area.

2.3.2. Main Material Parameters

Slightly different from the indoor pre-study, the asphalt used in the project is SBS I-C-modified asphalt. The technical indicators are shown in Table 5. The coarse and fine aggregates are all limestone from Xinhe, Xi'an, China. The aggregate density of each grade is shown in Table 6, and the other indicators meet the requirements of the specification. The fillers are slow-release snow-melting salts, and the replacement rate of limestone filler is 100%.

Table 5. Properties of SBS I-C for SSAM.

Index		Unit	Test	Standard
Penetration (100 g, 5 s, 25 °C)		0.1 mm	72	60–80
Penetration index		-	1.02	≥−0.4
Ductility (5 °C, 5 cm/min)		cm	51	≥30
Softening point		°C	82	≥55
Kinematic viscosity at 135 °C		Pa·s	1.763	≤3
Elastic recovery at 25 °C		%	92.5	≥65
After TFOT	Quality loss	%	−0.320	±1.0
	Penetration ratio at 25 °C	%	80.6	≥60
	Ductility (5 °C, 5 cm/min)	cm	30	≥20

Table 6. Density of aggregate and filler.

Particle Size/mm	0–3	3–6	6–11	11–17	17–22	Snow-Melting Salt
Apparent specific gravity/(g/cm^3)	2.663	2.723	2.725	2.722	2.737	2.170
Bulk specific gravity/(g/cm^3)	2.731	2.666	2.693	2.689	2.707	2.136

2.3.3. Mix Design for Construction

In the early stage, the performance of the SMA-10 salt-storage wear layer was studied indoors. For the consideration of construction costs, the project directly selected AC-16 as the wear layer of the pavement, that is, the upper layer. There are five stone silos in the mixing station, corresponding to limestone aggregates with particle sizes of 0–3, 3–6, 6–11, 11–17, and 17–22 mm, respectively. The gradation curve used for production is shown in Figure 2. Five groups of asphalt aggregate ratios of 3.8%, 4.3%, 4.8%, 5.3%, and 5.8% were proposed to determine the optimal asphalt aggregate ratio. The results of the Marshall test are shown in Table 7. The optimal asphalt aggregate ratio determined by calculation is 4.135%, and then the asphalt aggregate ratio is adjusted to 4.2%. The performance indicators of the AC-16 asphalt mixture under the best asphalt aggregate ratio are also shown in Table 7.

Table 7. Marshall test results.

Asphalt Aggregate Ratio/%	VFA/%	VV/%	Flow Value/mm	Stability/kN	Bulk Gravity/(g/cm^3)
3.8	61.97	5.16	2.24	16.70	2.394
4.3	71.71	3.76	2.97	15.68	2.412
4.8	78.62	2.91	2.92	14.43	2.416
5.3	84.58	2.16	3.08	11.55	2.419
5.8	85.88	2.12	4.19	10.68	2.403
4.2 (optimum)	71.90	3.61	3.28	15.75	2.418

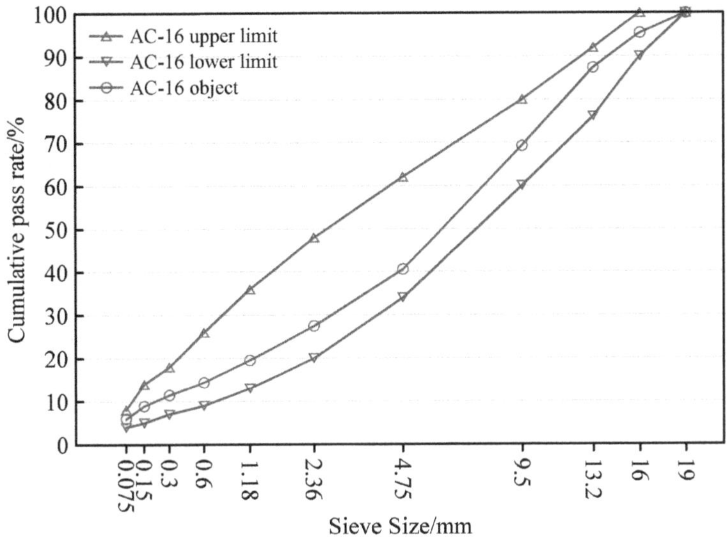

Figure 2. AC-16 grading curve for construction.

3. Results and Discussion

3.1. Mechanical Properties

3.1.1. High-Temperature Stability

Figure 3 shows the high-temperature performance of different types of SSAMs. Comparing the salt-storing asphalt mixture in the SBS group with SBS00 without snow-melting salt, it can be seen that the snow-melting salt reduces the high-temperature performance of the asphalt mixture. The dynamic stability of SBS50, SBS75, and SBS100 was 9.8%, 15.0%, and 15.1% lower than that of SBS00, respectively. This may be because the interaction between snow-melting salt and asphalt is not as good as the interaction between alkaline limestone filler and asphalt [33,34], which leads to poor adhesion between asphalt mortar and aggregate, and the shear resistance of SSAM decreases at high temperatures. When the content of snow-melting salt is the same, the high-temperature performance of SSAM in the HEA group is better than that in the SBS group. For example, when the replacement rates of snow-melting salt are 50%, 75%, and 100%, the dynamic stability of the HEA group is 15.7%, 19.7%, and 11.3% higher than those in the SBS group, respectively. This indicates HEA can significantly improve the high-temperature performance of the salt-storage wear layer. As a highly elastic polymer, the network structure formed by TPS after swelling in SBS-modified asphalt increases the elasticity of the asphalt [5], which may improve the high-temperature performance of the asphalt mixture. Whether in the SBS group or the HEA group, the snow-melting salt adversely affects the rutting resistance, and the higher the snow-melting salt content, the greater the decrease in the high-temperature performance. Compared with SBS00, the dynamic stability of HEA50 and HEA70 is even higher than that of SBS00. This shows that HEA can compensate for the negative impact of snow-melting salt on high-temperature performance to a certain extent. Among the six SSAMs, the dynamic stability of SBS75 and SBS100 with lower dynamic stability are 4469 and 4462 times/mm, respectively. The highest dynamic stability is HEA50. Even the SBS100 with the lowest dynamic stability, its high-temperature performance can meet the specification requirements of 3000 times/mm.

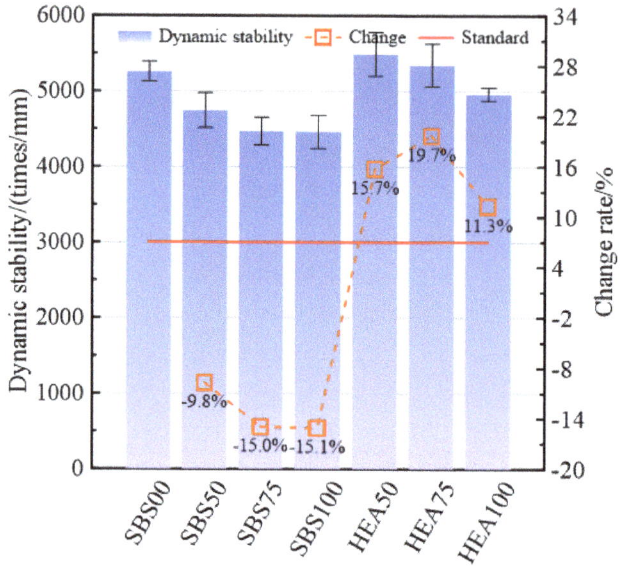

Figure 3. High-temperature stability performance of salt-storage asphalt wear layer.

3.1.2. Moisture Resistance

The results of the immersion Marshall test and the freeze–thaw split test are shown in Figure 4a,b, respectively. In Figure 4a, Compared with SBS00, the residual stability of SBS50, SBS75 and SBS100 decreased by 6.3%, 7.8% and 8.0%, respectively. Snow-melting salt is not good for moisture resistance. Snow-melting salts can lead to the reduction in structural asphalt in the asphalt, which in turn reduces the integrity of the asphalt mixture, resulting in a decrease in moisture resistance [16]. In addition, after the snow-melting salt is dissolved in water, the micropores of the asphalt mixture increase, and the water is more likely to penetrate into the interior of the asphalt mixture [33]. This may also be one of the reasons for the decrease in the residual stability ratio. In the HEA group, the addition of the high-elastic agent TPS significantly improved the residual stability ratio. HEA50, HEA75 and HEA100 were 7.9%, 8.8% and 7.4% higher than SBS50, SBS75 and SBS100, respectively. This shows that TPS increased the bond between the asphalt and the aggregate. The viscosity of HEA is higher than that of SBS-modified asphalt in Table 1, which supports this view. For two groups of SSAM, the higher the replacement rate of snow-melting salt, the greater the decrease in residual stability. When the replacement rate of snow-melting salt is lower than 75%, the moisture resistance of HEA50 and HEA75 is even higher than that of SBS00 without snow-melting salt. The high viscosity of HEA compensates for the weaker adhesion between snow-melting salt and asphalt, making the asphalt mixture more resistant to moisture damage. Among the SSAMs, the lowest and highest residual stability ratios are SBS100 and HEA50, respectively. Although snow-melting salts were detrimental to the residual stability ratios, the residual stability ratios of all SSAMs were higher than the 80% required by the specification.

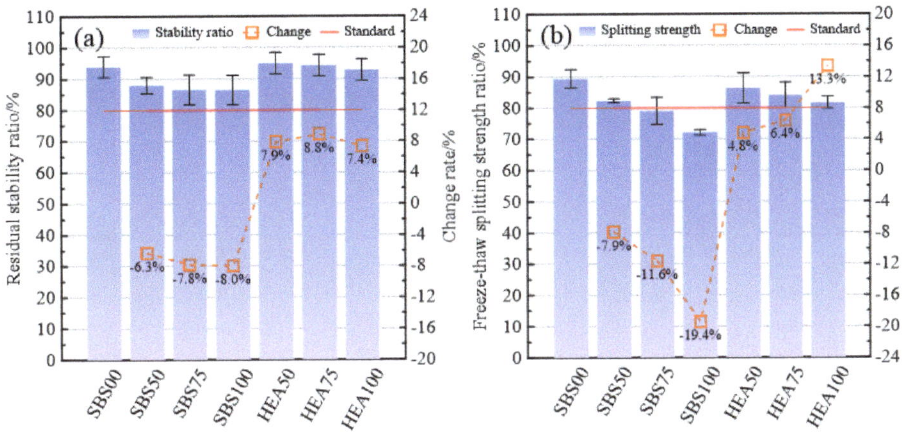

Figure 4. Moisture resistance performance of SSAM. (**a**) Residual stability; (**b**) splitting strength ratio.

In Figure 4b, the freeze–thaw splitting strength ratio of the SSAM in the SBS group is lower than that of the SBS00, and the decrease of the SBS100 even reaches 19.4%. In addition, to the void structure left by the precipitation of snow-melting salt and the reduction in asphalt cohesion by snow-melting salt, the corrosion effect of the salt solution may also be one of the reasons for the decline of moisture resistance [26,34]. Compared with the SBS group, the splitting strength ratio of the HEA group was increased to varying degrees, and the most obvious increase was HEA100, which was 13.3% higher than that of SBS100. The better viscosity and low-temperature properties of HEA are responsible for the improved splitting strength ratio. Similar to Figure 4a, for both the SBS group and the HEA group, as the content of snow-melting salt increases, the freeze–thaw splitting strength ratio of the mixture decreases linearly, and this is especially evident in the SBS group. Although HEA improves the moisture resistance of SSAM to a certain extent, the freeze–thaw splitting strength ratio of all SSAMs is lower than that of SBS00. It shows that the improvement effect of HEA is limited. Even the splitting strength ratio of HEA50 with the highest splitting strength ratio is about 3.0% lower than that of SBS00. According to the specification requirements [35], the freeze–thaw splitting strength ratio of SMA using modified asphalt should not be lower than 80%. The freeze–thaw splitting strength ratio of SSAM using HEA can meet this requirement, but SBS75 and SBS100 do not meet the requirement. The content of snow-melting salt should not be too much, especially when using SBS-modified asphalt.

3.1.3. Low-Temperature Crack Resistance

Figure 5a shows the tensile strain of the SSAM when it fails in bending. The larger the tensile strain, the higher the flexibility and crack resistance of the mixture at low temperatures. In the asphalt mixture of the SBS group, the maximum tensile strains of salt-asphalt mixtures are lower than SBS00. The maximum tensile strain decreased by 12.3% after all the limestone fillers were replaced with snow-melting salt. After using HEA high-elastic modified asphalt, the maximum tensile strain of the HEA group significantly increased compared with the SBS group. Compared with the SBS group, HEA50, HEA75, and HEA100 increased by 6.7%, 4.2%, and 12.3%, respectively. This is because HEA has better flexibility than SBS-modified asphalt at low temperatures, which is illustrated by the higher ductility of HEA than SBS-modified asphalt in Table 1. Regardless of the HEA group or the SBS group, the maximum tensile strain decreased with the snow-melting salt. The poor adhesion between snow-melting salt and asphalt may be the main reason for this problem. In addition, the snow-melting salt is mainly distributed in the asphalt, and its dispersibility characteristics may reduce the continuity of the asphalt on the fracture surface [20,36], and then reduce the toughness of the asphalt. Under the condition that the replacement rate does not exceed 75%, the tensile strain of the HEA group is higher than that of SBS00. Even with all limestone fillers replaced, the maximum tensile strain of HEA100 is only 1.5% lower than that of SBS00. HEA compensates for the adverse effects of snow-melting salts. Among the six SSAMs, HEA50 has the best low-temperature crack resistance, and SBS100 is the most prone to cracking. Nevertheless, the tensile strains of all SSAMs meet the specification requirements.

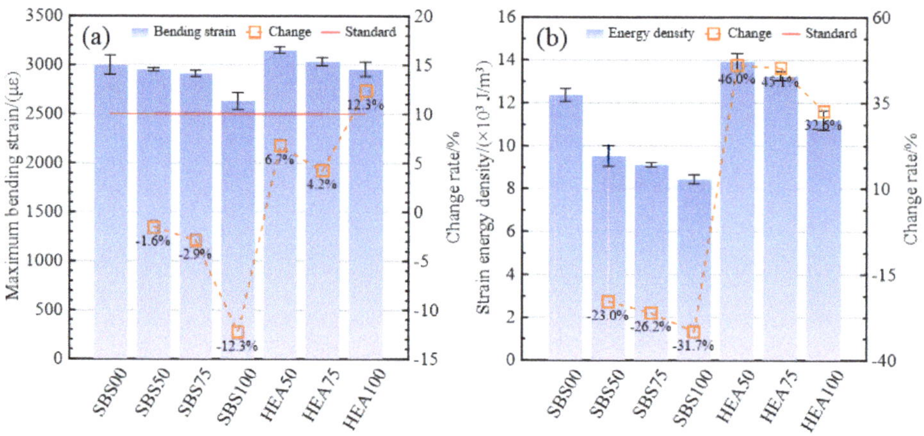

Figure 5. Low-temperature performance index of SSAM. (**a**) Maximum tensile strain; (**b**) strain energy density.

The strain energy densities of different types of salt-storage wear layers at −10 °C are shown in Figure 5b. The higher the strain energy density, the higher the energy required for the material to fracture, and the less likely it will fracture. The strain energy density of the SSAM in the SBS group was lower than that in the SBS00 group, and the strain energy density in the HEA group was much higher than that in the SBS group. Therefore, it can be concluded that snow-melting salt and HEA have negative and positive effects on the low-temperature crack resistance of salt-storing asphalt mixtures, respectively. The strain energy density of HEA50 and HEA75 is higher than that of SBS00, which is similar to Figure 5a. In addition, HEA50 has the highest strain energy density and SBS100 has the lowest strain energy density in the SSAM. Highly elastic asphalt is beneficial for promoting the application of salt-storage materials.

3.2. *Surface Properties*

3.2.1. Surface Anti-Skid Performance

As a surface functional layer, the SSAM must meet the anti-skid requirements to ensure driving safety while meeting the conventional mechanical properties. The anti-skid performance of different SSAMs is shown in Figure 6. The British pendulum number (BPN) of the SSAM in the SBS group was higher than that of the SBS00, and the BPN of the SBS100 increased most significantly by 6.6%. This shows that the snow-melting salt is beneficial in improving the anti-skid performance. There may be two reasons, one is that the snow-melting salt particles are coarser than the limestone ore powder [9,16]. Snow-melting salt may improve the surface roughness of salt-storage asphalt [37]. Second, before testing BPN, the sprayed water may have caused the precipitation of surface salt, leaving micro-pores that make the surface of the salt-storage asphalt rougher. Under the same replacement amount of snow-melting salt, the BPN value of the asphalt mixture in the HEA group is higher than that in the SBS group. For example, the BPN value of HEA100 is 8.6% higher than that of SBS100. TPS increases the elasticity of asphalt and improves the viscosity of asphalt; therefore, high-elastic asphalt can better improve the anti-skid ability of pavements during winter. With the increase in the replacement amount of snow-melting salt, the BPN of the two groups of SSAMs gradually increased, indicating that the snow-melting salt has an advantage over the limestone filler in improving the anti-skid performance of the mixture. Among the six SSAMs, the best anti-skid performance is HEA100. Although SBS50 has the worst skid resistance, it still meets the specification requirements.

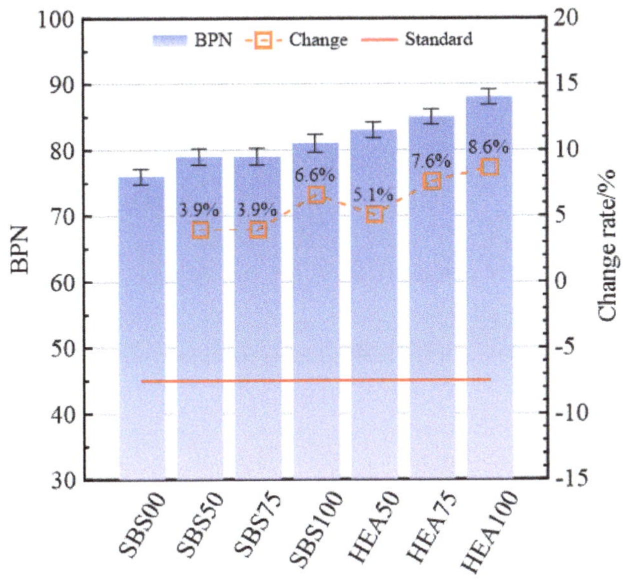

Figure 6. Anti-skid index of SSAM.

3.2.2. Surface Spalling Resistance

As shown in Figure 7a, the asphalt loss of the SSAM in the SBS group is higher than that in the SBS00. Among them, the asphalt loss of SBS100 is 26.6% higher than that of SBS00. This may be because the snow-melting salt increases the free asphalt in the asphalt mixture [16]. The asphalt loss exhibited by the HEA group was different from that of the SBS group, which may be because TPS increased the high-temperature performance of SBS-modified asphalt. This is consistent with the penetration and softening points of HEA in Table 1 being lower and higher than that of SBS-modified asphalt, respectively. With the increase in the replacement rate of snow-melting salt, the quality of asphalt adhering to the beaker increases, especially in the SBS group. Therefore, the content of free asphalt in the mixture may increase with the content of snow-melting salt. This may be because the addition of snow-melting salt reduces the optimum asphalt aggregate ratio of the mix. However, the fillers used in determining the optimal asphalt aggregate ratio are all limestone fillers, and the optimal asphalt aggregate ratio may slightly change after the snow-melting salt replaced part of the limestone filler. The lack of adjustment of the optimal asphalt aggregate ratio resulted in a larger asphalt loss of the specimens with snow-melting salt than that without snow-melting salts [38]. Compared with the SBS00 group, the SSAM in the HEA group has a higher asphalt loss rate. Therefore, although HEA is helpful to improve asphalt bonding, the gain effect of HEA on asphalt cannot completely cover the negative effect of snow-melting salt. In the SSAM, the least asphalt loss is HEA50, and the largest one is SBS100. Since the maximum asphalt loss required by the specification is 0.1%, the free asphalt content of the various SSAMs is satisfactory. This indirectly shows that the anti-spalling performance of the SSAM meets the requirements.

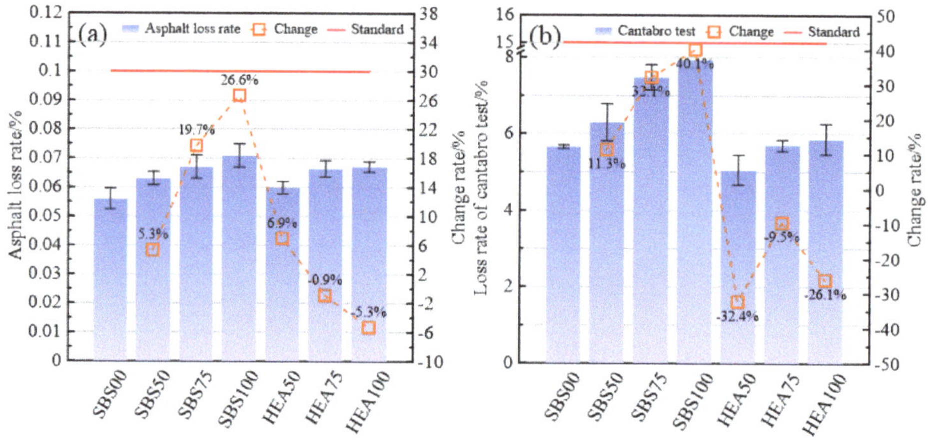

Figure 7. Anti-spalling index of SSAM. (**a**) Asphalt loss rate; (**b**) Cantabro loss rate.

In Figure 7b, the mass loss of the SSAM in the SBS group was significantly higher than that in the SBS00 group. The mass loss of SBS50, SBS75, and SBS100 is 11.3%, 32.1%, and 40.1% higher than that of SBS00, respectively. This may be because the coating effect of the mortar formed by asphalt and snow-melting salt may be weak, which cannot accurately ensure the adhesion of asphalt and aggregates, reducing the integrity of the mixture. Compared with the SBS group, the HEA group had a more significant decrease trend. The high viscosity and high elasticity of HEA can improve the adhesion of asphalt mortar. Whether it is the HEA group or the SBS group, with the increase in the replacement rate of snow-melting salts, the loss of mass also tended to increase. Snow-melting salt is not good for spalling resistance of SSAM. The mass loss of HEA50 is lower than that of SBS00, and HEA75 is on par with SBS00. HEA100 is higher than SBS00, which may be caused by the change in the asphalt-filler ratio [39]. HEA improves the spalling resistance of SSAM to a certain extent. In the specification for SMA using modified asphalt, the loss rate in the Cantabro test should not exceed 15% [35]. Therefore, the anti-loose performance of the SSAM meets the requirements.

3.3. *Determination of Optimal Dosage of Snow-Melting Salt*

From the analysis in Sections 3.1 and 3.2, it can be seen that the snow-melting salt is unfavorable to the mechanical properties of asphalt mixtures, including high-temperature performance, moisture resistance, and low-temperature performance. In terms of surface properties, snow-melting salt is beneficial to the anti-skid performance of the SSAM, but at the same time harms the anti-spalling performance. To determine the optimal dosage of snow-melting salt, the influence of snow-melting salt on mechanical properties and surface properties should be comprehensively considered. Here, the performance of various SSAMs is comprehensively evaluated by the grey target decision [40]. The indicators used in the evaluation are dynamic stability, residual stability ratio, freeze–thaw splitting strength ratio, maximum flexural–tensile strain, strain energy density, BPN, asphalt loss, and Cantabro loss. The raw data of the 8 indicators are shown in Table 8.

Table 8. Raw data for eight indicators.

SSAMs	Dynamic Stability/ (Times/mm)	Residual Stability Ratio/%	Freeze–Thaw Splitting Strength Ratio/%	Maximum Tensile Strain/με	Strain Energy Density/(J/m³)	BPN	Asphalt Loss/%	Cantabro Loss/%
SBS00	5258.4	93.9	89.4	2998.6	12,371.4	76	0.0560	5.7
SBS50	4744.9	88.0	82.3	2951.3	9531.6	79	0.0631	6.3
SBS75	4468.9	86.6	79.0	2910.8	9127.7	79	0.0670	7.5
SBS100	4462.1	86.4	72.1	2630.0	8448.5	81	0.0709	7.9
HEA50	5491.3	94.9	86.2	3149.8	13,912.6	83	0.0599	5.1
HEA75	5347.1	94.2	84.1	3033.4	13,246.8	85	0.0664	5.7
HEA100	4964.2	92.8	81.7	2954.7	11,204.5	88	0.0671	5.9

Among the eight indicators, except for the two economic indicators of asphalt loss and Cantabro loss, the other six indicators are benefit indicators. The lower the economic index, the better the performance of the asphalt mixture, while the benefit index is the opposite. To evaluate the comprehensive performance of SSAM, it is necessary to normalize each index [41]. The normalized values of each index are shown in Table 9. After normalization, the optimal data in each index is 1. If the data of an index are closer to 1, then the performance represented by this index is better. Here, [1,1,1,1,1,1,1,1] is used as the target vector to represent the best overall performance. The closer the distance of the vector composed of the comprehensive performance of various asphalt mixtures to the target vector, the better the comprehensive performance of the asphalt mixture.

Table 9. Normalized data.

SSAMs	Dynamic Stability	Residual Stability Ratio	Freeze–Thaw Splitting Strength Ratio	Maximum Tensile Strain	Strain Energy Density	BPN	Asphalt Loss	Cantabro Loss
SBS00	0.958	0.989	1.000	0.952	0.889	0.864	1.000	1.000
SBS50	0.864	0.927	0.921	0.937	0.685	0.898	0.888	0.899
SBS75	0.814	0.912	0.884	0.924	0.656	0.898	0.835	0.757
SBS100	0.813	0.910	0.806	0.835	0.607	0.920	0.790	0.714
HEA50	1.000	1.000	0.965	1.000	1.000	0.943	0.935	1.119
HEA75	0.974	0.992	0.940	0.963	0.952	0.966	0.843	0.993
HEA100	0.904	0.978	0.914	0.938	0.805	1.000	0.834	0.966

The distances between the comprehensive performance of the seven asphalt mixtures and the target vector are shown in Table 10. If the snow melting potential is not considered, the comprehensive performance of the seven asphalt mixtures is HEA50, HEA75, SBS00, HEA100, SBS50, SBS75, and SBS100 in descending order. It can be seen that the comprehensive performance of the HEA group is better than that of the SBS group, and with the increase of snow-melting salt content, the comprehensive performance of SSAM decreases. Based on the current results, it is suggested that HEA is used as the cement for SSAM, and the content of snow-melting salt should be 50%. In practical applications, the amount of snow-melting salt must be determined by the snow-melting performance.

Table 10. Comprehensive performance of SSAM.

SSAMs	Distances	Overall Performance Ranking
SBS00	0.187	3
SBS50	0.408	5
SBS75	0.526	6
SBS100	0.628	7
HEA50	0.151	1
HEA75	0.184	2
HEA100	0.296	4

3.4. On-Site Construction Performance

3.4.1. Performance Verification of Salt-Storage Wear Layer before Construction

- High-temperature performance verification

Referring to the specification [35], the climate zone where Xi'an is located is 1–3. A rutting tester was used to form three standard rutting plate specimens with a thickness of 5 cm. The rutting plate specimens were kept in an environment of 60 °C for 7 h, the rutting test was then carried out, and the tire ground pressure was 0.7 MPa. The dynamic stability data of the rutting plate are shown in Table 11. The average dynamic stability is 6368.6 times/mm, which is much higher than the maximum deformation resistance requirement of 2800 times/mm for the modified asphalt mixture in the specification.

Table 11. Dynamic stability of asphalt mixture.

Specimen Order	1	2	3	Mean	Standard
Dynamic stability/(times/mm)	6847.8	6326.7	5931.3	6368.6	1800–2800

- Moisture resistance verification

The moisture resistance test was carried out on Marshall specimens prepared from SSAMs. A group of specimens was kept at a constant temperature in a water tank at 60 °C for 48 h, and then the stability was measured. One group was kept at a constant temperature in a water tank at 60 °C for 30 min to test its stability. The stability test results are shown in Table 12. The residual stability is 87.84%, which meets the specification requirements of 80%.

Table 12. Marshall specimen residue stability.

Specimen Order	1	2	3	Mean	Standard
Stability before immersion/kN	15.86	15.92	16.69	/	/
Stability after immersion/kN	14.20	13.65	14.73	/	/
Residual stability ratio/%	89.56	85.71	88.25	87.84	≥80%

- Low-temperature performance verification

According to the specification requirements [35], the flexural and tensile strain of the modified asphalt mixture should not be less than 2500 με at low-temperature failure. In Table 13, the average failure strain of the salt-storage asphalt wear layer is 2668 με, which is slightly higher than the specification requirements, so the SSAM used in the project can meet the requirements of crack resistance in winter.

Table 13. Tensile strain at bending failure.

Specimen Order	1	2	3	Mean	Standard
Height/mm	35.5	35.4	35.2	/	/
Deflection/mm	0.4874	0.4991	0.5225	/	/
Span/mm	200	200	200	/	/
Tensile strain/με	2595	2650	2759	2668	≥2500

- Water permeability verification

In engineering applications, in addition to testing the conventional mechanical properties, it is also necessary to verify the water permeability of the mixture. First, the rutting plate specimen was formed with SSAM, and the annular sealing area was smeared on the surface of the rutting plate with grease. The timer was started when the water in the graduated cylinder decreased to 100 mL. The scale of the water in the measuring bucket was read every 60 s. Due to the slow falling speed of the water, the amount of water seepage at 3 min was measured. The permeability coefficient is shown in Table 14. The average value of the permeability coefficient is 44.7 mL/min, which is far less than the 120 mL/min required by the specification, so the water permeability of the SSAM meets the specification requirements.

Table 14. Water permeability of SSAM.

Specimen Order	1	2	3	Mean	Standard
Water level at 1 min/mL	140	165	155	/	/
Water level at 2 min/mL	172	230	189	/	/
Water level at 3 min/mL	205	280	217	/	/
Permeability coefficient/(mL/min)	35	60	39	44.7	≤120

3.4.2. Production and Construction of Salt-Storage Wear Layer

- Dosing of snow-melting salt

The asphalt mixing machine is HLB4000. The mixing method of snow-melting salt is the external mixing method, and the snow-melting salt required for one mixing was added to the mixture by temporary hoisting (Figure 8). The amount of each material required for one mixing is shown in Table 15.

Figure 8. Hoisting the snow-melting salt into the mixing plant.

Table 15. The amount of materials required for one mixing.

1#silo	2#silo	3#silo	4#silo	5#silo	Snow-Melting Salt	Asphalt	Asphalt Aggregate Ratio	Total Aggregate
620 kg	355 kg	682 kg	630 kg	112 kg	100 kg	105 kg	4.2%	2499 kg

- Mixing and transport

The manufacturing process of SSAM is generally carried out following the production process of ordinary asphalt mixture: raw material preparation, heating, stone consumption measurement, dry mixing, asphalt spraying and mixing for 10 s, adding snow-melting salt and mixing for more than 40 s, and discharging. The process temperature control of the SSAM at different stages in the mixing process is as follows [35]: the heating temperature of the mineral aggregate is 180–190 °C, the heating temperature of the modified asphalt is 165–170 °C, the mixing temperature is 165–170 °C, and the output temperature is 165–170 °C. The material temperature is 170 °C, the waste temperature is 195 °C, and the temperature on arrival at the construction site is 150 °C.

The journey from the mixing station to the paving site is about 30–40 min, and the transporter is covered with canvas for insulation. The laying of the AC-20 cushion was completed the day before the laying of the AC-16, and the top surface of the AC-20 was sprayed with tack coat oil.

- Paving and rolling

The receiving hopper of the paver is pre-painted with a release agent. The screed of the paver is preheated in advance and kept above 100 °C. The forward speed of the paver operation was maintained at 2 m/min.

On-site rolling is carried out in three stages: initial rolling, re-rolling, and final rolling using a double-drum roller and a tire roller. For the initial pressure, a double-drum roller was used for one static rolling and one vibration rolling at low speed. Double-drum and rubber rollers were used for cross-rolling in the re-rolling, and the rolling times were no less than four times. The final layer was rolled by a double-drum roller. To reduce the dissolution of snow-melting salts, a mop was used to apply light oil, such as soybean oil or rapeseed oil, on the steel wheel and rubber wheel. No water spray equipment was used throughout the rolling process.

- Active ingredient identification and pavement inspection

The on-site confirmation method of the active ingredient of the slow-release snow-melting salt is the silver nitrate solution titration method. Pavements laid on site were tested for the presence of snow-melting salts at intervals using a formulated silver nitrate solution. The milky white color of the solution in Figure 9 proves that the snow-melting salt has withstood the high-temperature mixing and paving process. The existence of active ingredients proves that the experiment road can melt snow and ice in winter.

Figure 9. Active ingredient test of snow-melting salt.

After the paving of the experiment road, the pavement was sampled by drilling cores. In addition, indexes, such as the permeability coefficient and flatness, were also tested. The test results are shown in Table 16. According to the test results, all the test indicators of the snow melting pavement meet the requirements of the specification; therefore, it is feasible to use the conventional hot mix asphalt pavement method for the salt-storage asphalt pavement.

Table 16. On-site inspection results of salt-storage pavement.

Test Order	1	2	3	4	5	6	7	Design
Compactness/%	98.2	98.1	98.5	98.3	98.2	98.3	98.1	≥98
Permeability coefficient/(mL/min)	109	110	108	112	113	110	110	≤300
Flatness/mm	1.51	1.52	1.53	1.5	1.51	1.52	1.52	1.5

4. Conclusions and Recommendations

(1) Snow-melting salt reduces the mechanical properties of SSAM. With the increase in the replacement amount of snow-melting salt, the mechanical properties of the asphalt mixture significantly decreased. Among them, the high-temperature performance of the SBS group SSAM decreased by 9.8–15.1%, the moisture resistance decreased by 6.3–19.4%, and the low-temperature performance decreased by 1.6–12.3%. All mechanical indexes of SBS50 meet the requirements, while the freeze–thaw splitting strength ratios of SBS75 and SBS100 do not meet the specification requirements.

(2) The mechanical properties of the HEA group SSAM with TPS are higher than those of the SBS group. Among them, the high-temperature, moisture resistance, and low-temperature performance of the SSAM in the HEA group are 11.3–19.7%, 4.8–13.3%, and 4.2–12.3% higher than those in the SBS group. This may be because the TPS not only improves the elasticity of SBS-modified asphalt but also improves the bonding performance between asphalt and aggregate. Therefore, the use of snow-melting salts with TPS is recommended.

(3) HEA can make up for the negative impact of snow-melting salt on the mechanical properties of SSAM to a certain extent, which makes the high-temperature performance, residual stability index of moisture resistance performance, and maximum bending tensile strain index of HEA50 and HEA75 not lower or even higher than SBS00 without snow-melting salt.

(4) In terms of surface properties, snow-melting salts can improve the skid resistance of SSAM, but they have a negative effect on surface spalling resistance. Whether it is a positive effect or a negative effect, the more snow-melting salt is added, the more obvious this effect is. Due to the high viscosity properties of HEA, HEA can both improve the anti-skid performance and surface anti-spalling performance.

(5) Considering both mechanical properties and surface properties, the comprehensive properties of the seven asphalt mixtures are HEA50, HEA75, SBS00, HEA100, SBS50, SBS75, and SBS100 from high to low. Therefore, HEA is suggested to be used as the cementing material of SSAM, and the content of snow-melting salt should be lower.

(6) In terms of engineering application, the mixture verified by indoor mechanics and surface properties can be used for salt-storage asphalt pavements. Moreover, the conventional paving method for hot asphalt mixtures can meet the acceptance requirements of salt-storage asphalt pavements.

In this paper, the mechanical properties, surface properties, and construction properties of SSAMs are studied from indoor and field perspectives. The durability such as fatigue performance of SSAM needs further exploration. It is worth noting that the comprehensive performance decision only considers the mechanical properties and surface properties, but in practical applications, the snow-melting salt content needs to be determined according to the snow-melting effect of the SSAM. In terms of on-site construction, the feeding method of snow-melting salt needs to be optimized.

Author Contributions: Conceptualization, Y.C., F.Z.; methodology, Y.C.; validation, B.S.; formal analysis, Y.C.; investigation, Y.C., J.L. and X.L.; resources, B.S.; data curation, B.S.; writing—original draft preparation, Y.C.; writing—review and editing, Y.C. and F.Z.; visualization, F.Z., J.L. and X.L.; supervision, Z.L.; project administration, Z.L.; funding acquisition, Z.L. All authors have read and agreed to the published version of the manuscript.

Funding: This research was funded by the Fundamental Research Funds for the Central University, CHD (300102212701) and the Scientific Innovation Practice Project of Postgraduates of Chang'an University (300103722016).

Institutional Review Board Statement: Not applicable.

Informed Consent Statement: Not applicable.

Data Availability Statement: All data used during the study appear in the published article.

Conflicts of Interest: The authors declare no conflict of interest.

References

1. Sha, A.; Liu, Z.; Jiang, W.; Qi, L.; Hu, L.; Jiao, W.; Barbieri, D. Advances and development trends in eco-friendly pavements. *J. Road Eng.* **2021**, *1*, 1–42.
2. Chen, L.; Wang, Y.; Wang, Z.; Chang, H.; Fan, F. Diffusion resisting performance of concrete modified with sodium methyl silicate in saline soil area. *Constr. Build. Mater.* **2022**, *350*, 128767.
3. Tan, Y.; Zhang, C.; Xu, H.; Tian, D. Snow melting and deicing characteristics and pavement performance of active deicing and snow melting pavement. *China J. Highw. Transp.* **2019**, *32*, 1–17.
4. Liu, Z.; Sha, A.; Jiang, W. Advances in asphalt pavements containing salts: Additives, mixtures, performances, and evaluation. *China J. Highw. Transp.* **2019**, *32*, 18–31+72.
5. Zhong, K.; Sun, M.; Chang, R. Performance evaluation of high-elastic/salt storage asphalt mixture modified with Mafilon and rubber particles. *Constr. Build. Mater.* **2018**, *193*, 153–161. [CrossRef]
6. Li, J.; Cao, Y.; Sha, A.; Song, R.; Li, C.; Wang, Z. Prospective application of coal gangue as filler in fracture-healing behavior of asphalt mixture. *J. Clean. Prod.* **2022**, *373*, 133738.
7. Cao, Y.; Li, J.; Sha, A.; Liu, Z.; Zhang, F.; Li, X. A power-intensive piezoelectric energy harvester with efficient load utilization for road energy collection: Design, testing, and application. *J. Clean. Prod.* **2022**, *369*, 133287.
8. Liu, K.; Huang, S.; Jin, C.; Xie, H.; Wang, F. Prediction models of the thermal field on ice-snow melting pavement with electric heating pipes. *Appl. Therm. Eng.* **2017**, *120*, 269–276. [CrossRef]
9. Zhang, Y.; Liu, Z.; Shi, X. Development and use of salt storage additives in asphalt pavement for anti-icing: Literature review. *J. Transp. Eng. Part B Pavements* **2021**, *147*, 03121002. [CrossRef]
10. Wu, S.; Zheng, M.; Wang, C.; Bi, S.; Wang, C.; Li, Y. Salt-dissolved regularity of the self-ice-melting pavement under rainfall. *Constr. Build. Mater.* **2019**, *204*, 371–383. [CrossRef]
11. Yao, T.; Han, S.; Men, C.; Zhang, J.; Luo, J.; Li, Y. Performance evaluation of asphalt pavement groove-filled with polyurethane-rubber particle elastomer. *Constr. Build. Mater.* **2021**, *292*, 123434. [CrossRef]
12. Zhu, X.; Zhang, Q.; Du, Z.; Wu, H.; Sun, Y. Snow melting pavement design strategy with electric cable heating system balancing snow melting, energy conservation, and mechanical performance. *Resour. Conserv. Recycl.* **2022**, *177*, 105970. [CrossRef]
13. Ho, I.; Li, S.; Abudureyimu, S. Alternative hydronic pavement heating system using deep direct use of geothermal hot water. *Cold Reg. Sci. Technol.* **2019**, *160*, 194–208. [CrossRef]
14. Zhao, Y.; Chen, C.; Xiang, Y.; Wang, J. Preparation, characterization, and anti-icing properties of sustained-release low-freezing-point asphalt mixture. *J. Mater. Civ. Eng.* **2022**, *34*, 04022171. [CrossRef]
15. Deng, S. Effect of salt storage material on performance of melting snow asphalt pavement material. *New Build. Mater.* **2020**, *47*, 64–67.
16. Liu, Z.; Chen, S.; He, R.; Xing, M.; Bai, Y.; Dou, H. Investigation on the properties of asphalt mixtures containing antifreeze fillers. *J. Mater. Civ. Eng.* **2015**, *27*, 04014180. [CrossRef]
17. Tan, Y.; Sun, R.; Guo, M.; Zhong, Y.; Zhou, S. Research on deicing performance of asphalt mixture containing salt. *China J. Highw. Transp.* **2013**, *26*, 23–29.
18. Liu, Z. Anti-Freezig Materials Used in Asphalt Pavement and Its Application Properties. Ph.D. Thesis, Chang'an University, Xi'an, China, 2013.
19. Tan, Y.; Hou, M.; Shan, L.; Sun, R. Development of sustained release complex salt filler for asphalt pavement included salt. *J. Build. Mater.* **2014**, *17*, 256–260.
20. Feng, D.; Yi, J.; Wang, D.; Chen, L. Impact of salt and freeze-thaw cycles on performance of asphalt mixtures in coastal frozen region of China. *Cold Reg. Sci. Technol.* **2010**, *62*, 34–41. [CrossRef]
21. Lysbakken, K.; Lalagüe, A. Accuracy of sobo 20 in the measurement of salt on winter pavements. *Transp. Res. Rec.* **2013**, *2329*, 24–30.
22. Borst, M.; Brown, R. Chloride released from three permeable pavement surfaces after winter salt application. *JAWRA J. Am. Water Resour. Assoc.* **2014**, *50*, 29–41. [CrossRef]
23. Hossain, S.; Fu, L.; Hosseini, F.; Muresan, M.; Donnelly, T.; Kabir, S. Optimum winter road maintenance: Effect of pavement types on snow melting performance of road salts. *Can. J. Civ. Eng.* **2016**, *43*, 802–811. [CrossRef]
24. Roseen, R.; Ballestero, T.; Houle, K.; Heath, D.; Houle, J.J. Assessment of winter maintenance of porous asphalt and its function for chloride source control. *J. Transp. Eng.* **2014**, *140*, 04013007. [CrossRef]

25. Derya, A.; Riza, K.; Ramazan, O.; Kizilel, S. Gelation-stabilized functional composite-modified bitumen for anti-icing purposes. *Ind. Eng. Chem. Res.* **2015**, *54*, 12587–12596.
26. Zheng, M.; Zhou, J.; Wu, S.; Yuan, H.; Meng, J. Evaluation of long-term performance of anti-icing asphalt pavement. *Constr. Build. Mater.* **2015**, *84*, 277–283. [CrossRef]
27. İkiz, N.; Galip, E. Computerized decision tree for anti-icing/pretreatment applications as a result of laboratory and field testings. *Cold Reg. Sci. Technol.* **2016**, *126*, 90–108. [CrossRef]
28. Zhang, Z.; Luo, Y.; Zhao, F. Review of research on the effect of salt storage deicing material on the preformance of asphalt mixture. *Chem. Ind. Eng. Prog.* **2018**, *37*, 2282–2294.
29. Liu, Z.; Xing, M.; Chen, S.; He, R.; Cong, P. Influence of the chloride-based anti-freeze filler on the properties of asphalt mixtures. *Constr. Build. Mater.* **2014**, *51*, 133–140. [CrossRef]
30. Shi, K.; Fu, Z.; Song, R.; Liu, F.; Ma, F.; Dai, J. Waste chicken fat oil as a biomass regenerator to restore the performance of aged asphalt: Rheological properties and regeneration mechanism. *Road Mater. Pavement.* **2021**, 1–25. [CrossRef]
31. *JTG E20-2011*; Standard Test Methods of Bitumen and Bituminous Mixtures for Highway Engineering. China Communications Press Co., Ltd.: Beijing, China, 2011.
32. *JTG E42-2005*; Test Methods of Aggregate for Highway Engineering. China Communications Press Co., Ltd.: Beijing, China, 2005.
33. Xu, O.; Xiang, S.; Yang, X.; Liu, Y. Estimation of the surface free energy and moisture susceptibility of asphalt mastic and aggregate system containing salt storage additive. *Constr. Build. Mater.* **2022**, *318*, 125814. [CrossRef]
34. Yuan, D.; Jiang, W.; Xiao, J.; Lu, H.; Wu, W. Thermal-oxygen aging effects on viscoelastic properties of high viscosity modified asphalt. *J. Chang. Univ. Nat. Sci. Ed.* **2020**, *40*, 1–11.
35. *JTG F40-2004*; Technical Specifications for Highway Asphalt Pavements. China Communications Press Co., Ltd.: Beijing, China, 2004.
36. Yuan, D.; Jiang, W.; Hou, Y.; Xiao, J.; Ling, X.; Xing, C. Fractional derivative viscoelastic response of high-viscosity modified asphalt. *Constr. Build. Mater.* **2022**, *350*, 128915. [CrossRef]
37. Zhang, X.; Chen, H.; Hoff, I. The mutual effect and reaction mechanism of bitumen and de-icing salt solution. *Constr. Build. Mater.* **2021**, *302*, 124213. [CrossRef]
38. Dou, H. Design and Pavement Performance of Ultra-Thin Snow Melt Salt Asphalt Mixture Pavement Overlay. Ph.D. Thesis, Chang'an University, Xi'an, China, 2015.
39. Liu, Z.; Sha, A.; Xing, M.; Li, Z. Low temperature property and salt releasing characteristics of antifreeze asphalt concrete under static and dynamic conditions. *Cold Reg. Sci. Technol.* **2015**, *114*, 9–14. [CrossRef]
40. Chen, Q.; Wang, C.; Yu, S.; Song, Z.; Fu, H.; An, T. Low-temperature mechanical properties of polyurethane-modified waterborne epoxy resin for pavement coating. *Int. J. Pavement Eng.* **2022**, 1–13. [CrossRef]
41. Chen, Q.; Wang, C.; Wen, P.; Wang, M.; Zhao, J. Comprehensive performance evaluation of low-carbon modified asphalt based on efficacy coefficient method. *J. Clean. Prod.* **2018**, *203*, 633–644. [CrossRef]

materials

Article

Influence of Fineness Levels and Dosages of Light-Burned Dolomite on Portland Cement Performance

Wenxiu Jiao [1,2], Aimin Sha [1,2,*], Zhuangzhuang Liu [1,2] and Shuo Li [2]

1 Key Laboratory for Special Area Highway Engineering of Ministry of Education, Chang'an University, Xi'an 710064, China
2 School of Highway, Chang'an University, Xi'an 710064, China
* Correspondence: ams@chd.edu.cn

Abstract: The paper aims to understand the effect of light-burned dolomite powders (LBD) on ordinary Portland cement (OPC) and evaluate the influence of LBD dosages and fineness levels on the mechanical properties and hydration properties of OPC. The LBD/OPC pastes were prepared by OPC blended with LBD at various replacement dosages and fineness levels. The mechanical properties were studied by flexural and compressive strength tests, while the hydration properties were investigated by X-ray diffraction (XRD), thermogravimetric analysis (TG), differential scanning calorimetry (DSC), and reaction degree of LBD. Experiment results indicated that the flexural and compressive strength of LBD/OPC samples were higher than reference sample at all ages. The fineness levels of LBD was C (C-LBD) with 0.5–1.5 wt% dosages, and the fineness levels of LBD was B (B-LBD) with 1.5–2.5 wt% dosages can significantly improve the strength of cement-based materials. The main mineral components of LBD are MgO and $CaCO_3$, of which MgO could react with water to form $Mg(OH)_2$ quickly, and $CaCO_3$ could hydrate with C_3A to from hydrated calcium carboaluminate ($C_3A \cdot CaCO_3 \cdot 11H_2O$), which prevents the conversion of AFt to AFm.

Keywords: light-burned dolomite powders; ordinary Portland cement; mechanical properties; hydration properties

1. Introduction

Limestone filler can be utilized as a replacement of cement clinker to improve the particle composition and adjust and optimize the structure of the pores [1,2]. Besides, it has good economic and environmental benefits, such as reducing carbon dioxide emissions during cement production process, which has been accepted and applied in most countries of the world [3–5]. As the main associated mineral of limestone, dolomite is rich in dosage and widely distributed in the world [6,7]. In China, the Shanxi, Hubei, Guizhou, and Liaoning provinces and the coastal areas of Shandong are the main sources of dolomite. Use of dolomite as one of cement component has been the focus of many research efforts in recent years [8–11].

The main chemical components of dolomite are calcium carbonate and magnesium carbonate ($CaCO_3 \cdot MgCO_3$). In addition, dolomite is also the main mineral component of dolostone and dolomitic limestone [12,13]. Dolomite can be used in cement in two ways. First, as aggregates for the production of concrete, and second as a mineral admixtures of cement which plays the role of active gel additions [14]. The application of dolomite as a mineral admixture in cement can improve some properties of cement-based materials. For example, it can compensate for the volume shrinkage of concrete to a certain extent, and improve the flowability and working performance. There is no negative impact on the volume stability of cement-based materials [14,15]. However, when dolomite is used as mineral admixture, the strength of OPC will be reduced [8].

Calcium carbonate ($CaCO_3$) and magnesium carbonate ($MgCO_3$) are the main components of dolomite; the decomposition temperature of them is different and the difference is more than 200 °C [16,17]. By controlling the calcination temperature and calcination time of dolomite, the $MgCO_3$ in dolomite can be fully decomposed without $CaCO_3$ participating in the decomposition. Thereby, an LBD containing active magnesium oxide (MgO) and inert $CaCO_3$ is obtained [17–19]. The decomposition process of dolomite was usually divided into two steps. Firstly, dolomite compound salt is decomposed to inert $CaCO_3$ and $MgCO_3$; secondly, $MgCO_3$ is further decomposed into MgO.

Massive reaction mechanisms and performance analyses of dolomite in cement have been studied with the progressive development and maturation of cement admixtures. Xu et al. studied the reaction mechanism of dolomite powder in the Portland cement by XRD, TGA, MIP and SEM. Results have shown that with the increase of curing temperature, the chemical reactivity of dolomite was significantly higher than that of limestone. Therefore, the pore structure of Portland—dolomite cement was refined and its long-term compressive strength was improved [12]. Xu and Chen et al. investigated the hydration and mechanical properties of dolomite powder on calcium sulfoaluminate (CSA) cements. They found that the hydration of CSA cement was accelerated, and the strength loss of the CSA mortar was alleviated with the addition of dolomite powder [20]. Nguyen investigated the effect of dolomite powder as part of fine particles on the engineering properties of slag—cement based self-compacting mortar. The workability of self-compacting mortar was enhanced and the microstructure was improved as the addition of dolomite substituting slag powder [21]. The stability of hydrate phase combination in Portland cement containing dolomite after leaching, carbonization, and chloride exposure was studied by Machner. The hydrotalcite was formed in the dolomite cement paste, which can withstand high degrees of leaching and carbonation [22].

Although notable progress has been achieved in the research and application of dolomite in cement, few works in the literature have reported the study of LBD as mineral admixtures for cement concrete in previous research. The reasons are as follows: (i) There is insufficient research on whether LBD has similar physical and chemical effects to dolomite and limestone in cement-based materials; (ii) MgO as one of the main mineral components of LBD was able to participate in hydration reaction quickly, which is significantly different from the almost inert limestone powder; and (iii) the effects of LBD on the composition, microstructure and long term stability of the pastes are also unclear [23]. However, it should be noted that, as a relatively economical limestone associated mineral, LBD can react with water fast to form calcium hydroxide ($Ca(OH)_2$), which can be used as an activator for cement-based materials. At the same time, a certain amount of LBD is beneficial to the development of pastes strength.

In order to clarify the mechanical properties of LBD in cement-based materials, and quantitatively analyze its active effect, this paper studies the influence of fineness levels, dosages and curing times of LBD/OPC pastes, respectively. The mechanical properties were studied by compressive and flexural strength tests, meanwhile the hydration properties were researched by XRD, TG, DSC and reaction degree of LBD. It is expected to provide a basis for the safety and efficient application of LBD on cement-based materials.

2. Materials and Methods

2.1. Raw Materials

OPC 42.5 was produced by Jidong Cement Company, Xi'an, China, and the LBD was supplied by Pengyuan New Material Company, Xi'an, Shaanxi province, China. The chemical composition of OPC and LBD were summarized in Table 1 and the main mineral components of LBD and OPC can be clearly observed in Figure 1. It should be noted that the technical properties of OPC 42.5 used in this study (provided by the producer) are as follows: specific surface area (Blaine) = 360 m^2/kg, density = 3.02 g/cm^3, initial setting time = 2.8 h, and final setting time = 4.7 h. Three types of LBD were considered according to its fineness levels: blaine specific surface area of LBD was 411 m^2/kg, 623 m^2/kg and 812 m^2/kg and noted as A-LBD, B-LBD, and C-LBD, respectively.

Table 1. Chemical composition of raw materials.

Raw Materials	CaO	SiO_2	Al_2O_3	Fe_2O_3	SO_3	MgO	K_2O	LOI
OPC 42.5	65.8	18.00	5.2	4.8	3.00	1.3	1.1	0.8
LBD	32.4	0.74	0.17	0.12	0.09	21.3	0.01	45.17

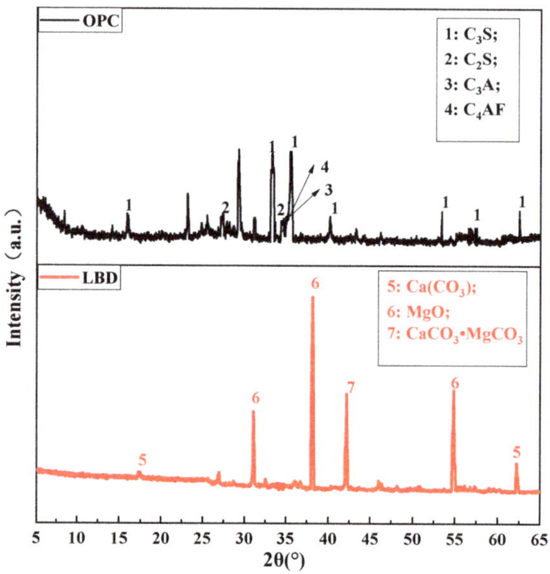

Figure 1. XRD patterns for OPC and LBD powders.

2.2. Experimental Design

The amount of LBD with different dosages mixed with OPC, and the dosages of LBD were 0, 0.5 wt%, 1.5 wt%, 2.5 wt%, and 4 wt%, respectively (LBD is equivalent to replace the OPC). The LBD/OPC pastes were prepared by manually mixing according to ASTM C305-14, typically 1000 g of solids with 450 g of clear water, and the water-to-solid weight ratio (w/s) of samples used for strength measurement was set as 0.45. The fresh LBD/OPC pastes were poured into molds (40 × 40 × 160 mm) and then they were placed in a curing chamber with a temperature of 20 ± 2 °C and relative humidity about 95% for 3, 7 and 28 d, respectively. The LBD/OPC samples were summarized in Table 2. When the specified curing age was reached, the samples were taken out from the curing chamber, and the molds were removed. The mechanical properties were evaluated by flexural strength test and compressive strength test. Three samples were prepared from each batch for flexural strength test, and six samples were prepared from each batch for compressive strength test. After curing 3, 7, and 28 d from the preparation, roughly 10 g of pastes were withdraw and the hydration reaction was stopped by solvent exchange (add to isopropanol). After two weeks, the isopropanol was drained and then placed in an oven at 40 °C for 7 d, until all of the water was evaporated. The samples were ground using an agate mortar to through 0.075 mm sieve completely, finally the prepared samples were put into a sealed bottle. The preparation process and samples were shown in Figure 2.

2.3. Test Methods

2.3.1. Preparing LBD/OPC Pastes

The preparation process of LBD/OPC pastes were as follows: first, the cement mixer should be wiped clean, and then the OPC was replaced by LBD with different dosages and different fineness levels, respectively. The LBD and OPC powder were poured into the mixer, meanwhile 450 g of clear water were slowly added to the mixer. Turn on the machine, the cement mixer was operated at a low speed of 140 rpm for 120 s, stop mixing for 15 s, and then continued to mix at a fast speed of 285 rpm for another 120 s. After mixing was completed, the mixer was stopped and both fresh pastes were poured and compacted into the triple molds (40 × 40 × 160 mm).

Table 2. Test samples of LBD/OPC pastes.

Marked Number	Dosages of LBD (%)	Time (Days)	W/S Ratio	Temperature and Humidity
A-LBD/OPC B-LBD/OPC C-LBD/OPC	0, 0.5, 1.5, 2.5, 4	3, 7, 28	0.45	20 °C, 95%

Where: A-LBD/OPC noted as fineness levels of LBD was A equivalent to replace the OPC; B-LBD/OPC noted as fineness levels of LBD was B equivalent to replace the OPC; C-LBD/OPC noted as fineness levels of LBD was C equivalent to replace the OPC.

Figure 2. The operation process of samples.

2.3.2. Flexural Strength Test

A three-point bending test (flexural) was conducted on the specimens. The flexural strength test standard (T0506 in JTG E30-2005) were used as a reference to determine the testing procedure. The specimen was placed on two supporting points with a certain distance, and a downward load was applied to the upper specimen of the midpoint of the two supporting points. When the two contact points of the specimen form two equal moments, the specimen will break at the midpoint, the flexural strength data were automatically recorded by three-point bending test machine. The loading rate = 50 ± 0.2 N/S, three samples were tested per batch and the results were averaged to ensure the accuracy of the test.

2.3.3. Compressive Strength Test

The compressive strength test standard (T0506 in JTG E30-2005) was used as a reference to determine the testing procedure. The specimen was placed on the bottom plate of the testing machine, and the supporting surface of the specimen should be perpendicular to the top surface. The center of the specimen should be aligned with the center of the bottom plate of the testing machine, and the load should be applied uniformly and continuously until the specimen was crushed. The loading rate = 2.4 ± 0.2 KN/S, six samples were tested per batch and the results were averaged.

2.3.4. X-ray Diffraction (XRD)

XRD (Bruker, Karlsruhe, Germany, D8 Advanced, Cu-Ka) was employed to qualitatively analyze the mineral composition of hydration products. Samples were scanned between 5° to 65° (2θ) and with an integrated step scan of 0.05°/step (2θ). The XRD test was used to determine the mineral components involved in the hydration reaction and resulting composition.

2.3.5. Thermogravimetric and Differential Scanning Calorimetry (TG/DSC)

TG/DSC (TA, SDT Q600 V8.0 Build 95, Nitrogen condition) was used to determine the variation of weight and heat quantity of LBD/OPC pastes with the temperature rise. The nitrogen was used as the shielding gas and the heating temperature ranges between 20 °C and 900 °C with a heating rate of 10 °C/min.

3. Results and Analysis

3.1. Mechanical Properties

3.1.1. Flexural Strength

The flexural strength of LBD/OPC samples under different fineness levels and dosages were studied systematically. As shown in Figure 3a, the strength of A-LBD/OPC samples reduced first and then increased with increasing of dosages of LBD at all ages. When the dosage of LBD was 0.5%, the strength of the sample was smallest, even lower than that of the control sample. The strength of the samples increased significantly when the dosages of LBD exceeded 0.5%, while the strength of the sample reached the maximum value at the dosage of LBD was 4%. There was no significant difference in the flexural strength between the 3rd and the 7th day. However, there was a significant increase in the strength of 28th day, suggesting that the addition of LBD to OPC had little effect on the early flexural strength, mainly for improve the post strength of the system.

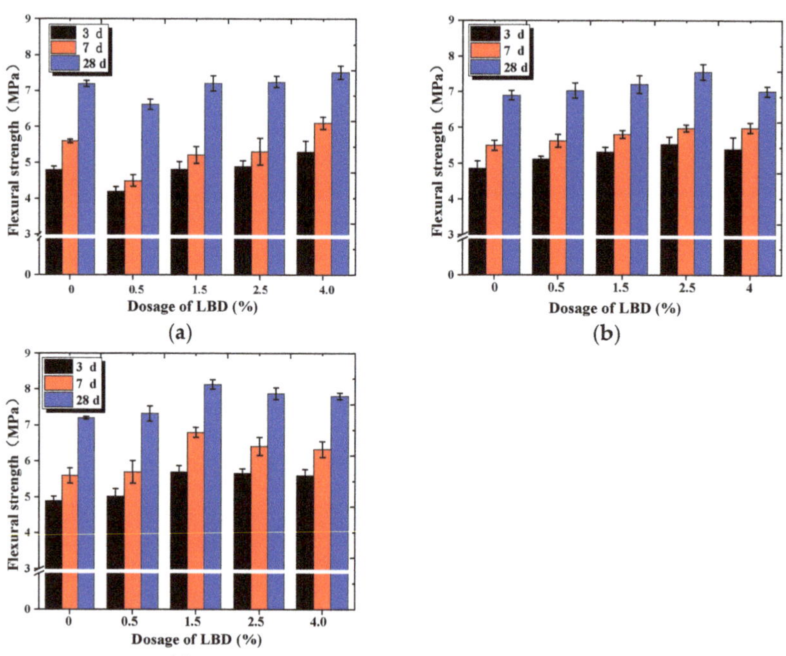

Figure 3. Flexural strength of LBD/OPC samples: (**a**) A-LBD/OPC samples; (**b**) B-LBD/OPC samples; (**c**) C-LBD/OPC samples.

Figure 3b,c present the flexural strength of the sample higher than reference sample (0% additives) at all ages, and the flexural strength increased first and then decreased slightly with the increasing LBD dosages. When the fineness levels of LBD were different, the optimal dosages of LBD were different. For the B-LBD/OPC sample, when the dosage was 2.5%, the strength reached the peak, and the strength of sample was 45.04%, 29.26%, and 24.06% higher than that of reference sample on the 3rd, 7th and 28th days, respectively. For the C-LBD/OPC sample, the optimal dosage was 1.5%, and the flexural strength was improved significantly on the 7th day, indicating that the addition of C-LBD has an impact on the early flexural strength of OPC.

Figure 4 shows the flexural strength contribution rate of the different dosages and fineness levels of LBD to cement-based materials. As shown in Figure 4, the strength contribution rate of LBD/OPC samples was greater than 0, which indicates that LBD has a certain hydration activity to OPC, and has a positive effect for the flexural strength of cement-based materials. From Figure 4a,c, the flexural strength contribution rate reached the maximum when the dosage of B-LBD was 2.5%, the strength contribution rate of the sample increased by about 14% after curing for 3 d. However, for C-LBD, the optimal dosage was 1.5%, and the strength contribution rate of sample increased by 22% after curing for 7 d. From Figure 4c, the strength contribution rate of the C-LBD/OPC samples was generally larger, and its maximum contribution rate can reach 13.5%, which was about 44% higher than the maximum contribution rate of the B-LBD/OPC samples. This indicates that a small amount of C-LBD can make a greater contribution to the flexural strength of cement-based materials due to its relatively larger specific surface area.

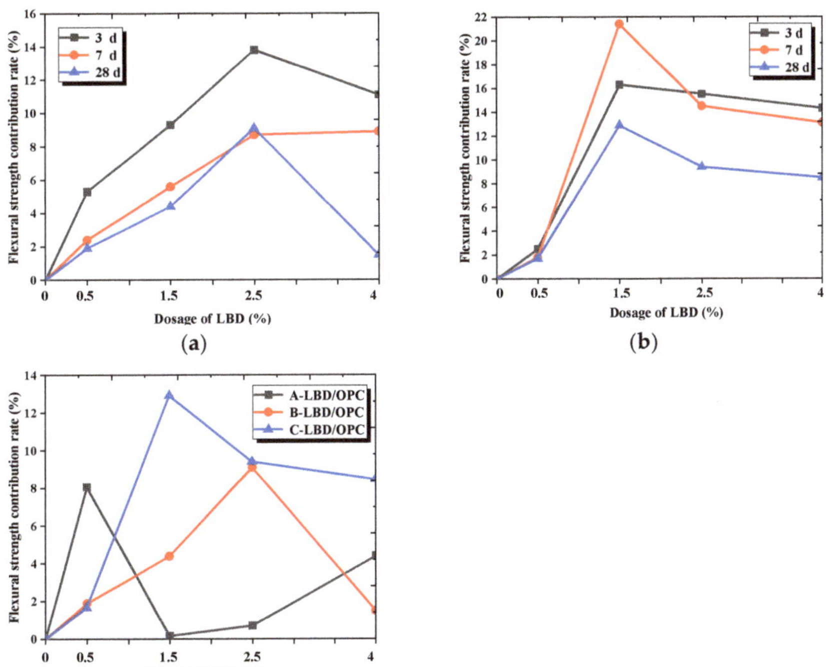

Figure 4. Flexural strength contribution rate of LBD to OPC materials: (**a**) Influence of B-LBD dosages; (**b**) influence of C-LBD dosages; (**c**) influence of LBD fineness levels (curing for 28 d).

3.1.2. Compressive Strength

Figure 5 shows the development law diagram of the compressive strength of the LBD/OPC samples under standard curing conditions. Comparing Figures 3 and 5, it can be seen that the compressive strength and flexural strength of the samples varied with the similar regulation, the fineness levels and the dosages of LBD have an effect on the strength of the samples. However, there were also some small differences between them. Figure 5a shows the compressive strength of A-LBD/OPC samples at ages of 3rd, 7th and 28th days. The experimental results indicate that with the increase of dosages of A-LBD, the strength of the samples increased first and then decreased slightly. When the dosage of LBD was 1.5%, the compressive strength of the sample was reached on peak, at this time, the compressive strength of LBD-OPC samples at the 3, 7, and 28 days was increased by 4.68%, 8.11% and 8.70%, respectively, compared with the reference samples. When the dosage of LBD was 0.5%, the strength of the sample was smallest, even lower than that of the reference sample, the strength of samples at the 3, 7, and 28 days was decreased by 2.5%, 5.08%, and 3.07%, respectively. It is speculated that the reason may be that the specific surface area of A-LBD is relatively small and its hydration activity is low.

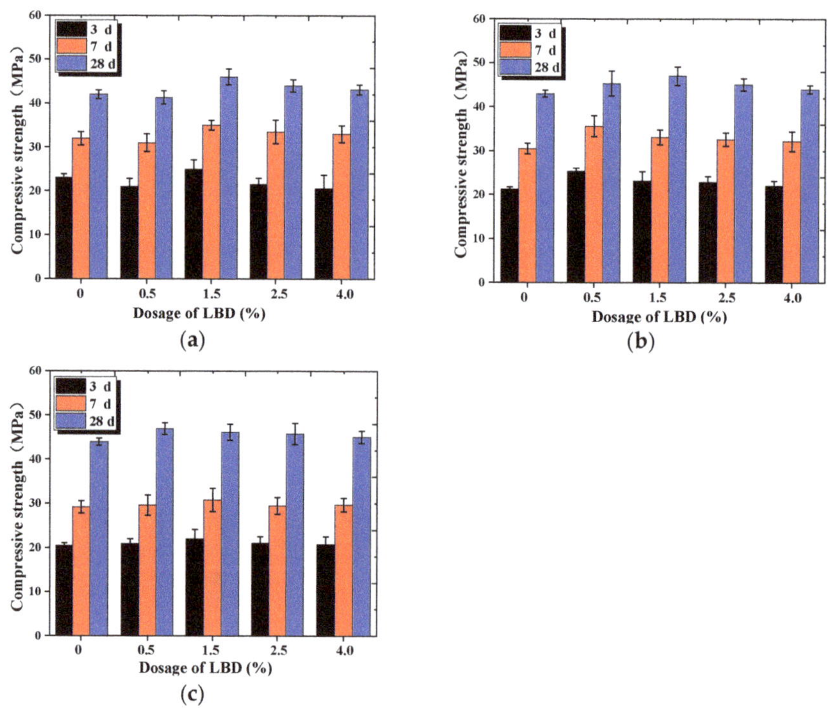

Figure 5. Compressive strength of LBD/OPC samples: (**a**) A-LBD/OPC samples; (**b**) B-LBD/OPC samples; (**c**) C-LBD/OPC samples.

Figure 5b,c present that the compressive strength of samples increased first and then decreased slightly with increasing of LBD dosages, and they were larger than the reference sample. The compressive strength of B-LBD/OPC samples reached the peak when the dosage was 1.5%, while for C-LBD/OPC sample reached the maximum value when the dosage was 0.5%. It indicates that the fineness levels of LBD have different effects on the optimal dosage of the compressive strength, and the early and late strength development of samples was also affected by the fineness levels of LBD.

Figure 6 shows the compressive strength contribution rate of the different dosages and fineness levels of LBD to cement-based materials. The strength contribution rate of cement-based materials was greater than 0 after adding LBD powder, besides A-LBD/OPC samples at dosage of 0.5%. Which indicates that LBD has a certain hydration activity to OPC, and it can improve the compressive strength of cement-based materials. From

Figure 6a,b, the compressive strength contribution rate reached the maximum when the dosage of B-LBD and C-LBD was 0.5% and 1.5%, respectively. The strength contribution rate of the samples increased by about 7.5%, and 20% for B-LBD and C-LBD after curing for 3 d. From Figure 6c, the strength contribution rate of the A-LBD/OPC sample was generally smaller, it mainly due to the relatively smaller specific surface area.

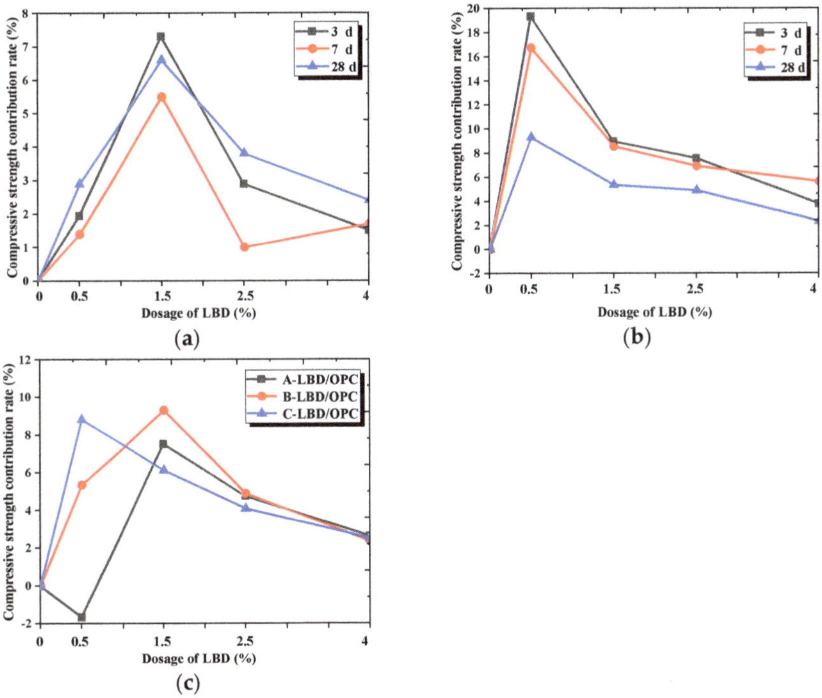

Figure 6. Compressive strength contribution rate of LBD to OPC materials: (**a**) Influence of B-LBD dosages; (**b**) Influence of C-LBD dosages; (**c**) Influence of LBD fineness levels (curing for 28 d).

Combined flexural strength and compressive strength test results, C-LBD powder with dosage 0.5–1.5 wt% is recommended for cement-based material in this study to improve its strength, besides, B-LBD powder with dosage 1.5–2.5 wt% is also a good choice for improving the strength of cement materials.

3.2. Hydration Properties

3.2.1. XRD

Figure 7a shows the XRD analysis of LBD/OPC samples under different curing times (A-LBD, with 1.5% dosage). There were diffraction peaks of hydration products appearing such as $Ca(OH)_2$, $CaCO_3$, MgO, $Mg(OH)_2$ and quartz. In addition, the diffraction peaks of secondary minerals such as AFt and $C_3A{\cdot}CaCO_3{\cdot}11H_2O$, as well as the diffraction peaks of C_3S and C_2S that are not involved in the hydration reaction can also be found. The diffraction peaks of $Ca(OH)_2$ was detected at about $2\theta = 18°$, $34°$ and $47°$, and the diffraction peaks of $Ca(OH)_2$ decreased gradually with the increase of curing time, it shows that more cement clinker participated in the hydration, which made the hydration more thorough. Meanwhile the diffraction peaks development law of C_3S and C_2S were similar to the $Ca(OH)_2$. MgO and $CaCO_3$ are main components of LBD, the diffraction peaks of MgO and $CaCO_3$ were detected at about $2\theta = 43°$ and $29°$, respectively. When curing for three days, the diffraction peaks of $Mg(OH)_2$ was obtained at $2\theta = 41°$, which indicates that once the MgO is mixed with water, the $Mg(OH)_2$ can be formed in a short time.

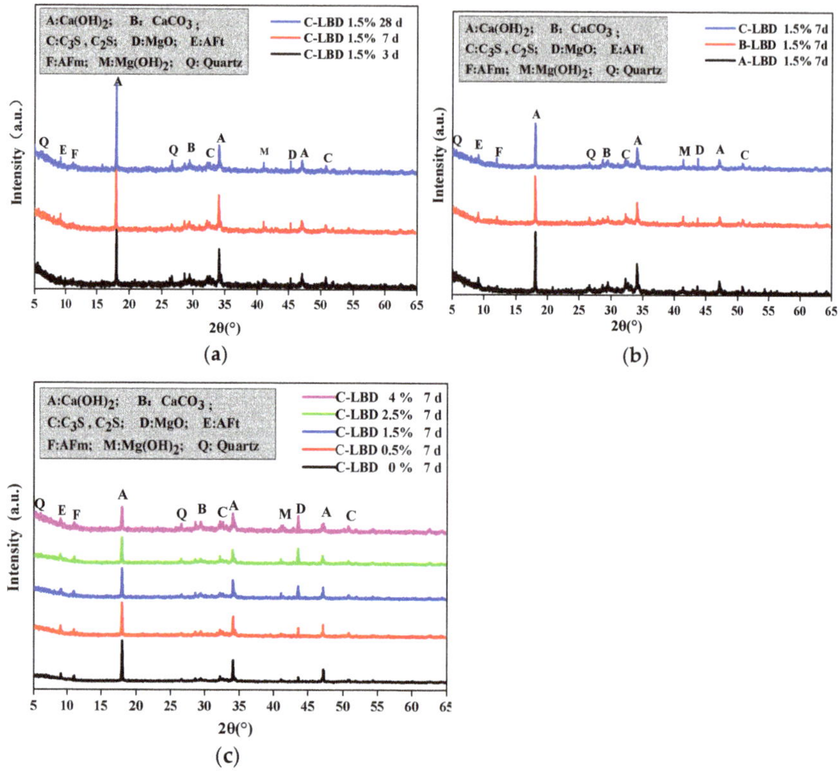

Figure 7. XRD diffractograms LBD/OPC pastes: (**a**) Influence of time; (**b**) influence of fineness; (**c**) influence of dosage.

The XRD of different fineness levels of LBD/OPC samples (7th day with 1.5% dosage) were arranged in Figure 7b. The diffraction peaks of $Ca(OH)_2$, C_3S and C_2S decreased gradually with the increasing of fineness levels. However the diffraction peaks development law of MgO was opposite compared with $Ca(OH)_2$, the main reason is that with the increase of LBD fineness levels, the specific surface area increases accordingly, and the contact area between LBD and OPC increases, and finally more hydration reactions is involved. It should be noted that the diffraction peaks of AFt decreased with increasing of LBD fineness levels, while diffraction peaks of $C_3A{\cdot}CaCO_3{\cdot}11H_2O$ show an increasing trend, this means that during the hydration reaction process, AFt was partially converted into AFm, while $CaCO_3$ could react with C_3A to form $C_3A{\cdot}CaCO_3{\cdot}11H_2O$.

Figure 7c shows the XRD results of LBD/OPC samples under different dosages (fineness level: C-LBD, curing time: 7th day). It can be seen that the dosages of MgO and $CaCO_3$ increased with the increasing of LBD dosages.

3.2.2. TG/DSC

The thermogravimetric results of LBD/OPC samples were shown in Figure 8 (7th day, C-LBD). The TG curve could quantitatively provide the mass loss of each individual phase. Three obvious mass losses could be observed from Figure 8a: one from 100 °C to about 300 °C with a larger span, the second in the 400–500 °C range; and the last from 600 °C to 700 °C. According to the related works in the literatures [7,24], the relative mass loss in the first weightlessness period was about 6.1%, which is mainly caused by the physical dehydration of hydration products. The physical dehydration mainly included the following parts: (i) the dehydration temperature of C-S-H gel was about

105 °C which produced by cement hydration; (ii) the dehydration temperature of AFt was about 125—140 °C; (iii) the dehydration temperature of AFm was about 175° C; (iv) and the dehydration temperature of $C_3A{\cdot}CaCO_3{\cdot}11H_2O$ was about 200—300 °C. The second loss was ascribed to the decomposition of $Ca(OH)_2$ with a relative mass of 4.5%. The third relative mass loss was smallest, only 2.7%, and the decomposition $CaCO_3$ or $CaMg(CO_3)_2$ to produce CO_2 was the main reason for the mass loss. This shows that the mass loss decreased first and then increased significantly with increasing LBD dosages. When the LBD dosage was 0.5%, the mass loss was the smallest, even lower than the reference sample. However, the mass loss was largest when the LBD dosage was 4%, which is due to the dosage of $CaCO_3$ increases with increasing of LBD, and more CO_2 is produced from the decomposition of $CaCO_3$.

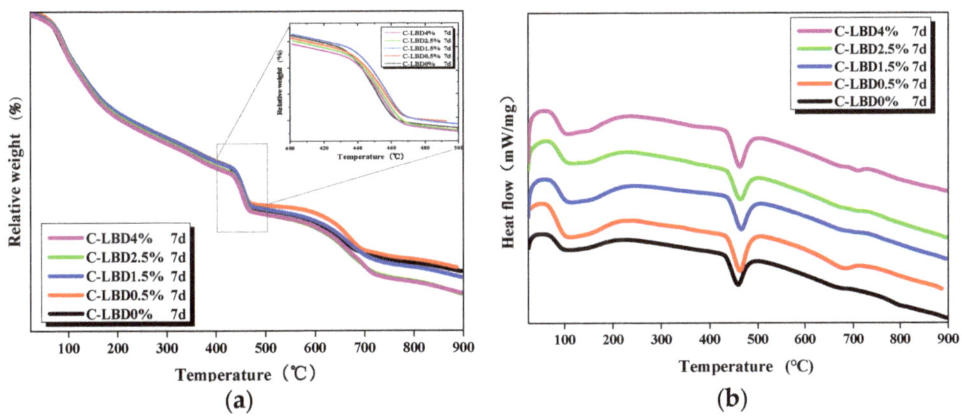

Figure 8. Thermal analysis curves of LBD/OPC pastes: (**a**) TG; (**b**) DSC.

Figure 8b shows the DSC results of samples at different dosages (7th day, C-LBD). It can be seen that the endothermic events occurred from 100 to 200 °C and 400—500 °C, while small endothermic events occurred around 650—750 °C. The first is mainly due to the physical dehydration of AFt and AFm; the second corresponds to $Ca(OH)_2$ decomposition to absorb heat; the last one is attributable to $CaCO_3$ or $CaMg(CO_3)_2$ decomposition producing CO_2.

The state of the water in the LBD/OPC samples was very complex, and it was difficult to strictly distinguish the physical combination or chemical combination of the water. For research convenience, dehydration below 105 °C is generally considered to be evaporable water and the dehydration at 105—500 °C is identified as non-evaporable water [25–27]. Table 3 shows the mass losses of C-LBD/OPC samples at different dosages (7th day, C-LBD). It was found that the mass losses percentage of the hydration products decreased first and then increased with the increase of C-LBD dosages at all of the different temperature ranges. When the LBD dosage was 1.5%, the mass losses was minimized, the loss rate of evaporable water, non-evaporable water and decarbonation of $CaCO_3$ or $CaMg(CO_3)_2$ in the samples were 3.16%, 10.99%, and 2.89%, respectively. The results also indicated that the mass loss rate of evaporable water of the samples was lower than reference sample when added to C-LBD, this means that C-LBD has a positive effect on cement hydration. The loss rate of non-evaporable water and decarbonation of $CaCO_3$ or $CaMg(CO_3)_2$ in the pastes were 11.41% and 3.63%, respectively. When the LBD dosage was 4%, the mass losses were larger than reference sample, mainly since when the fineness of LBD increases to a certain level, it has an adverse effect on the hydration of the LBD/OPC paste.

Table 3. Mass losses of the paste at different temperature ranges and corresponding processes.

Dosages of C-LBD	Evaporable Water (below 105 °C), %	Non-Evaporable Water (105–500 °C), %	Decarbonation of CaCO3 or CaMg(CO3)2, %
0%	3.525	11.17	3.24
0.5%	3.36	11.04	3.17
1.5%	3.16	10.99	2.89
2.5%	3.2	11.24	3.45
4%	3.4	11.41	3.63

3.3. The Reaction Degree of LBD

To describe the hydration process of LBD/OPC samples, the reactivity of LBD was determined by calculating the difference between the initial $CaCO_3$ dosage and the final $CaCO_3$ dosage, the initial $CaCO_3$ dosage was determined as Equation (1), and the mass loss at approximately 650—750 °C was used to measure the final $CaCO_3$ dosage as percentage by weight (wt%) using stoichiometry as presented in Equation (2) [28,29]

$$M = \frac{M_1}{M_2 + M_3 + M_4} \times 100\% \tag{1}$$

$$T = M_5 \cdot \frac{M_6 or M_7}{M_8 or 2M_8} \tag{2}$$

$$R = M - T \tag{3}$$

where M is the initial $CaCO_3$ dosage; T is the final $CaCO_3$ dosage; R is the reaction degree of LBD; M_1 is the mass of LBD (blended and interground), M_2 is the mass of OPC, M_3 is the mass of LBD, M_4 is the mass of water in the pastes, M_5 is the mass loss of pastes at 650—750 °C, M_6 is the mass of $CaCO_3$, M_7 is the mass of $CaMg(CO_3)_2$, M_8 is the mass of CO_2.

According to the results of TG curve, the reaction degree of LBD in OPC can be calculated by Equations (1)—(3) and shown in Figure 9 (7th day, C-LBD). Obviously, the dosages of C-LBD has a bigger effect on the degree of hydration reactivity, which is consistent with the conclusions of the literature [30]. When the dosage of LBD was 1.5%, the reaction degree was maximized, and the reaction degree is about 4.8%. This is consistent with the flexural strength results for C-LBD/OPC samples.

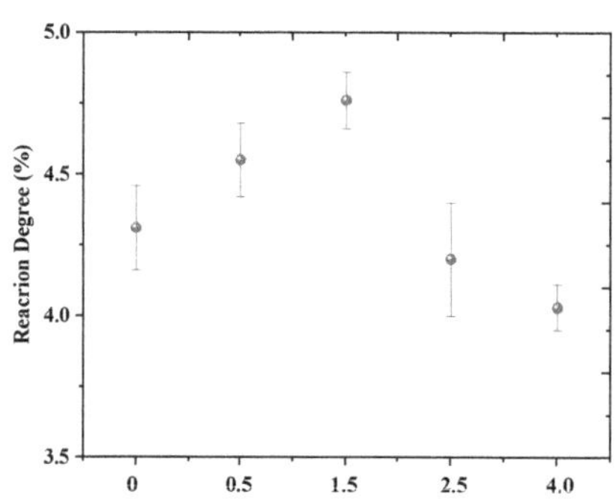

Figure 9. Reaction degree of LBD/OPC pastes.

4. Discussion

According to the data provided by flexural strength and compressive strength, the fineness levels and the dosages of LBD have an influence on the strength of the cement-based materials, this remarkable phenomenon can be observed at all curves. Typically, in a certain fineness levels range, the finer the LBD, the larger specific surface area, the greater contact area with the cement, and the more hydration reactions involved. In the early stages of cement hydration, the main hydration products were AFt, $Ca(OH)_2$ and C_2S, C_3S; with increasing of the curing time, part of the AFt translated into AFm. The main components of LBD are MgO and $CaCO_3$. Some studies have pointed out that $CaCO_3$ can react with C_3A to produce $C_3A{\cdot}CaCO_3{\cdot}11H_2O$, which prevents the conversion of AFt to AFm [24–26]. At the same time, the formation of $C_3A{\cdot}CaCO_3{\cdot}11H_2O$ could improve the pore structure of pastes and enhance the strength of the LBD/OPC samples [27,28]. However, when the fineness levels of LBD increased to a certain extent, the specific surface area was too large, and the agglomeration phenomenon was prone to occur. With the increase of LBD dosages, the aggregation phenomenon will affect the contact area between LBD and cement powder, thereby reducing the amount of LBD participating in the cement hydration reaction and reducing the strength of the paste. Therefore, the excessive LBD would have detrimental effect on the strength of the LBD/OPC pastes when it has sufficient fineness.

The presented results obtained by XRD, TG, DSC and reaction degree on LBD/OPC pastes hydration showed that the carbonated aluminate phase was formed. This new phase could improve the microstructure of the sample and make it more compact, which plays an important role in improving the strength of the LBD/OPC pastes [31]. Besides, LBD has physical effects such as nucleation, dilution and filling in OPC, the synergistic effect of the carbonated aluminate phase and this physical effect could improve the properties of OPC. The filling effect of the LBD is related to its fineness and particle size distribution, the dilution effect is related to the dosages of LBD, and the dilution effect mainly occurs in the early stage of cement hydration [32–34]. Some studies have pointed out that the carbonated aluminate phase was formed according to the following Equations (4)—(8) [35]. However, professor Zajac [36] held a different view on the formation of the carbonated aluminate; he believes that the formation of the carbonated aluminate was indirect (that is to say, the formation of the carbonated aluminate phase is divided into two steps: (i) the LBD was dissolved in the water; (ii) the $CaCO_3$ was reacted with aluminum-containing components in the pastes to form carbonated aluminate furtherly). At the same time, professor Zajac pointed out that $Mg(OH)_2$ is unstable and it can react with aluminum containing components in the pastes to form hydrotalcite. However, in this study, only the diffraction peak of $Mg(OH)_2$ were found in LBD/OPC pastes under different LBD dosages and curing time by XRD experiments, and no diffraction peak of hydrotalcite was found, it is different from professor Zajac idea. However, regardless of whether it is an indirect reaction mechanism or a direct reaction mechanism, the reaction occurs after the LBD was dissolved. When the fineness levels of LBD were appropriately increased, the dissolution rate was accelerated, and the reaction rate and degree were correspondingly improved.

LBD could release CO_3^{2-} ions and then combine with alumina to form a carboaluminates phase, which is mainly based on the dissolution rate of LBD. Some studies have shown that LBD has a lower dissolution [37–40], which indicates that the LBD reacts slowly in OPC system, while carboaluminates increased with curing time, which prevented the conversion of AFt to AFm. AFt can be combined with carboaluminates to ensure better filling of pore structure and well refined pore structure, and increasing the strength of Portland cement.

$$CaMg(CO_3)_2 \rightarrow CaCO_3 + MgCO_3 \tag{4}$$

$$Mg(CO_3)_2 \rightarrow MgO + CO_2 \tag{5}$$

$$MgO + H_2O \rightarrow Mg(OH)_2 \tag{6}$$

$$CaMg(CO_3)_2 + Ca\,(OH)_2 \rightarrow CaCO_3 + Mg(OH)_2 \tag{7}$$

$$CaCO_3 + C_3A + 11H_2O \rightarrow C_3A{\cdot}CaCO_3{\cdot}11H_2O \tag{8}$$

Limestone is the primary raw material for the production of clinker in the cement production process and has a positive effect on the strength and heat of hydration of Portland—limestone cement. However, dolomite and LBD are not to be used as raw materials for clinker production, as they are a potential carbonate source and could be utilized together with Portland cement. Compared with limestone mixed with cement, there was no negative effect after LBD mixed with cement. The performance of dolomite mixed with cement is comparable to that of limestone mixed with cement, and no negative effect is found, which is related to the clinker produced by limestone containing dolomite [14].

5. Conclusions

Based on the experimental results in this study, the mechanical properties and hydration properties of LBD/OPC sample with different LBD dosages (0, 0.5, 1.5, 2.5, 4.0 wt%) and fineness levels (A, B and C) were investigated at the 3rd, 7th, and 28th days. the following conclusions can be drawn:

(1) LBD powder has a certain hydration activity to OPC materials, and it can improve the post strength of the paste significantly. The flexural strength of LBD/OPC samples was higher than reference sample at all ages. For the B-LBD/OPC sample, the flexural strength contribution rate reached the maximum with dosage of 2.5%, and for the C-LBD/OPC sample, the optimal dosage was 1.5%.
(2) The compressive strength of LBD/OPC samples increased first and then decreased slightly with increasing of LBD dosages. The compressive strength reached its peak when the B-LBD dosage was 1.5%, and C-LBD dosage was 0.5%, and the strength contribution rate of the samples increased by about 7.5%, and 20% for B-LBD and C-LBD samples after curing for 3 days. In conclusion, C-LBD powder with dosages 0.5–1.5 wt% and B-LBD powder with dosages 1.5–2.5 wt% are recommended for cement-based material.
(3) The main components of LBD are MgO and $CaCO_3$, of which MgO could react with water rapidly to form $Mg(OH)_2$, and $CaCO_3$ could react with C_3A to produce $C_3A{\cdot}CaCO_3{\cdot}11H_2O$, which prevents the conversion of AFt to AFm.
(4) Three principle mass losses can be observed: one from 100 °C to about 300 °C with a larger span, the second approximately in the 400—500 °C range, and the last from 600 °C to 700 °C. Three endothermic events were found in the DSC curve, which were caused by the physical dehydration of AFt and AFm, $Ca(OH)_2$ decomposition to absorb heat and $CaCO_3$ or $CaMg(CO_3)_2$ decomposition to produce CO_2, respectively.

Author Contributions: Conceptualization, A.S. and Z.L.; methodology, W.J.; validation, W.J. and S.L.; investigation, S.L.; resources, S.L.; data curation, W.J.; writing—original draft preparation, W.J.; writing—review and editing, A.S.; supervision, Z.L.; project administration, A.S.; funding acquisition, A.S. All authors have read and agreed to the published version of the manuscript.

Funding: This work was financially supported by the National Key Research and Development Program, China (2018YFB1600200); Modern Urban Environmental Protection Intelligent Road Research Center, 20191222115KYPT012JC014.

Institutional Review Board Statement: Not applicable.

Informed Consent Statement: Not applicable.

Data Availability Statement: Not applicable.

Acknowledgments: The authors would like to acknowledge the Key Laboratory for Special Area Highway Engineering, MOE, China, and the Materials Analysis Center, School of Materials Science and Engineering, Chang'an University, Xi'an China. They gratefully acknowledge the support making these laboratories and their operation possible. Finally, the authors are grateful to the reviewers of this paper for their comments and suggestions.

Conflicts of Interest: The authors declare no conflict of interest.

References

1. Liu, Y.; Zhuge, Y.; Duan, W.; Huang, G.; Yao, Y. Modification of microstructure and physical properties of cement-based mortar made with limestone and alum sludge. *J. Build. Eng.* **2022**, *58*, 105000. [CrossRef]
2. Garcia, J.E.; Tiburzi, N.B.; Folliard, K.J.; Drimalas, T. Mechanical properties and electrical resistivity of portland limestone cement concrete systems containing greater than 15% limestone and supplementary cementitious materials. *Cement* **2022**, *8*, 100026. [CrossRef]
3. Nepomuceno, M.C.S.; Bernardo, L.F.A.; Pereira-de-Oliveira, L.A.; Timóteo, R.O. Cement-based grouts for masonry consolidation with high content of limestone filler, metakaolin, glass powder and ceramic waste. *Constr. Build. Mater.* **2021**, *306*, 124947. [CrossRef]
4. Güneyisi, E.; Gesoğlu, M. Properties of self-compacting portland pozzolana and limestone blended cement concretes containing different replacement levels of slag. *Mater. Struct.* **2011**, *44*, 1399–1410. [CrossRef]
5. Zajac, M.; Rossberg, A.; Le Saout, G.; Lothenbach, B. Influence of limestone and anhydrite on the hydration of Portland cements. *Cem. Concr. Compos.* **2014**, *46*, 99–108. [CrossRef]
6. Sharma, R.S.; Sharma, A. Metamorphism of Dolomites and Limestones. In *Encyclopedia of Geology*, 2nd ed.; Alderton, D., Elias, S.A., Eds.; Academic Press: Oxford, UK, 2021; pp. 479–491.
7. Zajac, M.; Bremseth, S.K.; Whitehead, M.; Ben Haha, M. Effect of $CaMg(CO_3)_2$ on hydrate assemblages and mechanical properties of hydrated cement pastes at 40 °C and 60 °C. *Cem. Concr. Res.* **2014**, *65*, 21–29. [CrossRef]
8. Mikhailova, O.; Yakovlev, G.; Maeva, I.; Senkov, S. Effect of Dolomite Limestone Powder on the Compressive Strength of Concrete. *Procedia Eng.* **2013**, *57*, 775–780. [CrossRef]
9. Nguyen, H.-A.; Chang, T.-P.; Shih, J.-Y.; Suryadi Djayaprabha, H. Enhancement of low-cement self-compacting concrete with dolomite powder. *Constr. Build. Mater.* **2018**, *161*, 539–546. [CrossRef]
10. Su, A.; Chen, H.; Feng, Y.-x.; Zhao, J.-x.; Wang, Z.; Hu, M.; Jiang, H.; Duc Nguyen, A. In situ U-Pb dating and geochemical characterization of multi-stage dolomite cementation in the Ediacaran Dengying Formation, Central Sichuan Basin, China: Constraints on diagenetic, hydrothermal and paleo-oil filling events. *Precambrian Res.* **2022**, *368*, 106481. [CrossRef]
11. Chen, X.; Chen, H.; Chen, Q.; Lawi, A.S.; Chen, J. Effect of partial substitution of cement with Dolomite powder on Glass-Fiber-Reinforced mortar. *Constr. Build. Mater.* **2022**, *344*, 128201. [CrossRef]
12. Xu, J.; Lu, D.; Zhang, S.; Xu, Z.; Hooton, R.D. Reaction mechanism of dolomite powder in Portland-dolomite cement. *Constr. Build. Mater.* **2021**, *270*, 121375. [CrossRef]
13. Baghriche, M.; Achour, S.; Baghriche, O. Combined effect of cement kiln dust and calcined dolomite raw on the properties of performance magnesium phosphate cement. *Case Stud. Constr. Mater.* **2020**, *13*, e00386. [CrossRef]
14. Krishnan, S.; Bishnoi, S. Understanding the hydration of dolomite in cementitious systems with reactive aluminosilicates such as calcined clay. *Cem. Concr. Res.* **2018**, *108*, 116–128. [CrossRef]
15. Barbhuiya, S. Effects of fly ash and dolomite powder on the properties of self-compacting concrete. *Constr. Build. Mater.* **2011**, *25*, 3301–3305. [CrossRef]
16. Ptáček, P.; Šoukal, F.; Opravil, T. Thermal decomposition of ferroan dolomite: A comparative study in nitrogen, carbon dioxide, air and oxygen. *Solid State Sci.* **2021**, *122*, 106778. [CrossRef]
17. Zhou, F.; Yang, L.; Zhang, L.; Cao, J.; Zhang, H. Investigation of decomposition of dolomite and distribution of iodine migration during the calcination-digestion process of phosphate ore. *Hydrometallurgy* **2019**, *188*, 174–181. [CrossRef]
18. Samtani, M.; Dollimore, D.; Alexander, K.S. Comparison of dolomite decomposition kinetics with related carbonates and the effect of procedural variables on its kinetic parameters. *Thermochim. Acta* **2002**, *392–393*, 135–145. [CrossRef]
19. Qian, H.; Kai, W.; Hongde, X. A novel perspective of dolomite decomposition: Elementary reactions analysis by thermogravimetric mass spectrometry. *Thermochim. Acta* **2019**, *676*, 47–51. [CrossRef]
20. Xu, J.; Chen, J.; Lu, D.; Xu, Z.; Hooton, R.D. Effect of dolomite powder on the hydration and properties of calcium sulfoaluminate cements with different gypsum contents. *Constr. Build. Mater.* **2019**, *225*, 302–310. [CrossRef]
21. Nguyen, H.-A. Enhancement of engineering properties of slag-cement based self-compacting mortar with dolomite powder. *J. Build. Eng.* **2019**, *24*, 100738. [CrossRef]
22. Machner, A.; Zajac, M.; Ben Haha, M.; Kjellsen, K.O.; Geiker, M.R.; De Weerdt, K. Stability of the hydrate phase assemblage in Portland composite cements containing dolomite and metakaolin after leaching, carbonation, and chloride exposure. *Cem. Concr. Compos.* **2018**, *89*, 89–106. [CrossRef]
23. García, E.; Alfonso, P.; Labrador, M.; Galí, S. Dedolomitization in different alkaline media: Application to Portland cement paste. *Cem. Concr. Res.* **2003**, *33*, 1443–1448. [CrossRef]
24. Thongsanitgarn, P.; Wongkeo, W.; Chaipanich, A. Hydration and Compressive Strength of Blended Cement Containing Fly Ash and Limestone as Cement Replacement. *J. Mater. Civ. Eng.* **2014**, *26*, 04014088. [CrossRef]
25. Chen, X.; Zhou, J.; Yan, Y. Hydration of ultrafine and ordinary Portland cement at early ages. *KSCE J. Civ. Eng.* **2014**, *18*, 1720–1725. [CrossRef]
26. Deschner, F.; Lothenbach, B.; Winnefeld, F.; Neubauer, J. Effect of temperature on the hydration of Portland cement blended with siliceous fly ash. *Cem. Concr. Res.* **2013**, *52*, 169–181. [CrossRef]
27. Zhang, S.; Lu, D.; Ling, K.; Xu, J.; Xu, Z. Effects of Dolomite and Limestone Powders on Strength and Hydration of Cement Mortars. *J. Chin. Ceram. Soc.* **2016**, *44*, 1126–1133.

28. Viczián, I.; Mária, F. Handbook of the thermogravimetric system of minerals and its use in geological practice. *Cent. Eur. Geol.* **2013**, *56*, 397–400. [CrossRef]
29. Aqel, M.; Panesar, D.K. Hydration kinetics and compressive strength of steam-cured cement pastes and mortars containing limestone filler. *Constr. Build. Mater.* **2016**, *113*, 359–368. [CrossRef]
30. Hooton, D.; Nokken, M.; Thomas, M.D.A. Portland-Limestone Cement State of the Art Report and Gap Analysis for CSA A3000. *Cem. Assoc. Can.* **2007**, *1*, 1–59.
31. Peiyu, L.S.Y. Infulence of limestone powder on filling effect of cement paste and pore structure of sand grout. *J. Chinest Ceram. Soc.* **2008**, *36*, 69.
32. Kakali, G.; Tsivilis, S.; Aggeli, E.; Bati, M. Hydration products of C3A, C3S and Portland cement in the presence of $CaCO_3$. *Cem. Concr. Res.* **2000**, *30*, 1073–1077. [CrossRef]
33. Cyr, M.; Lawrence, P.; Ringot, E. Mineral admixtures in mortars: Quantification of the physical effects of inert materials on short-term hydration. *Cem. Concr. Res.* **2005**, *35*, 719–730.
34. Matschei, T.; Lothenbach, B.; Glasser, F.P. Thermodynamic properties of Portland cement hydrates in the system $CaO–Al_2O_3–SiO_2–CaSO_4–CaCO_3–H_2O$. *Cem. Concr. Res.* **2007**, *37*, 1379–1410. [CrossRef]
35. Matschei, T.; Lothenbach, B.; Glasser, F.P. The role of calcium carbonate in cement hydration. *Cem. Concr. Res.* **2007**, *37*, 551–558. [CrossRef]
36. Zajac, M.; Dienemann, W.; Bolte, G. Comparative experimental and virtual investigation of the influence of calcim and magnesium carbonates on reacting cement. In Proceedings of the 13th International Congress on the Chemistry of Cements, Madrid, Spain, 2 July 2011.
37. Pokrovsky, O.S.; Schott, J. Kinetics and mechanism of dolomite dissolution in neutral to alkaline solutions revisited. *Geochim. Cosmochim. Acta* **2013**, *301*, 597–626. [CrossRef]
38. Morse, J.W. The dissolution kinetics of major sedimentary carbonate minerals. *Earth-Sci. Rev.* **2002**, *58*, 51–84. [CrossRef]
39. Mittermayr, F.; Höllen, D.; Klammer, D.; Leis, A.; Dietzel, M. Dissolution of dolomite in alkaline cementitious media. In Proceedings of the 13th International Congress on the Chemistry of Cement, Madrid, Spain, 3–8 July 2011; pp. 1–7.
40. Xu, P.; Sheng, M.; Lin, T.; Liu, Q.; Wang, X.; Khan, W.A.; Xu, Q. Influences of rock microstructure on acid dissolution at a dolomite surface. *Geothermics* **2022**, *100*, 102324. [CrossRef]

materials

MDPI

Article

Research on Mechanical Behavior of the Steel–Concrete–Steel Composite Structures Subjected to High Temperature of Fire

Peng Cao [1,2,3], Xuebing Hu [2,3], Enlong Liu [1], Jianzhong Chen [2,3], Shouchao Jiang [4] and Hao Ding [2,3,*]

[1] College of Water Resource & Hydropower, Sichuan University, Chengdu 610065, China; caopeng1@cmhk.com (P.C.); liuenlong@scu.edu.cn (E.L.)
[2] National Engineering Research Center of Road Tunnel, Ltd., Chongqing 400067, China; huxuebing@cmhk.com (X.H.); chenjianzhong@cmhk.com (J.C.)
[3] China Merchants Chongqing Communications Technology Research & Design Institute Co., Ltd., Chongqing 400067, China
[4] College of Civil Engineering, Tongji University, Shanghai 200092, China; scjiang@tongji.edu.cn
* Correspondence: dinghao@cmhk.com; Tel.: +86-23-62653121

Abstract: A new type of steel–concrete–steel composite structure was adopted and widely used in the immersed tunnel of the Shenzhen–Zhongshan access. The research on the mechanical behavior of the new composite structure under a high temperature of fire is of great engineering significance to the fire protection design of the structure. Both the model test and a numerical simulation were adopted to study the mechanical behavior and damage characteristics of the new composite structure under fire. The RABT standard temperature rise curve was used to simulate the temperature rising law under fire (it reflects the characteristics of temperature rise in case of fire in an enclosed environment: rapidly raised to 1200 °C within 5 min, maintained at 1200 °C for 120 min, then it is cooled to normal temperature after 110 min). The temperature distribution law inside the structure, the deformation development law of the roof and the crack distribution were analyzed based on the thermal–mechanical coupling analysis method. The results showed that the internal part of the composite structure close to the fire surface was directly affected by the high temperature, and the temperature presented a step distribution law, while the part far from the fire surface was affected by the lag effect of the temperature transfer, and the temperature presented a continuous growth law. The roof deformation presented a three-stage model of "rapid growth-deformation stability-deformation recovery" with time. The overall cracks of the composite structure showed a symmetrical distribution under fire. The composite structure presented a shear failure mode as a whole. The results could provide a reference for the study of fire resistance for the new composite structure and support the structural fire protection design of the immersed tunnel of the Shenzhen–Zhongshan access.

Keywords: tunnel engineering; mechanical behavior under fire; model test; the steel–concrete–steel composite structures; temperature distribution; failure mode

1. Introduction

China has now become the country with the largest scale, the largest number, the most challenging geological circumstance and structure forms, and the quickest development pace of highway tunnels in the world, thanks to the rapid development of China's economic construction [1–3]. China has a heavily populated river system, multiple straits and bays, and an 18,000 km coastline. Along the river and the sea, about 20 significant cities have been created. China constructed 245 undersea tunnels by the end of 2020, with 23 of them being immersed tunnels, accounting for approximately 10% of the total [4–6]). One of the most prevalent disasters in traffic tunnels is fire. The primary effect of fire on underwater tunnels is the damage and destruction of the main structure due to high temperatures, resulting in the damage and loss of its bearing capacity and function. Furthermore, due to the unique climatic conditions of underwater tunnels, fires are often more severe than those found in overland traffic tunnels, and repairs are extremely complex [7–10]. The steel–concrete–steel composite structure has considerable benefits over typical reinforced concrete schemes, such as a high bearing capacity, good waterproof performance, minimal prefabricated site needs, fast building speed, and so on. It is better suited to large immersed tunnels and will be the future development path for large immersed tunnels. As a result, future studies on the fire mechanical response of steel–concrete–steel composite structures will play a critical role in fire safety design, tunnel construction, and operation safety of large-scale submerged tunnels.

Steel–concrete composite structural components are commonly utilized in building constructions as vertical and horizontal load-bearing components, and they are exposed to fire directly. As a result, the majority of research on the fire resistance of composite structures focuses on the construction of composite beam structures. Lennon and Moore (2003) conducted seven fire experiments on eight-story full-scale composite structures and discovered that steel–concrete composite beams could preserve stability and integrity even when deflected significantly [11]. Wang et al. (2017) studied the effect of various shear connection degrees on partially shear-connected composite beams under fire and proposed an analytical method for predicting the fire response of composite beams with steel–concrete partial shear connections, taking into account the effect of temperature on material properties as well as the slip behavior of shear connections under high temperatures [12,13]. Zhou et al. (2005) used a finite element model verified by the test results of simply supported composite beams to analyze and study the relevant parameters such as concrete slabs, steel beams, studs, loads, and fire protection layers of steel beams that affect the fire resistance of composite beams. The load ratio and the thermal resistance coefficient (or thickness) of the fire protection layer of steel beams are two critical characteristics that influence the fire resistance performance of composite beams, according to the findings [14]. Wang et al. (2015) studied the fire resistance of composite steel beams with reinforced truss floor bearing plates considering the shear slip of bolts. The research results showed that the composite steel beams with a fully shear resistance design had better fire resistance at room temperature, but shear slip and lift displacement still occurred at high temperatures [15]. Ma (2014) conducted a finite element numerical simulation analysis of the fire resistance of the honeycomb composite beam and revealed the temperature field distribution and failure mode of the honeycomb composite beam under fire [16]. Han (2020) conducted an experimental study on the fire resistance of stainless steel and concrete composite beams [17]. The test found that the composite beams all showed bending-shear failure modes under fire, and the specimens had large rebound deformation after the test. Li et al. (2010) proposed the critical temperature method for the fire resistance design of simply supported composite beams according to the method of checking the fire resistance of simply supported composite beams in the Code for Fire Safety of Steel Building Structures (GB51249-2017), and they provided the corresponding critical temperature calculation method under different floor thicknesses, fire resistance limits and load ratios [18]. Song et al. (2008) summarized some research results obtained by domestic and foreign researchers on the fire resistance of composite structures and summarized the working mechanism and fire resistance design methods of composite structures under fire, as well as some new progress in fire resistance research. For an immersed tunnel with a steel–concrete–steel composite structure, the stability and durability of the composite structure under a high fire temperature are of great significance to the safety of the whole tunnel structure, because the external steel shell has both waterproof and load-bearing functions [19]. Song et al. (2019) carried out three flexural beam tests and three shear beam tests for the steel–concrete–steel composite structure. Based on the test results, a shear design method combining the shear mechanism of steel–concrete composite with the shear mechanism of the I-steel web was proposed [20]. At present, the research foundation for the mechanical response and failure mode under a high temperature of fire, and the coordination of stress and deformation under a high temperature of the steel–concrete–steel composite structure are weak.

In this paper, a study on the mechanical behavior of the steel–concrete–steel composite structure under a high temperature of fire was carried out based on the physical model test and numerical simulation analysis for the steel–concrete–steel composite structure used in the immersed tube tunnel of the Shenzhen–Zhongshan link, revealing the temperature transfer mechanism, component damage mode, and structural mechanical response characteristics in the internal structure of the composite structure under a high temperature of fire. Related results can provide a reference for the research on fire protection technology and fire mechanical behavior of steel–concrete–steel immersed tunnels.

2. Structural Form of Immersed Tube Tunnel Sections

At present, the main tube structure of an immersed tunnel is mainly divided into reinforced concrete and steel plate concrete. The initial immersed tunnel adopted the form of a double steel plate or single steel plate. The outer steel plate was mainly used for waterproofing and as concrete pouring formwork, and the inner side was a reinforced concrete structure, so the outer steel plate was often not used as a load-bearing structure. With the advancement of immersed tunnel technology, in the 1980s, immersed tunnels with steel plate–concrete composite structures as the main bearing structure achieved engineering applications. In Japan, a new type of steel–concrete–steel composite structure was adopted in the Hong Kong Island Tunnel of Kobe Port: the inner structure was equipped with horizontal and vertical steel partitions, and different types of stiffeners were installed in the horizontal and vertical directions of the tunnel. The vertical and horizontal partitions and the inner and outer steel plates formed a whole to bear the external load, so that the component of the composite structure could fully utilize their respective material characteristics, as shown in Figure 1.

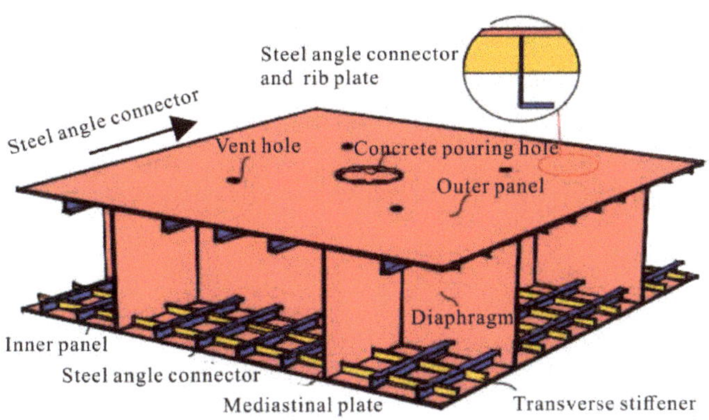

Figure 1. The steel–concrete–steel composite structure.

3. Experimental Scheme of Mechanical Behavior of the Steel–Concrete–Steel Composite Structure under a High Temperature of Fire

3.1. Processing of Components

The size of the steel–concrete–steel composite structural component was 2.9 m × 0.5 m × 0.5 m, the steel plate type was Q355B, the concrete label was C50, and there were horizontal and vertical steel plates, longitudinal T-ribs, welding studs, etc. The thickness of the steel plate at the top and bottom of the component was 10 mm, and the thickness of the transverse and longitudinal partitions was 4 mm. The specific structure of the steel structural component is shown in Figure 2.

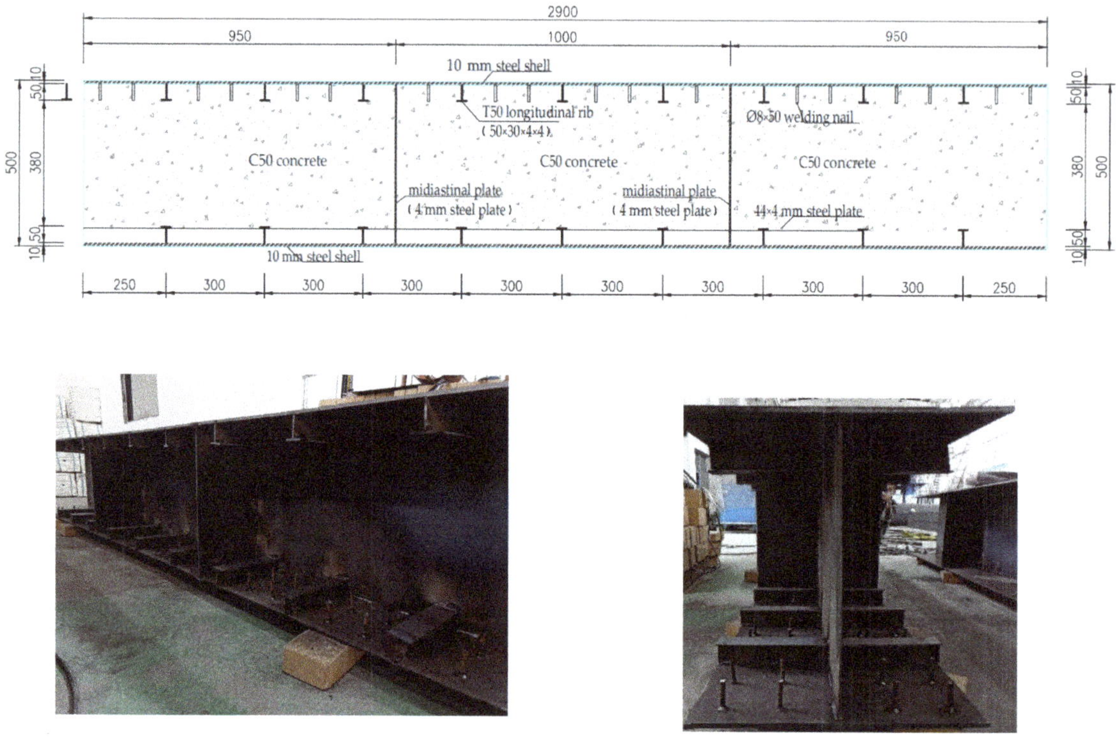

Figure 2. The specific construction of the steel structures.

After the steel structure was fabricated, thermocouples were arranged inside the component, the concrete was poured and cured, and the curing time was not less than 7 days.

3.2. *Arrangement of Measuring Points*

During the test, internal thermocouples were used to measure the temperature transfer law at different positions inside the component, and wire-pull displacement gauges were used to measure the vertical deformation of the steel plate at the top of the component.

3.2.1. Arrangement of Internal Temperature Measuring Points

The temperature measuring points inside the component were arranged nonuniformly along with the height of the section, and a total of four thermocouples were arranged on the measuring section. The distance between each measuring point and the upper surface of the steel plate at the bottom of the component was 0 mm (measuring point 1), 10 mm (measuring point 2), 240 mm (measuring point 3), and 470 mm (measuring point 4), respectively, as shown in Figure 3.

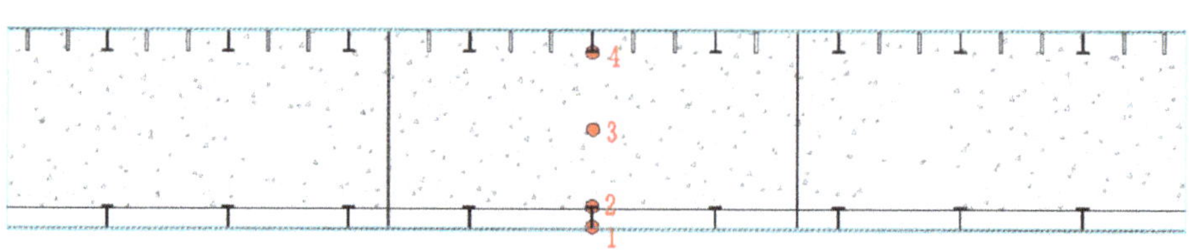

Figure 3. Arrangement of the measuring points of temperature in the structure.

3.2.2. Arrangement of the Measuring Points of the Roof Displacement

A total of five pull-wire displacement gauges were arranged on the top of the component, and the ends of the pull-wire were fixed on the top plate of the component. The specific arrangement positions were as follows: 1 in the middle of the component's clear span (measuring point 3), 1 at each of the 1/4 component's clear span (measuring point 2 and 4), and 1 at each end on both sides (measuring point 1 and 5), as shown in Figure 4.

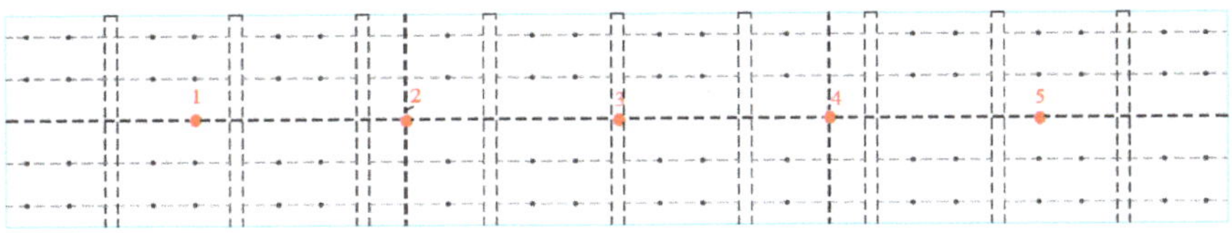

Figure 4. Arrangement of the measuring points of the roof displacement of the structure.

3.3. Test Implementation

The experiment was carried out with the test setup as shown in Figure 5. The roof components of the immersed tunnel were selected to conduct scale model testing under the most unfavorable condition. In this experiment, the geometric similarity scale was 1/3, and the elastic modulus similarity scale was 1. According to the similarity criterion, the similarity scale of concentrated force load was 1/9. According to the engineering design data, under the most unfavorable working condition, the design concentrated force load of the prototype structure was 1440 kN; therefore, the concentrated load of this experimental model was 1440/9 = 160 kN. The steel plate at the bottom of the component was directly affected by the fire, and the RABT standard temperature rise curve was adopted. A constant concentrated load was applied to the midspan position of the steel plate at the top of the component. The mechanical response and structural damage characteristics of the steel–concrete–steel composite structure under a high temperature of fire were analyzed. The specific implementation process was as follows: (1) lift the component to a predetermined position above the furnace with a crane, and place the component so that both ends of the component are on the support of the loading reaction beam; (2) carry out fire protection for the external wiring of the thermocouples, and install the loading system of the fire test furnace; (3) install the pull-wire displacement gauges, and the test data response is normal; (4) add fireproof cotton between the side of the component and the furnace cover to ensure that the component is under fire on one side; (5) apply a predetermined load to the top steel plate and turn on the burner for the test after the data are stable. The corresponding relationship between the temperature rise curve in the furnace and the RABT standard curve during the test is shown in Figure 6.

Figure 5. The picture of the test setup.

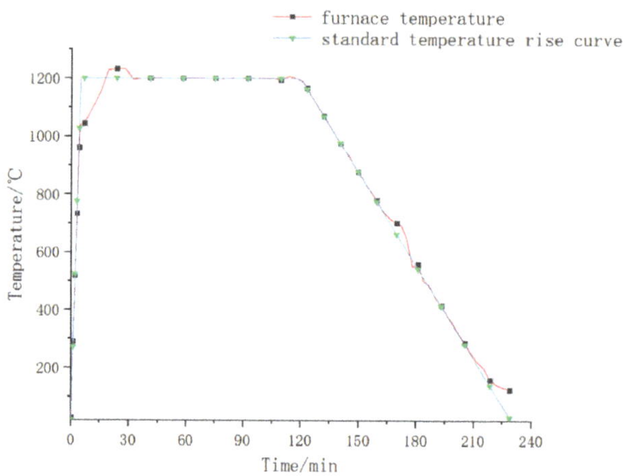

Figure 6. Temperature rise curve in the furnace during the test.

4. Analysis of Test Results

4.1. The Overall Test Phenomena

The damaged state of the component after the test is shown in Figure 6. It can be seen from the figure that due to the influence of the high temperature, the color of the concrete near the bottom steel plate has changed significantly, from blue-gray to gray-white. The color change area is about 10 cm in height, and the closer it is to the bottom steel plate, the more broken the concrete is; the concrete is powdery within 3 cm from the upper surface of the component bottom plate.

Due to the influence of the high temperature, the crack development on the side of the component is obvious, and the main distribution positions of the cracks are shown in Figure 7. It can be seen from the figure that under the action of thermal–mechanical coupling, the cracks couples, such as crack 1 and crack 4, crack 2 and crack 3, crack 6 and crack 8 distribute symmetrically according to the midspan line (dash-dot line in Figure 7). A shear failure model can be observed in the appearance of the specimen after the fire. An internal structure of the component comparison found that the side cracks of the component were mainly concentrated in the vicinity of the partition plates and the T-ribs. The crack statistics are shown in Table 1. This failure is mainly caused by the temperature internal force caused by the temperature gradient and thermal expansion difference in the component. When affected by the high temperature of the fire, a large temperature difference occurs at the interface between the steel plate and the concrete, resulting in a local thermal stress concentration, and then damage cracks. At the same time, the expansion and penetration of cracks are aggravated under the action of the external load.

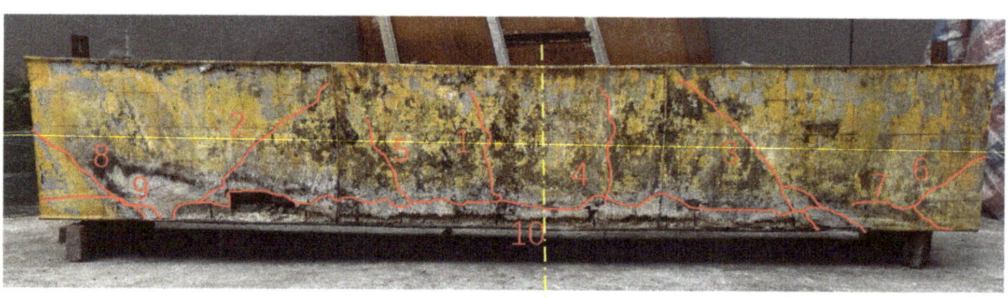

Figure 7. Distribution of cracks of the structure after test.

Table 1. Statistics of cracks of the structure after test.

Crack Number	Crack Type	Development Direction	Crack Length (cm)
1	Vertical crack	Vertical	34
2	Oblique crack	Slanted 45° inward	57
3	Oblique crack	Slanted 45° inward	57
4	Vertical crack	Vertical	34
5	Vertical crack	Vertical	34
6	Oblique crack	Obliquely outward	17
7	Oblique crack	Obliquely outward	9
8	Oblique crack	Obliquely outward	21
9	Oblique crack	Obliquely outward	11
10	Horizontal crack	Longitudinal along the component	210

4.2. Analysis of Temperature Transfer and Deformation Law of Composite Structures under a High Temperature of Fire

4.2.1. Temperature Distribution Law in Component

Firstly, the temperature distribution law in the steel–concrete–steel composite structure under the RABT heating curve was systematically analyzed. The temperature versus time curve for different sections in the component is shown in Figure 8. It can be seen from the figure that Section 1 (0 mm from the bottom plate) was directly affected by the high temperature of the fire during the test process. The temperature change law of this monitoring point was similar to the temperature rise curve in the furnace, and there were obvious heating and cooling stages: within the first 30 min, the temperature of the monitoring point was in the stage of rapid heating, and the maximum temperature reached 1007 °C; at 30~60 min, the temperature in the furnace was stable at 1200 °C, and the temperature of the monitoring point was stable at about 1050 °C, which was slightly less than 1200 °C, indicating that although the measuring point was directly affected by the high temperature, the temperature transfer in the component had a certain hysteresis; after 60 min, the temperature of the measuring point was abnormal, and the reason may have been that the range of the thermocouple embedded in the component was 800~1000 °C. After the thermocouple was continuously affected by the high temperature, the performance of the thermocouple was compromised, resulting in subsequent abnormalities in the temperature acquisition data.

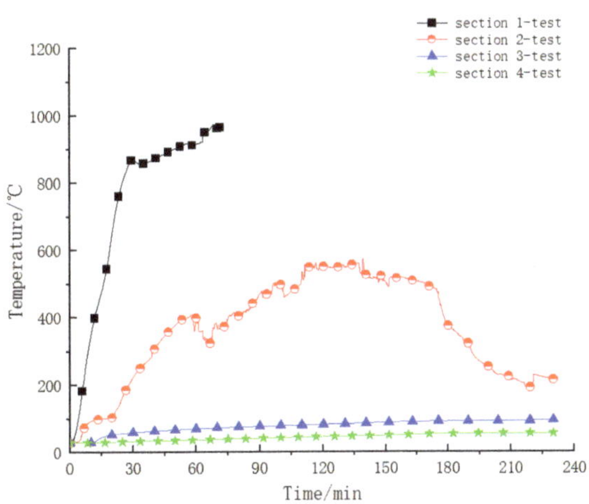

Figure 8. Temperature versus time curve for different sections during the test.

During the test, the temperature change rule of Section 2 (10 mm from the bottom plate) was similar to Section 1, but due to the hysteresis of the temperature transfer, in the initial heating stage, the temperature rise rate of Section 2 was significantly lower than that of Section 1, and the temperature reached the maximum value of 714 °C at 140 min; in the 20 min furnace cooling, the temperature of Section 2 entered the cooling stage, and the temperature of Section 2 was 321 °C when the test was completed.

The temperature of Section 3 (240 mm from the bottom plate) and Section 4 (470 mm from the bottom plate) was only in the growth stage, and there was no obvious decline stage, indicating that the hysteresis effect of the internal temperature transfer of the component was very obvious, with the highest temperature in Section 3 being 135 °C and the highest temperature in Section 4 being 61 °C.

4.2.2. Variation Rule of Roof Displacement

The variation law of the displacement of different measuring points on the top plate is shown in Figure 9. It can be seen from the figure that the variation law of the displacement of different measuring points during the test is very obvious. The three measuring points (measuring points 2, 3, and 4) located on the clear span of the component produced a vertical downward deformation; the measuring points (measuring points 1 and 5) at the end of the component produced a vertical upward deformation, indicating that under the effect of thermal–mechanical coupling, the overall deformation trend was that the middle sinks and the two ends rise. Moreover, the deformation law of the measuring points during the test was affected by the high temperature of the fire, showing three obvious stages: (1) in the rapid growth stage, at the beginning of the test, affected by the rapid heating stage in the furnace, the deformation of the component increased rapidly. The midspan position reached the maximum value of 11 mm; (2) in the stable deformation stage, the temperature in the furnace was stable at 1200 °C, and the deformation of the component was less affected by the temperature and remained basically stable; (3) in the deformation recovery stage, affected by the cooling stage in the furnace, the temperature in the component decreased, the deformation of the component was partially restored, and the residual deformation was 7 mm. This phenomenon showed that the deformation law of the steel–concrete–steel composite structure was closely related to its fire boundary conditions under the action of the thermal–mechanical coupling, and the effect of a high fire temperature on the mechanical mechanism of the components was very obvious.

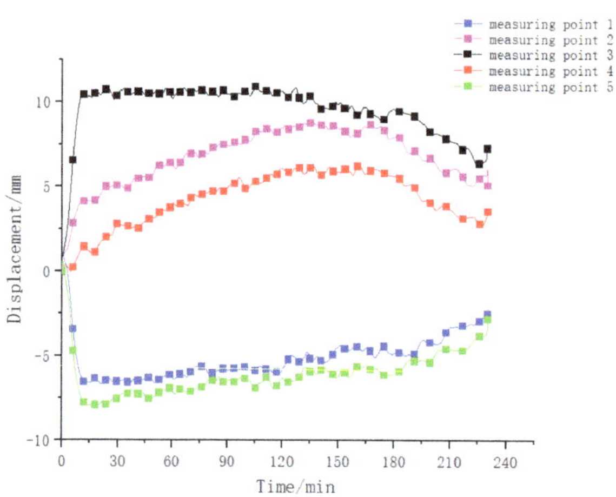

Figure 9. Displacement versus time curve of different monitoring points during the test.

5. Numerical Simulation Research on Mechanical Behavior of Steel–Concrete–Steel Composite Structure

In order to further analyze the mechanical behavior of the steel–concrete–steel composite structure under the action of the thermal–mechanical coupling, a comparative study was carried out between the numerical simulation and the physical model test results.

5.1. Material Parameters

5.1.1. Thermal Parameters

The thermal parameters of steel and concrete are shown in Table 2. The thermal parameters of ordinary concrete at high temperatures should be determined according to the following regulations.

Table 2. The thermal parameters for materials.

Material	Density kg/m^3	Heat Transfer Coefficient w/m/°C	Specific Heat Capacity J/m/°C
Steel	7850	45.0	600
Concrete	2300	With the change of temperature, Equation (1)	With the change of temperature, Equation (2)

The heat transfer coefficient of concrete λ_c should be calculated according to Equation (1):

$$\lambda_c = 1.68 - 0.19\frac{T_c}{100} + 0.0082\left(\frac{T_c}{100}\right)^2 \quad (1)$$

The specific heat capacity of concrete c_c should be calculated according to Equation (2):

$$C_c = 890 + 56.2\frac{T_c}{100} - 3.4\left(\frac{T_c}{100}\right)^2 \quad (2)$$

where: T_c—the temperature of the concrete (°C);

λ_c—the heat transfer coefficient of concrete (W/(m·°C));

c_c—specific heat capacity of concrete (J/(kg·°C)).

Considering the influence of moisture on the concrete's temperature, the main influence is the change of the concrete's specific heat capacity, so the concrete's specific heat capacity is corrected. Assuming that the vaporization of all water is completed between 90 °C and 110 °C, considering that the general concrete moisture content is 4%, and the vaporization heat of water (at 100 °C, one-atmosphere pressure) is 2257.2 kJ/kg, the density of concrete is taken as 2300 kg/m^3. The heat required for water vaporization in one cubic meter of concrete is:

$$2300 \times 0.04 \times 2257.2 = 207,662.4(\text{kJ}) \quad (3)$$

The specific heat of concrete at 100 °C is 942.8 J/(kg·°C) according to Equation (2), so the heat required for one cubic meter of concrete to heat up to 20 °C is:

$$2300 \times 20 \times 942.8 = 43,368.8(\text{J}) \quad (4)$$

Therefore, the specific heat value of concrete between 90 °C and 110 °C is:

$$\frac{207,662,400 + 43,368.8}{2300 \times 20} = 5103.2(\text{kJ}/(\text{kg °C})) \quad (5)$$

The specific heat of concrete in other temperature sections was taken according to Equation (2).

5.1.2. Heat Transfer Parameters

There are three basic ways of transferring heat: heat conduction, heat convection, and heat radiation. In this model, the heat transfer modes of the flue gas in the overfire area were thermal convection and thermal radiation.

The comprehensive emissivity was taken from the American standard ANSI/AISC360-10 "Specification for Structural Steel Buildings" (2010), which comprehensively considers the emissivity of the flue gas and the influence of the radiation angle coefficient [21]. The comprehensive emissivity of the selected steel shell concrete surface was taken as 0.7, and the heat dissipation coefficient of the film layer was taken as 25.0 $w/m^2/°C$.

5.1.3. Material Constitutive Relations at High Temperature

The tensile behavior of the concrete is linear elastic until the tensile strength. The stress–strain relationship of the compressive part of the high-temperature concrete adopts the high-temperature constitutive relationship proposed by Lie, as shown in Equation (6) [22].

$$
\begin{gathered}
\sigma_{c,T} = \begin{cases} f_{c,T}[1 - \left(\frac{\varepsilon_{cu,T}-\varepsilon_{c,T}}{\varepsilon_{cu,T}}\right)^2] & \varepsilon_{c,T} \leq \varepsilon_{cu,T} \\ f_{c,T}[1 - \left(\frac{\varepsilon_{cu,T}-\varepsilon_{c,T}}{3\varepsilon_{cu,T}}\right)^2] & \varepsilon_{c,T} > \varepsilon_{cu,T} \end{cases} \\
\varepsilon_{cu,T} = 0.0025 + (6T + 0.04T^2) \times 10^{-6} \\
f_{c,T} = \begin{cases} f_{c,20} & \varepsilon_{c,T} \leq \varepsilon_{cu,T} \\ f_{c,20}[2.011 - \frac{2.353(T-20)}{1000}] & \varepsilon_{c,T} > \varepsilon_{cu,T} \end{cases}
\end{gathered} \tag{6}
$$

where: $\sigma_{c,T}$—stress of concrete at high temperature;

$f_{c,T}$—compressive strength of concrete at high temperature;

$\varepsilon_{c,T}$—strain of concrete at high temperature;

$\varepsilon_{cu,T}$—strain of concrete corresponding to peak stress at high temperature

T—temperature.

The reduction coefficient of the concrete's tensile strength at high temperature was set according to the provisions of European specification EC2, as shown in Equation (7) [23].

$$
K_{c,T} = \begin{cases} 1.0 & T \leq 100\ °C \\ 1.0 - 1.0 \times \frac{T-100}{500} & 100\ °C < T \leq 600°C \end{cases} \tag{7}
$$

where $K_{c,T}$—the reduction coefficient of concrete tensile strength at high temperature.

In the analysis, the constitutive model of the steel at different temperatures adopted the ideal elastic–plastic model. The yield strength of steel at room temperature was 355 MPa, and that of the elastic model was $2.06 \times 10^5\ N/mm^2$. The elastic modulus and yield strength reduction coefficient at high temperatures were taken according to the provisions of the Code for Fire Safety of Steel Building Structures GB51249, as shown in Equation (8) [24]. The temperature has little effect on the Poisson's ratio of steel, and the Poisson's ratio takes a constant value of 0.3.

$$
\begin{gathered}
\chi_T = \begin{cases} \frac{7T_s-4780}{6T_s-4760} & 20\ °C \leq T_s < 600\ °C \\ \frac{1000-T_s}{6T_s-2800} & 600\ °C \leq T_s < 1000\ °C \end{cases} \\
\eta_T = \begin{cases} 1.0 & 20\ °C \leq T_s < 600\ °C \\ 1.24 \times 10^{-8}{T_s}^3 - 2.096 \times 10^{-5}{T_s}^2 & \\ +9.228 \times 10^{-3}T_s - 0.2168 & 300\ °C \leq T_s < 800\ °C \\ 0.5 - T_s/2000 & 800\ °C \leq T_s < 1000\ °C \end{cases}
\end{gathered} \tag{8}
$$

where: χ_T—the reduction coefficient of strength of steel at high temperature;

η_T—the reduction coefficient of Young's modulus of steel at high temperature.

5.2. Finite Element Model

There are many factors affecting the distribution of structural temperature fields under fire. In order to simplify the analysis process, the following assumptions were considered:

1. Steel and concrete are homogeneous and isotropic structural materials, and their thermal parameters such as thermal conductivity and specific heat capacity are the same in each direction at the same temperature;
2. A complete heat transfer between steel and concrete, ignoring the influence of contact thermal resistance;
3. The influence of structural high-temperature mechanical response on structure temperature field distribution is not considered;

The sequential thermal–mechanical coupling method was used to analyze the temperature field of the structure first, and then the results of the temperature analysis were applied to the mechanical model as loads to analyze the mechanical response of the model. The established finite element model is shown in Figure 10. The schematic diagram of the loading mode of the finite element model is shown in Figure 11. The load and the constraint configuration of the numerical model was consistent with the model test. As shown in Figure 11, using the loading surface to simulate the role of the distribution beam in the test, both ends of the model were placed on the pedestals and the displacements at the pedestals were constrained.

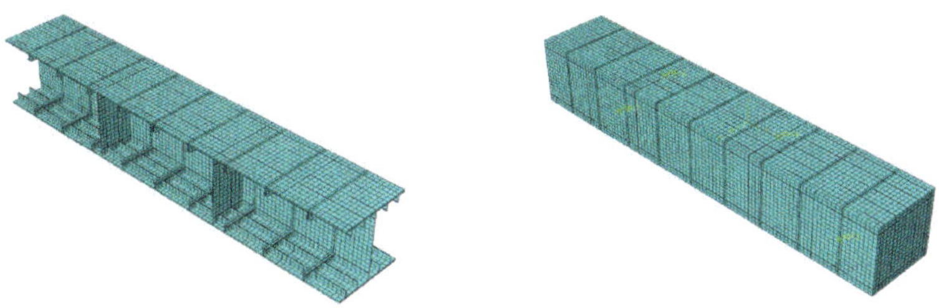

(**a**) (**b**)

Figure 10. The overall numerical model. (**a**) Model for the steel structure, (**b**) model for the concrete structure.

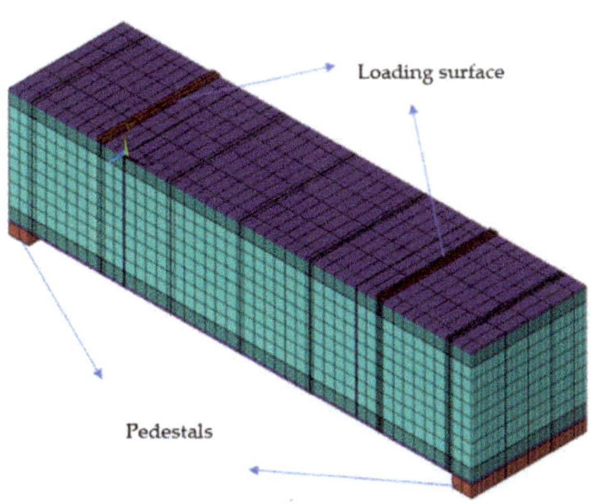

Figure 11. Schematic diagram of the loading mode of the finite element model.

5.3. Comparative Analysis of Physical Model Test and Numerical Simulation Results

(1) Comparative analysis of temperature transfer laws

The comparison results of the temperature–time curves are shown in Figure 12. It can be clearly seen from the figure that the numerical results are in good agreement with the physical test data as a whole, and the temperature distribution law of each section is obviously terraced. The larger the distance between the measurement points and the bottom steel plate, the lower the temperature. Due to the hysteresis effect of the temperature transfer, the heating period is longer. For Sections 1 and 2, which are closer to the fire surface, the numerical simulation results at the same time are slightly larger than the experimental data. The possible reasons are as follows: there is a certain difference between the simulation of boundary conditions and the actual situation; there are positional deviations due to pouring and other reasons during the production of the component; and the thermal parameters of the material are different from the actual ones.

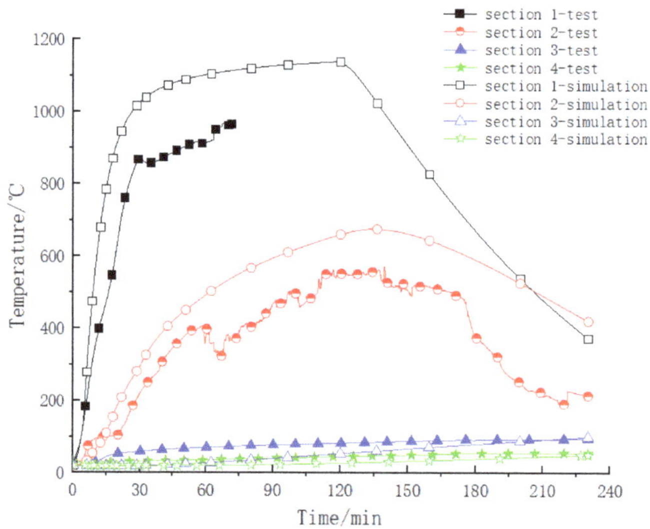

Figure 12. Comparative analysis of temperature transfer law of measuring points for each section.

(2) Contrastive analysis of roof deformation laws

The comparison results of the roof deformation laws are shown in Figure 13. It can be seen from the figures that the displacement obtained by numerical simulation and the displacement measured by the test are basically the same when the component is exposed to fire. Within half an hour before the fire, the difference between the two is very small, then there is a certain deviation where the displacement obtained by numerical simulation is larger than that measured by experiment. The overall displacement change has three stages: (1) A rapid growth stage: within 15 min of heating, the displacement rises faster. The main reason is that the temperature gradient of the section is large at this time, and the bending deformation of the section due to uneven thermal expansion increases rapidly. (2) A deformation stability stage: at this time, although the average temperature of the section increases, the temperature gradient of the section decreases. The displacement of the section increases due to the decrease of stiffness caused by the high temperature, but the decrease in temperature gradient reduces the bending caused by an uneven thermal expansion, and the overall displacement remains stable, or slightly increases. (3) A deformation recovery stage: after the temperature in the furnace decreases, the stiffness of the component is recovered somewhat, the temperature gradient in the section further decreases, and the displacement partially recovers, but the recovery speed is slow. The possible reason for the deviation is that in the numerical simulation, due to the large internal force of the temperature, the cracks in the concrete have a significant impact on the stiffness of the beam, and the deformation increases.

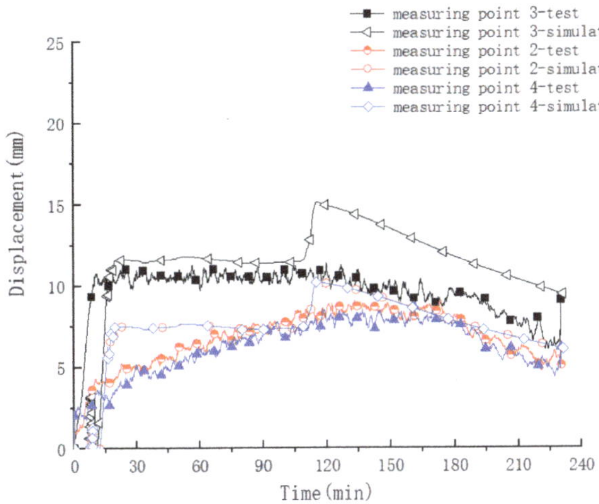

Figure 13. Comparative analysis of deformation development law of measuring points of clear span area of composite structure.

(3) Distribution analysis of structural plastic zone under a high temperature of fire

Based on the numerical simulation results, the plastic zone distribution of the steel–concrete–steel composite structure under a high temperature of fire was obtained, as shown in Figure 14. It can be seen from the figure that the lower flange is close to the fire surface and the plastic strain at the lower flange of the component is relatively large, mainly due to the fact that the higher the temperature, the greater the degradation of the material strength and elastic modulus; the thermal expansion deformation is constrained by the lower temperature concrete above, and the material enters a plastic state. The distribution of the plastic zone in other parts of the component is less, indicating that the high temperature of fire has a limited influence on the plastic development of the entire structure.

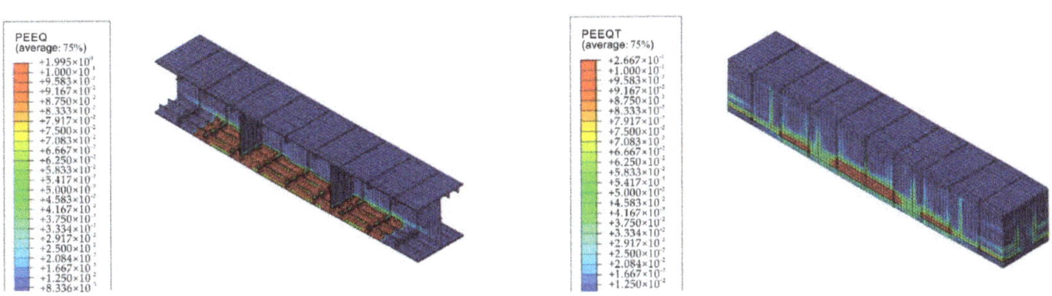

(**a**) Distribution of steel structure (**b**) Distribution of concrete structure

Figure 14. Distribution of the plastic zone of the composite structure under fire.

(4) Analysis of structural crack distribution under a high temperature of fire

Based on the numerical simulation results, the tensile failure cloud diagram of the component concrete was obtained, as shown in Figure 15, which was consistent with the physical model test results: the overall component crack distribution had a good symmetry. There was a longitudinal crack running through the component in parallel to the lower flange, which was consistent with the position of the test no. 10 crack. There was a 45° inclined crack on the connection line between the beam end support and the loading point.

The midspan crack developed vertically, and the crack distribution was in good agreement with the test results. It shows that the numerical simulation can well simulate the failure of the component concrete under a high temperature of fire.

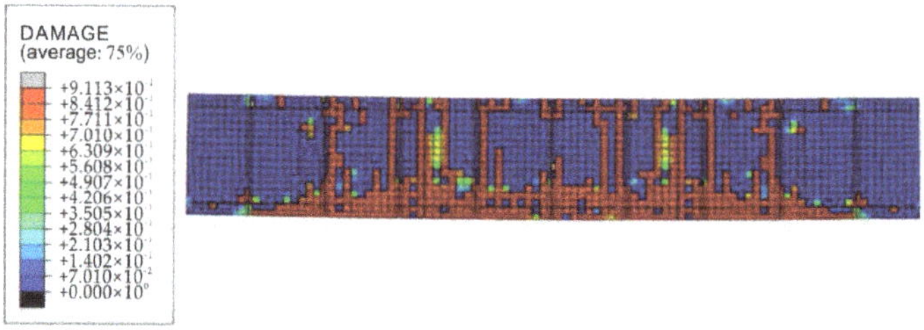

Figure 15. Distribution of failure of concrete in the composite structure caused by a high temperature under fire.

6. Conclusions

(1) Under the RABT standard temperature rise curve, the area directly affected by the high temperature under fire in the steel–concrete–steel composite structure was within a range of about 10 cm from the fire surface, and within 3 cm away from the fire surface, the concrete showed a powdered state; the crack distribution modes in the composite structures under thermal–mechanical coupling were given.

(2) The temperature transfer law of the measuring points at different depths of the combined structure were given: the temperature distribution of the measuring points closer to the fire surface presented a stepped distribution, and there was an obvious heating and cooling stage; the temperature of the far measuring points showed a continuous growth law due to the hysteresis effect of temperature transfer; the results of the numerical simulation and physical model test were compared to verify the temperature transfer law in the composite structure under a high temperature of fire.

(3) The evolution law of the roof displacement of the composite structure was given: the overall deformation trend of the structural component under the action of thermal–mechanical coupling was that the middle sank and the two ends upturned. Affected by the high temperature of the fire, the roof deformation presented three distinct stages with time: a rapid growth stage, a deformation stabilization stage, and a deformation recovery stage. The results of the numerical simulation and physical model test were compared to verify the mechanical response mechanism of composite structure under the high temperature of fire.

(4) The high temperature of fire had limited impact on the plastic development of the whole structure. The plastic zone was concentrated in the lower flange near the fire surface, and the plastic zone of the other parts was less distributed.

(5) The composite structure presented a shear failure mode as a whole: the crack distribution of the whole component had good symmetry, a 45° oblique crack appeared on the line connecting the beam end support and the loading point, and the midspan cracks developed vertically.

Author Contributions: Conceptualization, X.H., J.C., S.J. and H.D.; Data curation, P.C., J.C., S.J. and H.D.; Formal analysis, P.C. and H.D.; Investigation, P.C. and E.L.; Methodology, X.H., E.L. and H.D.; Resources, X.H., J.C. and S.J.; Supervision, E.L. All authors have read and agreed to the published version of the manuscript.

Funding: This study was funded by the National Key R&D Program of China (no. 2019YFB1600702), the National Natural Science Foundation of China (no. 52127814), the Chongqing Natural Science Foundation (no. cstc2021jcyj-msxmX0591).

Institutional Review Board Statement: Not applicable.

Informed Consent Statement: Not applicable.

Data Availability Statement: Data available on request.

Conflicts of Interest: The authors declare no conflict of interest.

References

1. Hong, K.R.; Feng, H.H. Development and thinking of tunnels and underground engineering in China in recent 2 years (from 2019 to 2020). *Tunn. Constr.* **2021**, *41*, 1259–1280.
2. Hong, K.R.; Feng, H.H. Development trends and views of highway tunnels in China over the past decade. *China J. Highw. Transp.* **2020**, *33*, 62–76.
3. Chen, X.S.; Xu, Z.H.; Bao, X.H.; Wang, X.T.; Fu, Y.B. Challenges and technological breakthroughs in tunnel construction in China. *China J. Highw. Transp.* **2020**, *33*, 1–14.
4. Lin, W.; Liang, J.Z.; Liu, L.F.; Zou, W.; Lin, M. Development and prospect of immersed tunnel and artificial technology. *Tunn. Constr.* **2020**, *41*, 2029–2036.
5. Liu, L.F.; Lin, W.; Yin, H.Q.; Liu, Y.P.; Zou, W.; Xu, S.Y.; Ji, Y.N. Construction of immersed tunnel engineering in the world and development status of immersed tunnel technology in China. *Cina Harb. Eng.* **2021**, *41*, 71–79.
6. Chen, J.Q.; Feng, X.Y.; Wei, H.; Feng, H.H. Statistics of underwater tunnels in China. *Tunn. Constr.* **2021**, *41*, 483–516.
7. Chen, Y. Application and developing trends of immersed tunnel. *Tunn. Constr.* **2017**, *37*, 387–393.
8. Zhang, E.Q. *Technical Study on Fire Protection for Joints and Segments of Undersea Immersed Tube Tunnel*; Chongqing Jiaotong University: Chongqing, China, 2013.
9. Zhang, M.X.; Huang, J.; He, X.Q. Coupling analysis of thermal field and stress field for immersed tunnels under fire load. *China Civ. Eng. J.* **2007**, *40*, 83–87.
10. Zhang, Z.Y. *Research of Mechanics Behavior for Immersed Tunnel under Fire Load*; Chongqing Jiaotong University: Chongqing, China, 2013.
11. Lennon, T.; Moore, D. The natural fire safety concept—full-scale tests at Cardington. *Fire Saf. J.* **2003**, *38*, 623–643. [CrossRef]
12. Wang, W.Y.; Engelhardt, M.D.; Li, G.Q.; Huang, G.S. Behavior of Steel-Concrete Partially Composite Beams Subjected to Fire-Part 1: Experimental Study. *Fire Technol.* **2017**, *53*, 1039–1058. [CrossRef]
13. Wang, W.Y.; Wang, K.; Engelhardt, M.D.; Li, G. Behavior of Steel-Concrete artially Composite Beams Subjected to Fire-Part 2: Analytical Study. *Fire Technol.* **2017**, *53*, 1147–1170. [CrossRef]
14. Zhou, H.Y.; Li, G.Q.; Wang, Y.Z. Parametric studies on fire-resistance of simaply-supported steel-concrete composite beams. *Steel Struct.* **2005**, *06*, 97–101.
15. Wang, W.Y.; Li, G.Q.; Chen, L.Z.; Jiang, S.; Huang, G. Experimental study on fire resistance of steel bar truss slab and steel composite beams. *China Civ. Eng. J.* **2015**, *48*, 67–75.
16. Ma, Q.J. *Investigation on the Fire Resistance of Composite Cellular Beams*; Shandong University: Jinan, China, 2014.
17. Han, Y.L. *Experimental Research on Mechanical Property of Stainless Steel-Concrete Composite Beams in Fire*; Southeast University: Nanjing, China, 2020.
18. Li, G.Q.; Wang, W.Y.; Zhou, H.Y. Simplified approach for fire-resistance design of simple-supported composite beams. *Build. Struct.* **2010**, *40*, 35–38.
19. Song, T.Y.; Han, L.H. Some new developments of the fire performance research on composite structures. *Eng. Mech.* **2008**, *25*, 230–253.
20. Song, S.Y.; Nie, J.G.; Xu, G.P.; Fan, J.S.; Tang, L.; Guo, Y.T. Development and application of steel-concrete-steel composite structures in immersed tunnels. *China Civ. Eng. J.* **2019**, *52*, 109–120.
21. *ANSI/AISC 360-10*; Specification for Structural Steel Buildings. American Institute of Steel Construction: Chicago, IL, USA, 2010.
22. Lie, T.T. A Procedure to calculate fire resistance of structural members. *Fire. Mater.* **1984**, *8*, 40–48. [CrossRef]
23. *BS EN 1992-1-2:2004*; Eurocode 2: Design of Concrete Structures, Part 1-2: General Rules—Structural Fire Design. British Standards Institution: London, UK, 2004.
24. *National Standarad of the People's Republic of China GB 51249-2017*; Code for Fire Safety of Steel Structures in Buildings. Ministry of Housing and Urban-Rural Development of the Peoples's Republic of China: Beijing, China, 2017.

materials

MDPI

Article

Fracture Behavior of Steel Slag Powder-Cement-Based Concrete with Different Steel-Slag-Powder Replacement Ratios

Ke-Xian Zhuo [1], Guo-Tao Liu [2], Xue-Wei Lan [3], Dong-Ping Zheng [1,4], Si-Quan Wu [1,4], Pei-Zong Wu [1,4], Yong-Chang Guo [1] and Jia-Xiang Lin [1,*]

[1] School of Civil and Transportation Engineering, Guangdong University of Technology, Guangzhou 510006, China; kexzhuo@163.com (K.-X.Z.); cosonz@163.com (D.-P.Z.); a623633772@163.com (S.-Q.W.); a623633772@126.com (P.-Z.W.); guoyc@gdut.edu.cn (Y.-C.G.)
[2] Guangdong GW Metal Industry Group Co., Ltd., Guangzhou 510030, China; gwjsim@gwjscyjt.com
[3] Guangzhou Zengcheng Zhengyuan Construction Engineering Testing Center Co., Ltd., Guangzhou 511300, China; zyjc06@163.com
[4] Guangzhou Hongchang Construction Technology Co., Ltd., Guangzhou 510006, China
* Correspondence: jxiang.lin@gdut.edu.cn; Tel.:+86-13710810754

Abstract: The influence of different replacement ratios of steel-slag powder as cement-replacement material on the fracture performance of concrete is studied in this paper. A three-point bending fracture test is carried out on slag powder-cement-based concrete (SPC)-notched beams with steel-slag powder as cementitious materials, partially replacing cement (0%, 5%, 10%, 15%, and 20%). Load-deflection curves and load-crack-opening displacement curves of SPC-notched beams with five different replacement ratios of steel-slag powder were obtained. The effects of different steel-slag-powder replacement ratios on the fracture properties (fracture energy, fracture toughness, and double-K fracture parameters) of the SPC were analyzed and discussed. The results showed that the incorporation of appropriate steel-slag powder can affect the fracture performance of SPC. Compared with concrete without steel-slag powder, adding appropriate steel-slag powder can effectively improve the bond performance between aggregate and matrix because the steel-slag powder contains hydration activity substances such as calcium oxide and aluminium trioxide. The fracture energy and fracture toughness of SPC increased and then decreased with the increase in steel-slag-powder replacement ratios, and the SPC concrete showed best fracture performance with a 5% steel slag powder replacement ratio. Its fracture energy increases by 13.63% and fracture toughness increases by 53.22% compared with NC.

Keywords: steel-slag powder; replacement ratio; fracture performance; concrete

1. Introduction

With the continuous development of civil-engineering technology, many new building materials are emerging, such as high-strength aluminum alloy, fiber concrete, sea-sand concrete, FRP materials, recycled concrete, steel-slag concrete, and so on [1–4]. Among them, steel-slag concrete is an environmental protection material [1]. Steel slag is a by-product of the steelmaking industry, accounting for emissions of about 10% to 20% in steel production [5,6]. China is a major industrial country with yearly production of steel consumption about 50% of the world's, resulting in an annual production of steel slag of approximately 10 million tons [7,8]. In recent years, how to turn waste steel slag into valuables has become a hotspot. Steel slag contains tricalcium silicate and dicalcium silicate mineral phases with similar composition to cement, as well as a slight amount of free calcium oxide and metallic iron [4,9]. Some researchers have proposed the use of steel slag as a coarse and fine aggregate for concrete, or finely ground into powder as cement admixture, as an effective way to improve the comprehensive utilization of steel slag [10–13]. Many studies have shown that when steel-slag powder is used as a cementitious material instead of cement, the steel-slag powder has a cementing activity that occurs under the mixture system with continuous hydration, which not only effectively reduces the amount of cement, but also improves the mechanical properties of concrete [12,14,15]. However, it is not that the higher steel-slag-powder replacement ratio is better. It was found that when the steel-slag-powder replacement ratio is 10%, the compressive strength of slag powder-cement-based concrete (SPC) is better than that of ordinary concrete (NC) and has good compatibility [14,16]. The steel-slag powder makes the concrete denser and effectively improves the impermeability of concrete, but the compressive strength gradually decreases with the increase in the steel-slag-powder replacement ratio. In addition, steel slag significantly improves the compressive strength, water absorption, and durability of concrete when used either as a coarse and fine aggregate in concrete [17–20]. Steel-slag concrete has promising advantages, not only for improving the workability of concrete, but also for being widely used in seismic-resistant buildings [21–23]. At present, several researchers have carried out various mechanical-property studies on steel-slag concrete [6,24–26]. The effect of expanding the range of steel-slag admixture on the compressive strength of normal and high-strength concrete has been studied [26]. The results showed that the appropriate amount of steel-slag incorporation not only improved the compressive strength of concrete, but also had a positive effect on the energy-dissipation capacity, especially more pronounced in normal-strength concrete. Palod, R. et al. [27] studied the properties of concrete with steel-slag powder as a substitute for cement with 20% replacement, which showed a promising and excellent performance on compressive strength, splitting-tensile strength, early-shrinkage properties, and resistivity. When steel slag was used as a partial replacement of cement in the UHPC mix [28], although the compressive strength decreased slightly, the self-shrinkage performance was improved with the increase in slag content, which was beneficial to the utilization of slag in UHPC. Steel-slag-concrete mixes have shown durability performance and mechanical stability as good as sea-sand concrete in acid and alkaline, freeze-thaw, and marine environments [29–31]. Saxena, S. et al. [18] used steel-slag aggregate to replace natural coarse aggregate at 15%, 25%, 50%, 75%, and 100% to study the compressive strength and flexural strength of concrete. The results indicated that the replacement of 50% basalt aggregate by steel-slag aggregate increased the compressive strength, flexural strength, and modulus of elasticity of concrete by 33% and 9.8%, respectively. Steel-slag concrete also exhibits better strain-ratio effects in terms of dynamics. Guo, Y.C. et al. [6] investigated the axial-impact compression properties of steel-slag-sand concrete (SSC) using 0%, 10%, 20%, 30%, and 40% steel-slag fine aggregates, respectively. The results of the study showed that the addition of steel slag as a fine aggregate can improve the dynamic properties of steel-slag concrete. In a further study, Rooholamini, H. et al. [32] examined the effect of partial and total replacement of natural aggregates by steel slag on the mechanical and fracture properties of compacted concrete, which were improved due to the high angularity and roughness of the coarse steel slag. It can be seen that the field of steel-slag-concrete research is mainly focused on use of steel-slag coarse and fine aggregates in concrete mixes. Steel slag has good promotion and environmental significance as a cement material with active material, as a partial replacement of cement in the preparation of concrete. At present, extensive research has been devoted to the properties of slag powder-cement-based concrete, and the results indicate that SPC is advantageous in terms of the mechanical strength and durability of the hardened mortars and concretes [33–35]. The addition of 10% steel-slag powder in SPC resulted in 6.5% and 11.9% higher compressive and tensile strength, respectively. When the dosage of the steel-slag powder was more than 10%, the strength clearly decreased [36]. In contrast to the incorporated steel slag in the15–30 μm and 30–45 μm ranges, the steel-slag particles with sizes of less than 15 μm and 45–80 μm were advantageous for the 7- and 28-day compressive-strength development of the concrete, especially at a low replacement ratio of up to 10% [37]. In conclusion, the mechanical properties of SPC are mainly focused on axial compressive and tensile behaviors, while the fracture properties of steel-slag-powder concrete are not clear.

Fracture-mechanics theory has been developed for decades. In 1961, Kaplan, M.F. et al. [38] applied the theory of fracture mechanics to concrete problems for the first time and conducted tests on the fracture toughness of concrete. Hillerborg, A. et al. [39] established a fracture model suitable for materials on concrete and deduced many valuable research results. Many studies have investigated the feasibility and effectiveness of fracture-mechanics theory in the study of different concrete mixes to further advance the research process of concrete fracture properties [40–42]. Since studies on fracture-mechanical properties of steel-slag powder-concrete have not been reported in the literature, and fracture properties are important parameters for structural-safety assessment and crack control, it is necessary to investigate the fracture properties of SPC. In this paper, based on the current developments in concrete fracture-mechanics theory, beam specimens of five groups of SPC mix, with different steel-slag-powder replacement ratios, were subjected to three-point bending fracture tests to explore the effect of substitution of steel-slag powder for cement, on the fracture properties of concrete, and to provide a basis for promoting the use of steel slag as an environmentally compatible building material.

2. Experimental Design

2.1. Test Materials

The composition of the specimens is mainly cement, steel-slag powder, river sand, granite crushed stone, water-reducing agent, and water. Among them, the cement is P.0 42.5 R ordinary silicate cement with the apparent density of 3100.0 kg/m^3. The apparent density of steel-slag powder is 2941.2 kg/m^3. The 9.7% particle size distribution is 0.09~0.30 mm and the 90.3% particle size distribution is 0.30~0.60 mm, produced by Lingshou Haotai mine processing plant in Hebei Province of China. The chemical compositions of the cement and steel-slag powder used in this study were determined by X-ray fluorescence method are as shown in Table 1. The crystalline phase of the steel slag was determined by powder X-ray diffraction (XRD) as shown in Figure 1. As can be seen from Table 1, the steel-slag powder has high CaO and Al_2O_3 content and low SiO_2 content, and has high activity. From Figure 1, it can be seen that the crystalline phase of steel-slag powder has more gel components, such as C_2S and C_3S. The fineness modulus of 2.52 of river sand, with an apparent density of 2678.6 kg/m^3 and with a water absorption ratio of 0.8%, are utilized. The natural coarse aggregate is granite crushed stone with apparent density of 2740.0 kg/m^3, water absorption ratio of 0.75%, and continuous gradation of grain size 4.75~25 mm. The additive, a reddish-brown liquid, is QL-pc5 polycarboxylic acid high-efficiency water-reducing agent with 20% water-reduction ratio and air content 1–5%, which was provided by the manufacturer. The amount of superplasticizer used in the concrete mixtures was set at 0.5% by weight of cement to meet the slump of 150 mm, which was based on the slump tests in accordance with Chinese standard [43].

Table 1. Chemical composition of raw materials (Mass fraction %).

Name	SiO_2	Al_2O_3	Fe_2O_3	CaO	MgO	SO_3	Na_2O	K_2O	R_2O	Loss
Cement	25.44	7.06	2.89	55.32	2.25	2.77	0.44	0.67	0.88	2.28
Steel slag	15.42	4.45	26.79	38.48	8.08	0.14	0.17	0.12	0.25	6.10

2.2. Specimens and Mix Proportions

In order to explore the effect of different replacement ratios of steel-slag powder on the fracture properties of the SPC, beam specimens of a total of five groups of specimens with 0%, 5%, 10%, 15%, and 20% replacement ratios of steel-slag powder were tested, respectively. Mix proportions are shown in Table 2. According to the recommendation of the International Commission on Fracture Mechanics, the three-point bending experimental method is used in this paper to determine the fracture properties of slag-concrete beams [44] with 15 specimens with dimensions of 150 mm × 150 mm × 550 mm (height × width × length), span of 450 mm, prefabricated crack depth of 45 mm, and thickness of 2 mm, as shown in Figure 2. Before the test, the specimens were placed at room temperature for 24 h and sealed with plastic film to maintain wet conditions, and the specimens were stored in a cool place and cured for 28 days by sprinkling water every day after demolding.

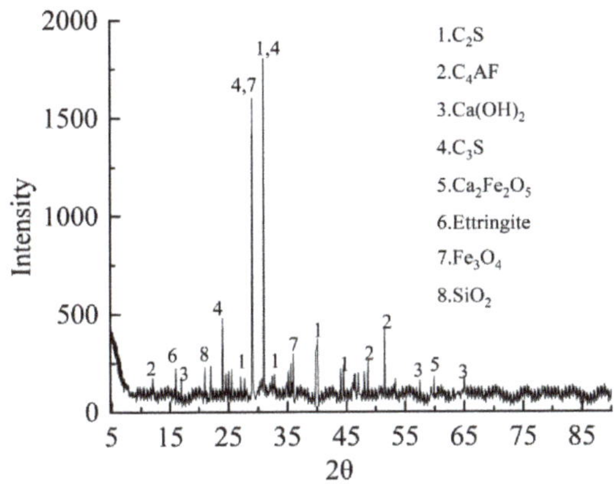

Figure 1. XRD pattern of steel-slag powder.

Table 2. Mix proportions of SPC (kg/m^3) and the corresponding compressive strength (MPa).

Specimens	Water	Cement	River Sand	Coarse Aggregate	Steel-Slag Powder	Superplasticizer	Compressive Strength
NC	168.8	287.7	530.3	990.3	0	1.42	34.17
SPC5	168.8	273.3	530.3	990.3	14.4	1.42	35.57
SPC10	168.8	258.9	530.3	990.3	28.7	1.42	32.47
SPC15	168.8	244.6	530.3	990.3	43.1	1.42	29.25
SPC20	168.8	230.2	530.3	990.3	57.5	1.42	28.82

Note: Each group has 3 specimens. NC represents ordinary concrete; SPC represents steel-slag-powder concrete, with the number at the end of the specimens representing the replacement ratios of steel-slag powder with equal gravity.

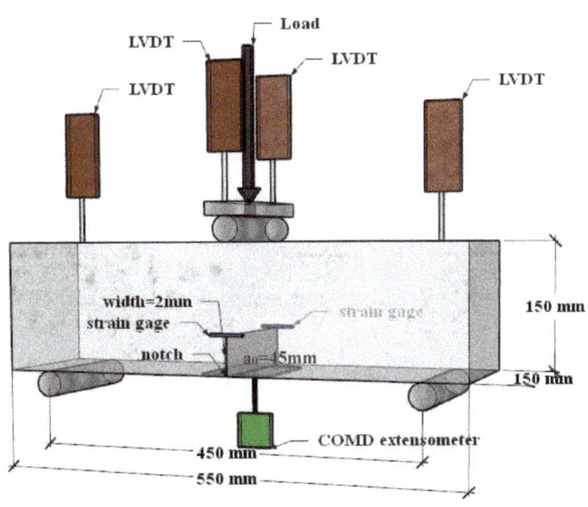

Figure 2. Experimental setup.

2.3. Experimental Setup

According to the recommendation of the International Commission on Fracture Mechanics, a three-point bending fracture test is carried out in this paper to investigate the fracture properties of SPC-notched beams [44–46]. The experimental setup is shown in Figure 2. A universal material-testing machine was adopted to perform the three-point bending fracture test. Displacement-loading mode was selected, with a loading ratio of 0.1 mm/min [47]. Mid-span load, mid-span deflection, and crack-tension displacement were collected. The loads in the span were measured by the load cell of the testing machine. The mid-span deflection measurement was measured by a displacement gauge (LVDT) with an accuracy of 0.01 mm. The crack mouth-opening displacement-measurement (CMOD) instrument is a high-precision clip-on extensometer. The load and deformation data were collected simultaneously by the high-speed static strain-acquisition system TDS-530.

3. Results and Analysis

3.1. Fracture Mode

The fracture morphology of the specimens with different steel-slag-powder replacement ratios is shown in Figure 3. Three kinds of fracture morphology of coarse-aggregate fracture, cement-matrix fracture, and coarse-aggregate surface were captured, as can be seen from Figure 3. Different fracture-morphology distributions are observed in fractures of the specimens with different steel-slag-powder replacement ratios. In order to analyze the fracture morphology of specimens under different steel-slag-powder replacement ratio, an image-recognition method is utilized to divide the fracture morphology based on the different color-gamut ranges of coarse-aggregate fracture, coarse-aggregate surface, and cement-matrix fracture by setting the color-interval range from 30 to 230, as shown in Figure 3, where black is the coarse-aggregate fracture, blue is the coarse-aggregate surface, and the white part is the cement matrix. Based on this, in order to analyze the fracture performance of the specimen, the ratios of fracture area of coarse aggregate to the total fracture area of the specimen (referred to as coarse-aggregate fracture ratio R) under different steel-slag-powder replacement ratios are analyzed, as shown in Figure 4. As can be seen from Figure 4, within the range of steel-slag-powder replacement ratios set in this paper, coarse-aggregate fracture ratio R first increases and then decreases with the increase in steel-slag-powder replacement ratios, in which the R of SPC5 concrete reaches a peak percentage of 32.46%, which is 35.53% higher than the NC percentage of 23.95%. The concrete of specimens SPC10, SPC15, and SPC20 have 24.29%, 15.43%, and 8.03% of R, which are 1.42% higher and 35.57% and 66.47% lower than NC, respectively. Obviously, the larger the R is, the greater the fracture-energy dissipation of the specimens. This is mainly because the steel-slag powder contains hydration-activity substances such as calcium oxide and aluminium trioxide, which strengthen the bonding interface between the aggregate and cement paste to a certain extent. At a lower replacement ratio of SPC, the effect of steel-slag powder and steel-slag powder particles fill the concrete interior to improve the concrete-fracture strength [15], so the crack extension proceeds based on the lower fracture strength of the coarse aggregate. With the increase in steel-slag-powder replacement ratio, the fractures of the specimens are rougher and the area of the coarse-aggregate-cement matrix-debonding interface increases, while the coarse-aggregate fracture ratio R decreases. The reason may be that higher levels of steel-slag-powder replacement ratio lead to lower amounts of cement, which affects the overall cement material effect. In addition, too large a dose of steel-slag powder cannot improve the interfacial properties between the cement and the aggregate, producing more internal voids, so the strength will be reduced.

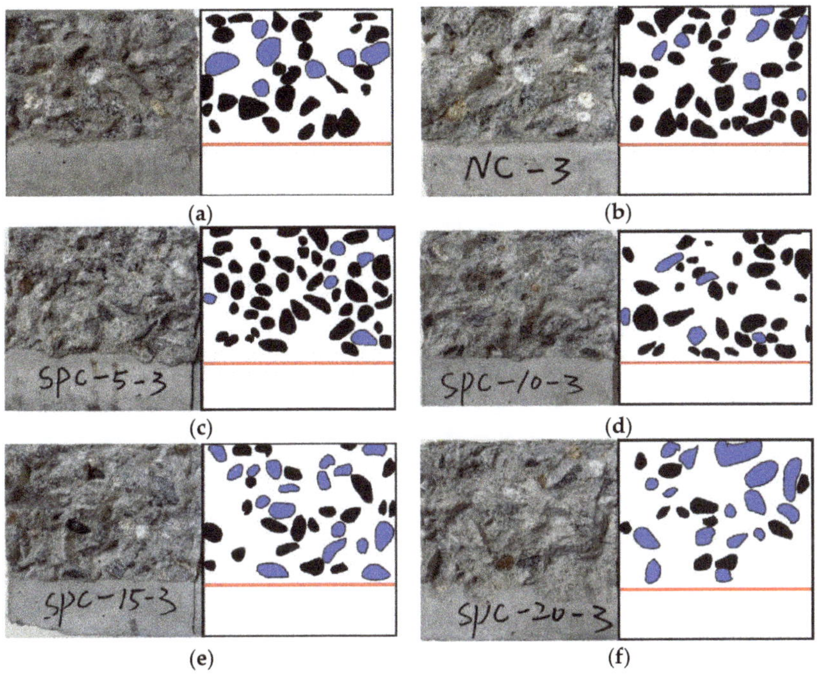

Figure 3. Fracture morphology of notched beam. (**a**) NC; (**b**) NC; (**c**) SPC5; (**d**) SPC10; (**e**) SPC15; (**f**) SPC20.

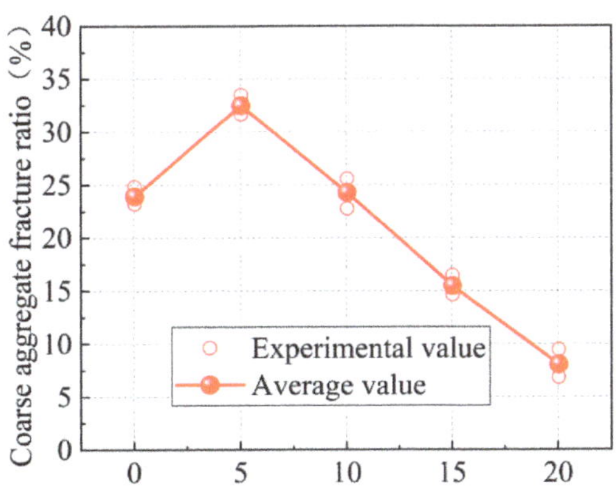

Figure 4. Relationship between coarse-aggregate fracture surface ratio R and steel-slag-powder replacement ratios.

3.2. Fracture Energy

Fracture energy is the energy consumed to form a unit fracture surface, which is a major parameter reflecting the fracture performance and energy-consumption capacity of concrete. For the three-point bending notched beam, the fracture energy includes four parts, which can be expressed as

$$W = W_0 + W_1 + W_2 + W_3 \tag{1}$$

W_0 is the work performed by the external force P, which is the envelope area under the load-deflection curve obtained in the test; W_1 is the work performed by the self-weight of the specimen before the external force P applied, which is small enough to be neglected compared with W_1; W_2 is the additional work performed by the self-weight of the specimen during the loading process [48], with $W_2 = 0.5\ mg\delta_0$. W_3 is calculated using the fracture-energy-calculation method proposed in the literature [49,50], after fitting a reasonable tail curve to the data measured by the test, The pertinent calculations are performed as suggested in Equations (2)–(5).

$$P = \beta\delta^{-\lambda}(\beta, \lambda{>}0) \tag{2}$$

$$\ln P = \beta - \lambda \ln \delta \tag{3}$$

$$W_3 = \int_{\delta_0}^{\infty} \frac{\beta}{\delta^{\lambda}} d\delta = \frac{\beta}{(\lambda - 1){\delta_0}^{(\lambda-1)}} \tag{4}$$

$$G_F = (W_0 + W_1 + W_2 + W_3)/A_{lig} \tag{5}$$

$$A_{lig} = t(h - a_0) \tag{6}$$

The energy of each of the above components is calculated. The true fracture energy G_F, where A_{lig} is the fracture area as shown in Equation (6). The concrete material at various mix ratios is obtained by Equation (5).

The load-deflection curves of the specimens are shown in Figure 5. The fracture energy of SPC is shown in Table 3. The fracture energy of specimens with different steel-slag-powder replacement ratios is shown in Figure 6. Compared with NC, it can be seen that the fracture energy of SPC5 and SPC10 concrete increases by 13.63% and 7.30%, respectively. The fracture energy of SPC5 is highest. With the increase in steel-slag-powder replacement ratios, the fracture energy of concrete gradually decreases. Therefore, SPC15 and SPC20, compared with NC, decreases by 2.04% and 9.62%, respectively. Steel-slag powder enhances the fracture energy of concrete for two reasons: On the one hand, it is related to the fracture-energy-absorptive capacity of concrete. During crack development, concrete needs to absorb a large amount of energy to destroy the internal cement-slurry-bonding force and aggregate shear stress. The hydration reaction of the active mineral of steel-slag powder can improve the fracture strength of the concrete matrix, effectively slowing the generation and propagation of cracks, and improving the energy-dissipation capacity of concrete. On the other hand, the steel-slag-powder particles enhance the compactness of the specimens. The addition of steel slag powder fills the cracks inside the concrete [51]. When the fracture surface expands further, the concrete needs to consume more energy to break. However, as the replacement ratio of steel-slag powder increases, the role of steel-slag powder becomes less obvious, which gradually reduces the strength of the concrete. At this time, the internal failure form of concrete changes the aggregate shear failure to interface peel failure, resulting in the gradual reduction in fracture energy.

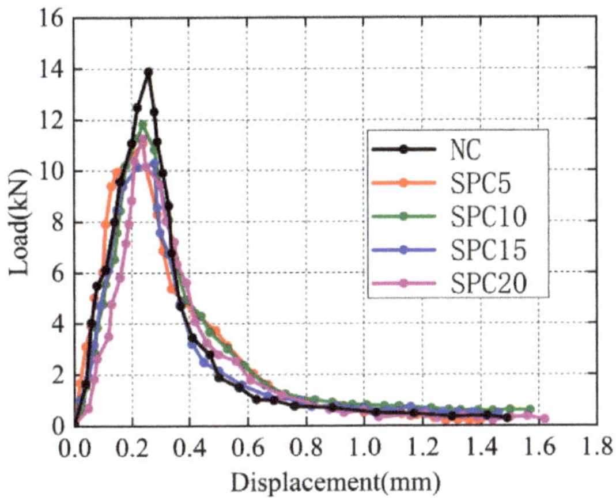

Figure 5. Load-deflection (P-δ) curves.

Table 3. Values of fracture energy of concrete of SFC.

Specimens	a_0 (mm)	Mg (N)	δ_0 (mm)	β	λ	R^2	W_0 (N·m)	W_2 (N·m)	W_3 (N·m)	A_{lig} (m^2)	G_f (J/m^2)	G_f (Average) (J/m^2)
N C-1	46.5	287.6						0.532			382.140	
N C-2	46.2	289.1	0.37	0.4	2.44	0.9830	4.236	0.535	1.163	0.01552	350.698	371.600
N C-3	46.8	286.1						0.529			381.961	
SPC5-1	45.7	288.1						0.677			456.938	
SPC5-2	45.5	289.9	0.47	0.99	1.77	0.9695	4.17	0.681	2.299	0.01564	353.150	422.262
SPC5-3	45.4	286.5						0.673			456.698	
SPC10-1	45.2	287.6						0.676			423.930	
SPC10-2	45.1	289.1	0.47	0.85	1.90	0.9653	4.125	0.679	1.863	0.01572	348.371	398.744
SPC10-3	44.8	287.6						0.676			423.930	
SPC15-1	44.8	286.2						0.644			363.999	
SPC15-2	44.7	288.0	0.45	0.61	1.81	0.9878	3.662	0.648	1.438	0.01578	364.255	364.008
SPC15-3	44.4	284.6						0.640			363.770	
SPC20-1	45	285.2						0.670			347.308	
SPC20-2	44.8	286.7	0.47	0.59	2.25	0.9819	3.587	0.674	1.213	0.01575	313.190	335.861
SPC20-3	44.7	283.7						0.667			347.084	

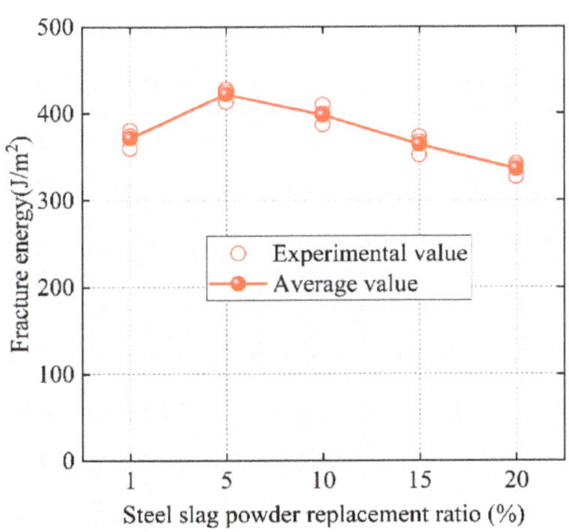

Figure 6. Fracture energy of SPC.

During the propagation of concrete cracks, tensile failure of the coarse aggregate occurs, absorbing a large amount of energy, so there is a correlation between the fracture area of coarse aggregate and fracture energy. In order to analyze the correlation between fracture energy G_F and the coarse-aggregate fracture ratio R and steel-slag-powder replacement ratio C, C and R are treated as independent variables in a regression analysis of the test data, finding the correlation coefficient r = 0.9987, through regression analysis to obtain the C, R regression equation as follows (7):

$$G_F = 253.66 + 222.04C + 492.65R \tag{7}$$

Equation (7) can better reflect the relationship between SPC fracture energy and steel-slag-powder replacement ratio, coarse-aggregate fracture ratio.

3.3. Fracture Toughness

The fracture toughness reflects the ability of the material to resist crack-instability propagation. In order to investigate the effect of different steel-slag-powder replacement ratio on the fracture toughness of SPC, the fracture toughness of the specimens in this paper is calculated and analyzed. For this paper, the fracture toughness K_{IC} of the SPC material can be calculated according to Equation (8):

$$K_{IC} = \frac{P_{\max}}{th^{\frac{3}{2}}} f\left(\frac{a_0}{h}\right) \tag{8}$$

where $P_{\max}$ is the peak vertical load; hts are the height, width, and span of the three-point bending specimen; a_0 is the depth of cut of the prefabricated specimen; $f(\frac{a_0}{h})$ is the geometric shape factor, whose expression is shown in Equation (9).

$$f(\frac{a}{h}) = 2.9(\frac{a}{h})^{1/2} - 4.6(\frac{a}{h})^{3/2} + 21.8(\frac{a}{h})^{5/2} - 37.6(\frac{a}{h})^{7/2} + 38.7(\frac{a}{h})^{9/2} \tag{9}$$

A fracture-process zone is formed at the front end of the precast crack in the concrete notched-beam specimen under load, while a critical crack is formed along the tip of the crack. The ASTM-recommended equation (Equation (8)) for fracture-toughness calculation is derived based on ordinary concrete materials. However, a new effective crack length a_c needs to be used instead of a in Equation (8) for calculation of fracture toughness of SPC, and the same can be obtained using Equation (10) developed by Chaudhuri [52].

$$a_c = \frac{2}{\pi}(h + h_0)\arctan\sqrt{\frac{tE(CMOD)_c}{32.6P_{\max}} - 0.1135} - h_0 \tag{10}$$

where h_0 is the thickness of the thin steel plate at the blade of the device clamped extensometer (mm); $CMOD_c$ is the critical value of the crack-opening displacement (mm); E is the calculated modulus of elasticity (GPa). Among them, the calculated modulus of elasticity E of the concrete mix is calculated as follows:

$$E = \frac{1}{tc_i}\left[3.70 + 32.60\tan^2\left(\frac{\pi}{2}\frac{a_0 + h_0}{h + h_0}\right)\right] \tag{11}$$

where $c_i = \frac{(CMOD)_i}{P_i}$ is the initial value of the specimen (mm/KN), calculated from the slope of the rising section of the specimen P-CMOD curve in Figure 7.

The load-crack-opening displacement curves (P-CMOD) of the five groups of specimens are shown in Figure 7. The details of calculations towards determining the fracture-toughness parameters and fracture-toughness values of SPC are shown in Table 4. The variations in fracture-toughness law of concrete mixes with different steel-slag-powder replacement ratios are shown in Figure 8.

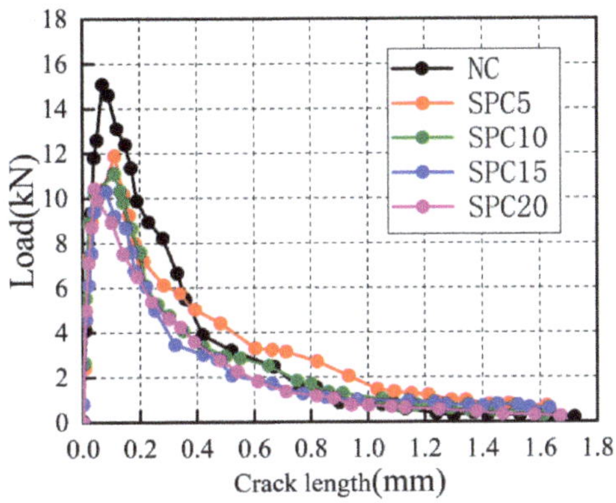

Figure 7. Load-crack-opening displacement curves (P-CMOD).

Table 4. Parameters and values of stress-intensity factor of the specimens.

Specimens	a_0 (mm)	c_i (mm/KN)	h_0 (mm)	mg (N)	Pmax (kN)	δ_0 (mm)	$CMOD_c$ (mm)	E (GPa)	A_c (mm)	K_{IC} ($MPa \cdot mm^{(1/2)}$)	K_{IC} (Average) ($MPa \cdot mm^{(1/2)}$)
N C-1	46.5	0.0033		287.6	15.08			26.86	61.90	1.663	
N C-2	46.2	0.0034	1.45	289.1	15.24	0.26	0.07	24.88	59.75	1.612	1.66
N C-3	46.8	0.0032		286.1	14.94			27.02	63.27	1.693	
SPC5-1	45.7	0.0026		288.1	11.88			33.13	90.79	2.565	
SPC5-2	45.5	0.0027	1.45	289.9	12.04	0.24	0.11	30.55	88.81	2.465	2.54
SPC5-3	45.4	0.0025		286.5	11.74			32.88	91.47	2.582	
SPC10-1	45.2	0.0031		287.6	11.08			27.29	77.55	1.711	
SPC10-2	45.1	0.0033	1.45	286.1	11.24	0.24	0.08	24.64	74.40	1.614	1.65
SPC10-3	44.8	0.0032		289.6	10.94			25.14	75.96	1.628	
SPC15-1	44.8	0.0041		286.2	10.30			20.34	70.11	1.345	
SPC15-2	44.7	0.0038	1.45	287.7	10.21	0.28	0.08	21.09	72.40	1.402	1.36
SPC15-3	44.4	0.0040		284.7	10.38			19.82	69.73	1.345	
SPC20-1	45.0	0.0038		285.2	10.41			22.11	48.73	0.898	
SPC20-2	44.8	0.0037	1.45	283.7	10.24	0.24	0.04	21.74	49.69	0.899	0.89
SPC20-3	44.7	0.0039		287.2	10.51			20.55	46.96	0.879	

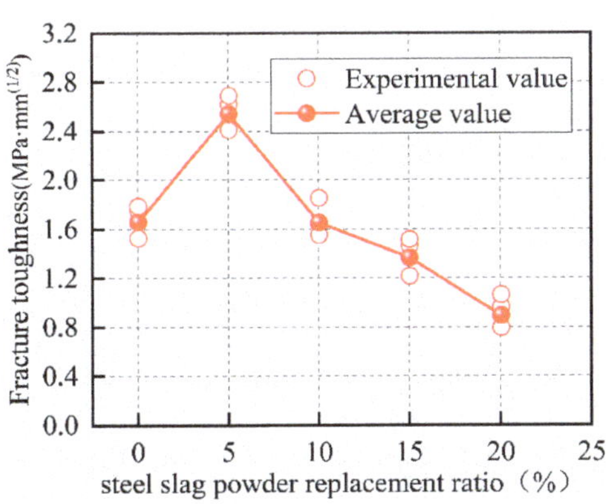

Figure 8. Fracture toughness of concrete at different steel-slag-powder replacement ratios.

It can be seen that the fracture toughness of the specimens increases and then decreases with the increase in the steel-slag powder-replacement ratio within the test range of this paper in Figure 8, which is consistent with the damage-mode pattern of the specimens.

The fracture toughness of SPC5 increases by 53.22% as compared to NC, while the values for SPC10, SPC15 and SPC20 decrease by 0.31%, 17.65%, and 46.14%, respectively. It can be concluded that the appropriate amount of steel-slag powder can improve the fracture toughness of concrete. The reason can be illustrated as follows: On the one hand, the steel slag has cement properties to produce more calcium silicate hydrates in the concrete matrix with cement, which increases the bond strength of cement paste and aggregate. On the other hand, with smaller-sized particles, the addition of steel-slag powder optimizes the fine-aggregate gradation, and enhances the compactness of concrete and increases its mechanical strengths to different degrees. When the strength of the coarse aggregate is lower than the bond strength of concrete, cracks will propagate along the coarse aggregate, resulting in the concrete-crack propagation requiring more energy consumption. However, as the replacement ratio of steel-slag powder increases, the mechanical strengths begin to decrease, which leads to a continued decrease in the ability of concrete to resist the stable propagation of cracks. The main reason is that steel-slag powder cannot replace cement as the main cement material. Reducing the amount of cement makes the concrete bond strength decrease gradually. At this time, the failure form of concrete is changed from the shear failure of coarse aggregate to the failure of the concrete-bonding layer, causing the fracture energy to decrease. This indicates that an appropriate content of steel-slag powder improves the fracture performance of SPC.

3.4. Double K Fracture Model

In this paper, the fracture model describing the development of cracks in SPC using the double-fracture criterion is also established [53], which is calculated by the following Equation (12):

$$K_{IC}^{ini} = \frac{3P_{ini}S}{2h^2t}\sqrt{a_0}F\left(\frac{a_0}{h}\right) \quad (12)$$

where P_{ini} is the cracking load measured by the three-point bending test; S is the span between the beam supports in the three-point bending tests; t is the width of the notched-beam specimen (cm); h is the height of the notched-beam specimen; a_0 is the initial crack length (cm); $F\left(\frac{a_0}{h}\right)$ is the geometric shape factor calculated according to Equation (13). The unstable fracture toughness K_{uni} of concrete is also calculated by the above two equations. The peak load $P_{\max}$ and the critical effective crack length a_c are used instead of P_{ini} and a_0 in the equation.

$$F\left(\frac{a_0}{h}\right) = \frac{1.99 - \left(\frac{a_0}{h}\right)\left(1 - \frac{a_0}{h}\right)\left(2.15 - 3.93\frac{a_0}{h} + 2.7\left(\frac{a_0}{h}\right)^2\right)}{\left(1 + 2\frac{a_0}{h}\right)\left(1 - \frac{a_0}{h}\right)^{\frac{3}{2}}} \quad (13)$$

Figure 9 shows the change law of double-fracture parameters of SPC under different steel-slag-powder replacement ratios of five groups of specimens, and the results of the double-fracture parameters of SPC specimens are shown in Table 5. The initial fracture toughness of the specimens decreases with the increasing steel-slag-powder replacement ratio, as observed in Table 5. Figure 9 shows the effect of increasing the steel-slag-powder replacement ratio on the calculated values of double-fracture parameters of SPC mixes. Among them, the initial fracture toughness of SPC20 is reduced by 41.22% compared with NC. There are two main reasons affecting the initial fracture toughness. On the one hand, the fracture strength of concrete depends on the concrete interfacial transition zone (ITZ) which is affected by the water-cement ratio and aggregate-cement matrix interface strength. On the other hand, when the modulus of elasticity of concrete increases, it is easier for a small deformation to produce initial cracks. The elastic modulus of concrete is increased by adding steel-slag powder appropriately. However, adding too much steel-slag powder will reduce the elastic modulus [54] and fracture toughness. In the process of fracture propagation of SPC with a low replacement ratio, the steel-slag powder can effectively prevent the rapid development and propagation of cracks. Until the unstable fracture toughness is reached, the crack will develop steadily when the load increases. The steel-slag powder has a positive effect on the unstable fracture toughness of concrete, as shown in Figure 9. It can be seen that the unstable fracture toughness of SPC5 increases by 46.79% compared with NC. With the increase in steel slag powder replacement ratio, the unstable fracture toughness of SPC decreases gradually, where SPC20 is 45.62% lower than NC. From above, the lower replacement ratio of steel-slag powder effectively improves the unstable fracture toughness of SPC. Moreover, it influences the unstable fracture toughness of the concrete with a high replacement ratio of steel-slag powder, which makes concrete beams more likely to enter into the unstable fracture-propagation phase. The reason is that the hydration reactions of steel-slag powder, along with those of cement, increase the strength of the cement matrix, resulting in the crack propagation along the aggregate cross section instead of the aggregate-cement-matrix interface, which takes more fracture energy. Therefore, the arrival of the crack-unstable fracture is delayed, which affects the speed of crack propagation. In addition, the reduction in cement is an important factor affecting the cementing property of the concrete. Adding too much steel-slag powder does not enhance the cementing property of the concrete. Cracks will rapidly propagate by passing aggregates, leading to the decline in the crack resistance of concrete, as similarly reported in a previous study [55]. Based on the results of the present set of experiments, the optimal fracture property of SPC is obtained when steel-slag powder replaces 5% of cement, when the cementing property can be significantly improved.

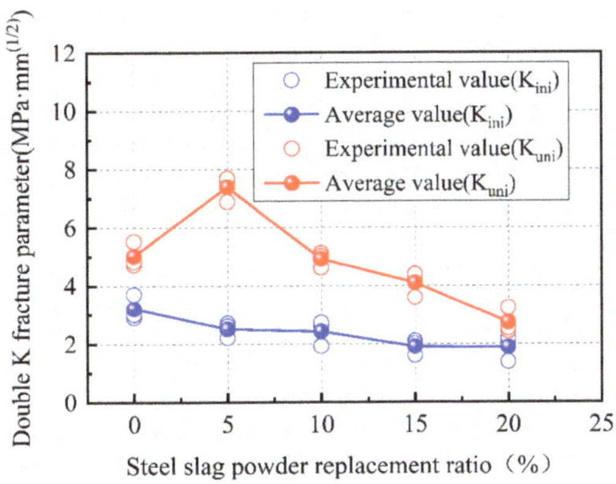

Figure 9. Double-K fracture parameters of concrete with different steel-slag-powder replacement ratios.

Table 5. The calculated results of double-K fracture parameters of SFC.

Test Piece	a_0 (mm)	P_{ini} (kN)	A_c (mm)	Pmax (kN)	K_{ini} (MPa·mm$^{(1/2)}$)	K_{uni} (MPa·mm$^{(1/2)}$)	K_{ini} (Average) (MPa·mm$^{(1/2)}$)	K_{uni} (Average) (MPa·mm$^{(1/2)}$)
N C-1	46.5	12.67	61.9	15.08	3.229	5.056		
N C-2	46.2	12.87	59.8	15.24	3.263	4.910	3.21	5.04
N C-3	46.8	12.27	63.3	14.94	3.143	5.139		
SPC5-1	45.7	10.04	85.9	11.88	2.524	6.601		
SPC5-2	45.5	11.03	84.8	12.04	2.764	6.523	2.64	6.58
SPC5-3	45.4	10.56	86.4	11.74	2.641	6.611		
SPC10-1	45.2	9.76	77.5	11.08	2.433	5.088		
SPC10-2	45.1	9.46	74.4	11.24	2.354	4.825	2.43	4.92
SPC10-3	44.8	10.06	76.0	10.94	2.491	4.855		
SPC15-1	44.8	7.80	70.1	10.30	1.931	4.049		
SPC15-2	44.7	8.00	72.4	10.21	1.977	4.205	1.91	4.10
SPC15-3	44.4	7.40	69.7	10.38	1.820	4.049		
SPC20-1	45.0	7.62	48.7	10.41	1.893	2.757		
SPC20-2	44.8	7.32	49.7	10.24	1.812	2.757	1.89	2.74
SPC20-3	44.7	7.92	47.0	10.51	1.957	2.700		

4. Conclusions

The influence of steel-slag-powder replacement ratio on the fracture performance of concrete is studied and concluded as follows:

(1) The image-recognition method is used to analyze the fracture surface of the specimens, and the coarse-aggregate fracture ratio representing the ratios of fracture area of coarse aggregate to the total fracture area of the specimen is proposed to provide a quantitative-analysis method for the fracture-failure mode of the specimens. The results show that the coarse-aggregate fracture ratio R is consistent with the changing law of fracture-performance parameters under different steel-slag-powder replacement ratios which can imply the fracture mechanism.

(2) Compared with concrete without steel-slag powder, a steel-slag-powder replacement ratio under 10% in weight can improve the fracture performance of the concrete, while a steel-slag-powder replacement ratio beyond 10% in weight shows a decreased effect, which implies positive and negative effects of steel-slag powder as a replacement to cement. The positive effect may be the filling effect of steel-slag powder, which strengthens the bond properties between the aggregate and matrix; and the negative effect may be the lower cementitious activity of steel slag compared with cement.

(3) The fracture energy and fracture toughness of SPC first increases and then decreases with the increasing steel-slag-powder replacement ratio. The SPC with a 5% steel-slag-powder replacement ratio in weight shows the best fracture performance, while the fracture energy increases by 13.63%, fracture toughness increases by 53.22%, and compressive strength increases by 4.1% compared with NC, which may due to the addition of steel-slag powder strengthening the bond properties between the aggregate and matrix to a certain extent.

Author Contributions: Investigation, Data curation, Writing—original draft, K.-X.Z.; Conceptualization, Funding acquisition, Methodology, Writing, Review and editing, J.-X.L.; Conceptualization, Funding acquisition, Supervision, Project administration, Y.-C.G.; Conceptualization, Supervision, Project administration, G.-T.L.; Conceptualization, Supervision, Review and editing, X.-W.L.; Conceptualization, Funding acquisition, Supervision, D.-P.Z.; Writing—review and editing, S.-Q.W.; Conceptualization, Supervision, Project administration, P.-Z.W. All authors have read and agreed to the published version of the manuscript.

Funding: This research received no external funding.

Institutional Review Board Statement: Not applicable.

Informed Consent Statement: Not applicable.

Data Availability Statement: Not applicable.

Acknowledgments: The authors gratefully acknowledge the financial support provided by the Guangdong Basic and Applied Basic Research Foundation (2019A1515110808), National Natural Science Foundation of China (No. 12002090).

Conflicts of Interest: The authors declare no conflict of interest.

References

1. Sha, F.; Liu, P.; Ding, Y. Application investigation of high-phosphorus steel slag in cementitious material and ordinary concrete. *J. Mater. Res. Technol.* **2021**, *11*, 2074–2091. [CrossRef]
2. Xiong, Z.; Mai, G.; Qiao, S.; He, S.; Zhang, B.; Wang, H.; Zhou, K.; Li, L. Fatigue bond behaviour between basalt fibre-reinforced polymer bars and seawater sea-sand concrete. *Ocean Coast. Manag.* **2022**, *218*, 106038. [CrossRef]
3. Lin, J.-X.; Song, Y.; Xie, Z.-H.; Guo, Y.-C.; Yuan, B.; Zeng, J.-J.; Wei, X. Static and dynamic mechanical behavior of engineered cementitious composites with PP and PVA fibers. *J. Build. Eng.* **2020**, *29*, 101097. [CrossRef]
4. Xiong, Z.; Lin, L.; Qiao, S.; Li, L.; Li, Y.; He, S.; Li, Z.; Liu, F.; Chen, Y. Axial performance of seawater sea-sand concrete columns reinforced with basalt fibre-reinforced polymer bars under concentric compressive load. *J. Build. Eng.* **2022**, *47*, 103828. [CrossRef]
5. Jiao, W.; Sha, A.; Liu, Z.; Jiang, W.; Hu, L.; Li, X. Utilization of steel slags to produce thermal conductive asphalt concretes for snow melting pavements. *J. Clean. Prod.* **2020**, *261*, 121197. [CrossRef]

6. Guo, Y.C.; Xie, J.; Zheng, W.; Li, J. Effects of steel slag as fine aggregate on static and impact behaviours of concrete. *Constr. Build. Mater.* **2018**, *192*, 194–201. [CrossRef]
7. Guo, J.; Bao, Y.; Wang, M. Steel slag in China: Treatment, recycling, and management. *Waste Manag.* **2018**, *78*, 318–330. [CrossRef]
8. Humbert, P.S.; Castro-Gomes, J. CO_2 activated steel slag-based materials: A review. *J. Clean. Prod.* **2019**, *208*, 448–457. [CrossRef]
9. Liu, W.; Li, H.; Zhu, H.; Xu, P. Effects of Steel-Slag Components on Interfacial-Reaction Characteristics of Permeable Steel-Slag-Bitumen Mixture. *Materials* **2020**, *13*, 3885. [CrossRef]
10. Parron-Rubio, M.E.; Perez-Garcia, F.; Gonzalez-Herrera, A.; Rubio-Cintas, M.D. Concrete Properties Comparison When Substituting a 25% Cement with Slag from Different Provenances. *Materials* **2018**, *11*, 1029. [CrossRef]
11. Shen, W.; Liu, Y.; Wu, M.; Zhang, D.; Du, X.; Zhao, D.; Xu, G.; Zhang, B.; Xiong, X. Ecological carbonated steel slag pervious concrete prepared as a key material of sponge city. *J. Clean. Prod.* **2020**, *256*, 120244. [CrossRef]
12. Roychand, R.; Kumar Pramanik, B.; Zhang, G.; Setunge, S. Recycling steel slag from municipal wastewater treatment plants into concrete applications—A step towards circular economy. *Resour. Conserv. Recycl.* **2020**, *152*, 104533. [CrossRef]
13. Miah, M.J.; Patoary, M.M.H.; Paul, S.C.; Babafemi, A.J.; Panda, B. Enhancement of Mechanical Properties and Porosity of Concrete Using Steel Slag Coarse Aggregate. *Materials* **2020**, *13*, 2865. [CrossRef]
14. Roslan, N.H.; Ismail, M.; Khalid, N.H.A.; Muhammad, B. Properties of concrete containing electric arc furnace steel slag and steel sludge. *J. Build. Eng.* **2020**, *28*, 101060. [CrossRef]
15. Peng, Y.; Hu, S.; Ding, Q. Preparation of reactive powder concrete using fly ash and steel slag powder. *J. Wuhan Univ. Technol.-Mater. Sci. Ed.* **2010**, *25*, 349–354. [CrossRef]
16. Roslan, N.H.; Ismail, M.; Abdul-Majid, Z.; Ghoreishiamiri, S.; Muhammad, B. Performance of steel slag and steel sludge in concrete. *Constr. Build. Mater.* **2016**, *104*, 16–24. [CrossRef]
17. Maslehuddin, M.; Sharif, A.M.; Shameem, M.; Ibrahim, M.; Barry, M.S. Comparison of properties of steel slag and crushed limestone aggregate concretes. *Constr. Build. Mater.* **2003**, *17*, 105–112. [CrossRef]
18. Saxena, S.; Tembhurkar, A.R. Impact of use of steel slag as coarse aggregate and wastewater on fresh and hardened properties of concrete. *Constr. Build. Mater.* **2018**, *165*, 126–137. [CrossRef]
19. Abu-Eishah, S.I.; El-Dieb, A.S.; Bedir, M.S. Performance of concrete mixtures made with electric arc furnace (EAF) steel slag aggregate produced in the Arabian Gulf region. *Constr. Build. Mater.* **2012**, *34*, 249–256. [CrossRef]
20. Khalaf, M.A.; Ban, C.C.; Ramli, M.; Ahmed, N.M.; Sern, L.J.; Khaleel, H.A. Physicomechanical and gamma-ray shielding properties of high-strength heavyweight concrete containing steel furnace slag aggregate. *J. Build. Eng.* **2020**, *30*, 101306. [CrossRef]
21. Pellegrino, C.; Cavagnis, P.; Faleschini, F.; Brunelli, K. Properties of concretes with Black/Oxidizing Electric Arc Furnace slag aggregate. *Cem. Concr. Compos.* **2013**, *37*, 232–240. [CrossRef]
22. Faleschini, F.; Zanini, M.A.; Toska, K. Seismic reliability assessment of code-conforming reinforced concrete buildings made with electric arc furnace slag aggregates. *Eng. Struct.* **2019**, *195*, 324–339. [CrossRef]
23. Zanini, M.A. Structural reliability of bridges realized with reinforced concretes containing electric arc furnace slag aggregates. *Eng. Struct.* **2019**, *188*, 305–319. [CrossRef]
24. Muhmood, L.; Vitta, S.; Venkateswaran, D. Cementitious and pozzolanic behavior of electric arc furnace steel slags. *Cem. Concr. Res.* **2009**, *39*, 102–109. [CrossRef]
25. Papachristoforou, M.; Anastasiou, E.K.; Papayianni, I. Durability of steel fiber reinforced concrete with coarse steel slag aggregates including performance at elevated temperatures. *Constr. Build. Mater.* **2020**, *262*, 120569. [CrossRef]
26. Guo, Y.; Xie, J.; Zhao, J.; Zuo, K. Utilization of unprocessed steel slag as fine aggregate in normal- and high-strength concrete. *Constr. Build. Mater.* **2019**, *204*, 41–49. [CrossRef]
27. Palod, R.; Deo, S.V.; Ramtekkar, G.D. Effect on mechanical performance, early age shrinkage and electrical resistivity of ternary blended concrete containing blast furnace slag and steel slag. *Mater. Today Proc.* **2020**, *32*, 917–922. [CrossRef]
28. Zhang, X.; Zhao, S.; Liu, Z.; Wang, F. Utilization of steel slag in ultra-high performance concrete with enhanced eco-friendliness. *Constr. Build. Mater.* **2019**, *214*, 28–36. [CrossRef]
29. Santamaría, A.; Orbe, A.; San José, J.T.; González, J.J. A study on the durability of structural concrete incorporating electric steelmaking slags. *Constr. Build. Mater.* **2018**, *161*, 94–111. [CrossRef]
30. Arribas, I.; Vegas, I.; San-José, J.T.; Manso, J.M. Durability studies on steelmaking slag concretes. *Mater. Eng.* **2014**, *63*, 168–176. [CrossRef]
31. Kim, S.; Kim, Y.; Usman, M.; Park, C.; Hanif, A. Durability of slag waste incorporated steel fiber-reinforced concrete in marine environment. *J. Build. Eng.* **2021**, *33*, 101641. [CrossRef]
32. Rooholamini, H.; Sedghi, R.; Ghobadipour, B.; Adresi, M. Effect of electric arc furnace steel slag on the mechanical and fracture properties of roller-compacted concrete. *Constr. Build. Mater.* **2019**, *211*, 88–98. [CrossRef]
33. Peng, Y.Z.; Chen, K.; Hu, S.G. Durability and Microstructure of Ultra-High Performance Concrete Having High Volume of Steel Slag Powder and Ultra-Fine Fly Ash. *Adv. Mater. Res.* **2011**, *255–260*, 452–456.
34. Qiang, W.; Mengxiao, S.; Jun, Y. Influence of classified steel slag with particle sizes smaller than 20 μm on the properties of cement and concrete. *Constr. Build. Mater.* **2016**, *123*, 601–610. [CrossRef]
35. Yang, J.; Lu, J.; Wu, Q.; Xia, M.F.; Li, X. Influence of steel slag powders on the properties of MKPC paste. *Constr. Build. Mater.* **2018**, *159*, 137–146. [CrossRef]

36. Fu, Q.; Ma, Q.; Zhu, Y. Experiment and Analysis on Compressive and Tensile Performance of Steel Slag Powder Concrete with Basalt Fiber. *Sci. Technol. Eng.* **2015**, *35*, 224–227.
37. Zang, J. The Influence of Steel Slag with Variable Particle Size Distribution on the Workability and Mechanical Properties of Concrete. *Ceram.-Silik.* **2018**, *63*, 67–75. [CrossRef]
38. Kaplan, M.F. Crack propagation and the fracture of concrete. *J. ACI* **1961**, *11*, 591–610.
39. Hillerborg, A.; Modeer, M.; Peterson, P.E. Analysis of crack formation and crack growth in concrete by means of fracture. *Cem. Concr. Res.* **1976**, *6*, 773–782. [CrossRef]
40. Guo, Y.C.; Zhang, J.H.; Chen, G.; Chen, G.M.; Xie, Z.H. Fracture behaviors of a new steel fiber reinforced recycled aggregate concrete with crumb rubber. *Constr. Build. Mater.* **2014**, *53*, 32–39. [CrossRef]
41. Xie, Z.H.; Xie, J.H.; Huang, P.Y.; Guo, Y.C. Effects of High Temperature Damage on Compressive and Flexural Behavior of High Strength Recycled Aggregate Concrete with Steel Fiber and Crumb Rubber. *China J. Highw. Transp.* **2016**, *29*, 17–24.
42. Hu, R.L.; Huang, P.Y.; Zhou, X.P. Determination of Double-K Fracture Toughness of Concrete via Size Effect Law. *Sci. Technol. Eng.* **2008**, *8*, 334–337.
43. *GB/T 50080-2016*; Standard for Test Method of Performance on Ordinary Fresh Concrete. China Architecture & Building Press: Beijing, China, 2016; 1–90.
44. Rilem, T.C. Draft recommendation: "Determination of the fracture energy of mortar and concrete by means of three-point bend tests on notched beams". *Mater. Struct.* **1985**, *18*, 484.
45. Domenico, D.; Urso, S.; Borsellino, C.; Spinella, N.; Recupero, A. Bond behavior and ultimate capacity of notched concrete beams with externally-bonded FRP and PBO-FRCM systems under different environmental conditions. *Constr. Build. Mater.* **2020**, *265*, 121208. [CrossRef]
46. Zhang, B.; Zhu, H.; Lu, F. Fracture properties of slag-based alkali-activated seawater coral aggregate concrete. *Theor. Appl. Fract. Mech.* **2021**, *115*, 103071. [CrossRef]
47. Kitsutaka, Y.; Kanakubo, T. Outline of JCI standard (JCI-S-001-003). *Concr. J.* **2006**, *44*, 10–15. [CrossRef]
48. Su-rong, L.; Hao, L. Experimental investigation on the fracture energy of fiber reinforced self-compacting concrete. *Eng. Mech.* **2010**, *27*, 119–123.
49. Zhang, D.; Liu, J.L.; Chen, B. Analysis on determination of concrete fracture energy by three-point bending method. *J. Build. Mater.* **1999**, *2*, 206–211.
50. Menou, A.; Mounajed, G.; Boussa, H.; Pineaud, A.; Carre, H. Residual fracture energy of cement paste, mortar and concrete subject to high temperature. *Theor. Appl. Fract. Mech.* **2006**, *45*, 64–71. [CrossRef]
51. Pan, Z.; Zhou, J.; Jiang, X.; Xu, Y.; Jin, R.; Ma, J.; Zhuang, Y.; Diao, Z.; Zhang, S.; Si, Q.; et al. Investigating the effects of steel slag powder on the properties of self-compacting concrete with recycled aggregates. *Constr. Build. Mater.* **2019**, *200*, 570–577. [CrossRef]
52. Chaudhuri, B.B.; Sarkar, N. Texture segmentation using fractal dimension. *IEEE Trans. Pattern Anal. Mach. Intell.* **1995**, *17*, 72–77. [CrossRef]
53. Xu, S.; Xiong, S.; Li, H.; Yao, L. Quantitative characterization and mechanism analysis on thickness-dependent size effect of concrete fracture. *China Civ. Eng. J.* **2017**, *50*, 57–71.
54. Zhuo, K.X. Experimental Study on Axial Compressive Properties of Environmental Steel Slag Concrete. Master's Thesis, Guangdong University of Technology, Guangzhou, China, 2016.
55. Hu, Y.Z.; Ge, Z. The effect of composite mineral admixtures containing steel slag on the properties of cementitious materials and concrete. *Bull. Chin. Ceram. Soc.* **2015**, *34*, 2737–2742.

MDPI AG
Grosspeteranlage 5
4052 Basel
Switzerland
Tel.: +41 61 683 77 34

Materials Editorial Office
E-mail: materials@mdpi.com
www.mdpi.com/journal/materials

www.ingramcontent.com/pod-product-compliance
Lightning Source LLC
Chambersburg PA
CBHW061837210926
51850CB00006B/915
* 9 7 8 3 7 2 5 8 5 6 6 5 7 *